Management of Ageing in Graphite Reactor Cores

Management of Ageing in Graphite Reactor Cores

Edited by

Gareth B Neighbour
Department of Engineering, University of Hull, Hull, UK

RSC Publishing

The proceedings of the Ageing Management of Graphite Reactor Cores held at the University of Cardiff on 28-30 Novermber 2005.

Special Publication No. 309

ISBN: 978-0-85404-345-3

A catalogue record for this book is available from the British Library

Published by The Royal Society of Chemistry,
Thomas Graham House, Science Park, Milton Road,
Cambridge CB4 0WF, UK

Registered Charity Number 207890

For further information see our web site at www.rsc.org

Printed by Henry Ling Ltd, Dorchester, Dorset. UK

Foreword

Nuclear power currently contributes some 23% of the electricity needs of the UK. Much of the power generated by nuclear fission is based on gas cooled reactor technology developed some fifty years ago. The first Magnox stations were commissioned in the 1960s whilst the first Advanced Gas Cooled Reactors came into service in 1976. Although the remaining Magnox reactors are scheduled to close over the next few years through to 2012 both they, and the fourteen AGRs in service, are essential to the power generation needs of the country.

The gas cooled reactors use graphite as a moderator in the form of layers of interlocked graphite bricks that provide channels for gas flow past fuel elements and for insertion of control rods to enable the reactor to be shut down when required. Graphite undergoes complex changes when exposed over long periods to the effects of irradiation and oxidation in the core of the reactor. It is essential that the ageing and deterioration of any complex and safety critical plant is managed effectively. The management of the effects of ageing of graphite in gas cooled reactor cores provides a significant challenge both to the industry and to the associated research community. This book provides an opportunity to review the latest developments in understanding of the behaviour of graphite as a material, as components in a reactor and in its behaviour in the complete reactor core. As well as presenting the current position with respect to ageing behaviour it is hoped that the papers presented will help to identify gaps in understanding which need to be filled to assist the management of the ageing behaviour of graphite and continued safe operation of the reactors concerned.

Michael Burdekin

(Chairman of the Conference Organising Committee)

Preface

The first international conference on the management of ageing processes in graphite reactor cores was held during 28th-30th November at the University of Cardiff, United Kingdom. The conference represented the gathering of the most eminent specialists in the field from around the world. The issues associated with polygranular graphite as a moderator have remained prominent in the forefront of nuclear energy since on December 2nd 1942, when Enrico Fermi initiated the first self-sustaining nuclear chain reaction. It is astounding to think that the pile of graphite blocks, known CP-1, in a squash court at the University of Chicago would mark the start of an amazing story in which many scientists and engineers would devote theirs life's work.

The United Kingdom has a number of gas-cooled, graphite-moderated nuclear reactors that are either nearing the end of their generating life, seeking life extension, or in the latter period of their design lives. As the graphite components of the reactor cores cannot be replaced, it is important to ensure that effective ageing-management strategies are in place to secure safe and reliable operation to the end of their generating life. This text, resulting from the proceedings, represents the accumulation of knowledge and presents the latest understanding and understanding in field that will remain significant for years to come in which graphite moderated reactors will remain centre stage in reactor technology, particularly Generation IV designs. The conference and this text had the objective of placing the next generation of scientists and engineers in good stead to improve the performance of the new generation of civil nuclear reactors and to share the UK experience with the rest of the world. I believe that the conference was very successful in achieving its aims particularly in the following topics:

- The understanding of the effects of irradiation on the properties and behaviour of graphite components
- Materials properties test procedures
- Modelling of graphite behaviour
- Brick and component behaviour
- Component integrity assessment methodologies
- Whole core modelling
- Inspection techniques
- Monitoring core condition
- Ageing-management strategies
- Regulatory requirements

Gareth B. Neighbour

University of Hull

June 2006

Acknowledgements

The editor would like to express his heartfelt thanks to the many authors who contributed to this volume. He would also like to thank his wife, Angela, for her love and support. Finally, but not least, the editor would like to recognize the support of the British Carbon Group (registered charity 207890, which itself is affiliated to The Royal Society of Chemistry, The Institute of Physics and The Society of the Chemical Industry) and the Organising Committee of the conference without which this book would not exist.

Organising Committee

Prof. F.M. Burdekin		*(GTAC & Conference Chair)*
Dr James Reed		*British Energy*
Dr Martin Metcalfe		*BNFL NSTS (Berkeley)*
Dr Nick Blackburn		*HSE NSD*
Dr Graham B. Heys		*HSE NSD*
Prof. Barry Marsden		*University of Manchester*
Dr Gareth Neighbour		*The University of Hull*
Dr A.J. (Tony) Wickham	*(Secretary)*	*Conference Manager*
Prof Malcolm Heggie		*Chair, British Carbon Group*

On behalf of the British Carbon Group, the editor would also like to express thanks to the following organisations for financial support for the initial meeting which led to this book.: The Nuclear Safety Directorate of The United Kingdom Health and Safety Executive (which is the body tasked with the licensing of nuclear installations in the UK), British Energy Ltd (reactor operators), British Nuclear Group Ltd, Nexia Solutions Ltd, and Serco Assurance plc.

Contents

Part A – Reactor Designs & Operation

Part B – Advances in the Fundamental Knowledge of Graphite Behaviour under Irradiation

PART A

Reactor Designs & Operation

The Management of Magnox Graphite Reactor Cores to Underwrite Continued Safe Operation

A T Ellis[1] and K M Staples
British Nuclear Group, Reactor Sites, Berkeley, Gloucestershire, UK.
Email: alun.t.ellis@magnox.co.uk

Abstract
The first generation of UK Nuclear Power generating plant consists of Magnox reactors that are graphite moderated and CO_2 gas cooled. Of the tranche of these power stations built over the period from the late 1950's to the early 1970's four sites remain in operation. It was recognised within the original design that there would be a loss of graphite, due to oxidation, during periods of operation. This oxidation of the PGA graphite occurs by a radiolytic process induced by ionising radiation and results in a degradation of physical, chemical and mechanical properties of the graphite. It is these changes that have to be accommodated in the arguments to demonstrate safe operation to the end of the declared life of these remaining reactors.

Following a review of the background this paper will describe the key challenges that are addressed and the processes adopted within British Nuclear Group, Reactor Sites, to provide this demonstration via three inter-related safety cases. These three safety cases consider core integrity, the long term graphite transient following a depressurisation fault and reactor shutdown reactivity margins. Attention will be given to the reactor core integrity issues where the safety case invokes a multi-legged approach to demonstrate the required high integrity.

Keywords
Magnox reactors, graphite reactor core, safety case

INTRODUCTION

The UK nuclear industry developed rapidly with twenty six Magnox reactors, at eleven power station sites commissioned between 1956 and 1972. All Magnox reactors are graphite moderated, and CO_2 gas cooled. All except the final four reactors were constructed with steel reactor pressure vessels; the final four featuring pre-stressed concrete vessels (Figure 1). In general these reactors have provided reliable electrical power generation over their operating lives. However they have now entered a planned closure and decommissioning phase. Eight (of the twenty six) reactors, managed by British Nuclear Group for the Nuclear Decommissioning Authority, are still operational although each has a planned date between 2006 and 2010 when generation will cease.

It was recognised during the original reactor design that graphite cores would be subject to weight loss due to the interaction of the graphite with oxidising species produced by radiolysis of the carbon dioxide coolant, leading to degradation of physical, chemical and mechanical properties. The effects of such degradation on continued safe operation of the reactors, until cessation of generation at each plant, are considered on a regular basis by plant monitoring and review of the associated safety cases.

---

[1] Now with Serco Assurance, Berkeley Centre, Gloucestershire.

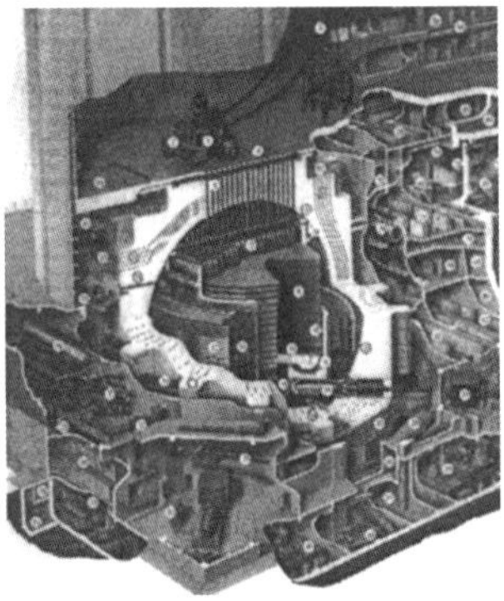

FIGURE 1: Cross section of Wylfa pre-stressed concrete pressure vessel reactor showing graphite core.

This paper will explain a typical reactor core design, describe the basis of each of the three individual, but inter-related, safety cases underwriting continued safe operation and conclude with comments on the key challenges.

REACTOR CORE DESIGN

The core of each reactor is an assembly of graphite bricks containing vertical channels for fuel, control devices, specimens and passages for coolant flow. The Oldbury core build dimensions are; overall height: 9.75m and radius: 6.8 to 7.2m (depending on reflector thickness) with an active core height: 8.6m and radius: 6.4m. Brick height is 81.3cm and either 17.1cm or 22.1 cm width. The structure is carried on a steel diagrid via support plates and a steel restraint structure assists in maintaining it in position and shape around the circumference (Figure 2). Disruption to the arrangement of the graphite structure during service could lead to channel flow impairment (and ultimately fuel melt) and restriction of control rod entry.

FIGURE 2: Construction of the Dungeness A core showing graphite brick keys.

The distribution of graphite bricks in the core structure is divided into two parts; the central moderator, or active core incorporating fuel and interstitial channels and, surrounding this, the reflector. The structure comprises top reflector bricks and a number of moderator bricks; the bottom bricks also acting partly as a bottom reflector. The cross section of the bricks varies with each station design but vertical keys were commonly used to interlock the columns in the later station designs (Figure 3). The keys allow relative movement, radially and vertically, between adjacent columns to accommodate the effects of thermal transients and irradiation induced dimensional changes in graphite bricks. Each core, dependent on design, is provided with a number of thermocouples to measure graphite material and channel outlet gas flow temperatures. The core restraint structure is also provided with monitoring thermocouples. Typically, core inlet gas temperatures are of the order of 225 °C whilst outlet gas temperatures are approximately 370 °C.

FIGURE 3: Oldbury graphite core showing keyed bricks and core restraint mechanism.

Two types of graphite were used in core construction; Pile Grade A (PGA) and Pile Grade B (PGB). PGA has a higher density and a lower neutron capture cross-section than the PGB material and forms the moderator. Top, bottom and side reflector blocks are a mixture of PGA and PGB materials dependent on the specific reactor design. Both materials were manufactured by a similar method using petroleum coke, a by-product produced during petroleum processing, for the filler material. This coke was ground, mixed with a coal-based binder pitch and then heated and extruded in a hydraulic press to form 'green bricks' which were baked, at ~1000 °C for several days, to a hard brittle material. The extrusion process created a material with anisotropic mechanical properties. To reduce the porosity and thereby increase the density of PGA material, these bricks were then impregnated with coal-tar at a high temperature (~1000 °C) and pressure before the final baking process at ~2800 °C. The resulting PGA material has a virgin porosity of ~20%. The individual bricks were finally machined before being used in the construction of the reactor core.

REACTOR CORE SAFETY CASES

Continued safe operation of each reactor core requires that; core structural integrity, the ability to manage reactivity for 'shut down' and long term 'hold down' purposes, and the effects of reactor pressure vessel depressurisation and subsequent air ingress to the core are addressed. British Nuclear Group addresses these issues in three inter-related site specific safety cases for each power station site as described below.

Core Integrity Safety Case

Graphite ageing arises from both radiolytic oxidation and fast neutron damage. Best, Stephen and Wickham (1985) reviewed graphite oxidation and presented a revised model to describe the available data. Their model explains that irradiation of carbon dioxide gas contained in graphite pores produces an 'active species'. The 'active species' entities may either recombine or diffuse to the walls of the pores in the graphite and react with it to produce carbon monoxide, thus enlarging the pores and reducing the graphite density (Figure 4).

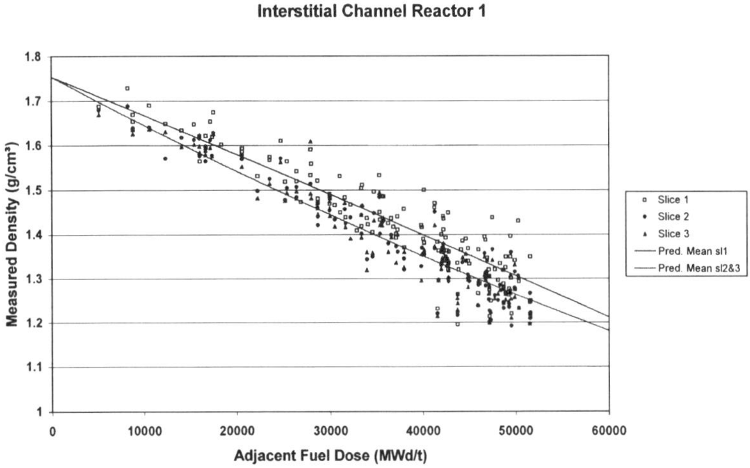

FIGURE 4: Plot of the measured and predicted values of density against adjacent fuel dose for interstitial channels at Oldbury.

The purpose of the graphite is to moderate high energy neutrons and in doing so some carbon atoms are displaced from their lattice position. This leads to graphite shrinkage (and the consequent development of internal stresses because of differential effects), changes in material properties and the accumulation of Wigner, or stored, energy which can be released when the core is heated above its normal operational temperatures.

In addition to its role as a moderator, the functional requirements for the graphite reactor core at the Magnox power stations is maintenance of both the geometry of the control rod channels and reactor core flow geometry. If the geometric integrity of the graphite core were to be seriously challenged, by distortion or widespread brick cracking and displacement of major fragments, insertion of the control rods could be affected. Additionally, through wall brick cracking, if accompanied by significant opening of the crack, could permit substantial flow by-pass from the associated fuel channel, degrading the cooling of the fuel in normal operation and during faults. To ensure a robust high integrity argument is provided for continued operation the safety case for each reactor is constructed around four legs. The four legs are inspection, monitoring, structural integrity assessment and consequences.

Inspection Leg – Forewarning of Severe Cracking

Each reactor is shutdown on a biennial basis for maintenance and inspection. During these outages all reasonably practicable steps are taken to confirm the integrity of the core graphite, and to provide either warning of brick cracking or other significant deterioration in the core. The activities undertaken during each reactor outage, including Norebore and TV viewing campaigns, are termed inspections. Channels to be inspected, which are a small portion of the total core complement, are selected on the basis of a targeted (higher risk), repeat (previous indications) and speculative inspection strategy and the results reviewed against defined acceptance criteria.

In a Norebore inspection, a device is passed down a fuel channel, providing continuous measurements of two perpendicular channel diameters, and of the departure of the channel from being straight and vertical. TV visual inspection of fuel and interstitial channels are undertaken using a sophisticated conical mirror camera enabling almost the entire wall area of a channel to be viewed at a single pass. This unit also incorporates a side viewing camera capable of providing high resolution crack gape images to 0.1 mm, for investigative purposes.

Monitoring Leg – Deterioration of the Graphite and Impact on the Overall Condition
On-line monitoring provides information about the reactor conditions at power and gives forewarning of significant brick damage. Material condition monitoring is provided by the removal of graphite samples from the reactor during outages. Graphite samples, typically 12 mm diameter x 20 mm in depth, are trepanned from the brick walls during the inspection programme. Additional capability is provided by special graphite samples installed in the reactor at start of life. These samples, some of which are removed periodically from the reactor, are used for both mechanical and chemical property measurements and provide a monitor of the condition of the graphite following in-service exposure. The data are incorporated into the overall data set and are used for safety case review and inputs to the revised structural integrity argument.

Confirmatory evidence of the retention of the design core geometry is routinely provided by monitoring for (i) free entry and movement of the control rods, (ii) the absence of anomalous channel gas outlet temperatures or graphite moderator temperatures and (iii) the absence of abnormal restrictions during refuelling. During normal plant operation 'freedom of movement' checks are undertaken on certain control rods to detect core distortion. Control rod checks are taken following shutdown, unplanned or planned and during start- up and during power operation. Following unplanned/planned shutdowns the operator immediately confirms that all control rods have fully inserted into the core. At statutory outage periods all control rods are tested during the first few hours of the shutdown and again on completion of maintenance. These tests are analysed for evidence of adverse trends in insertion time.

At power a daily calculation of maximum assessed fuel can temperature includes a full temperature survey. Any temperature indications which do not behave as expected will prompt an investigation. All fuel element, channel gas outlet and graphite temperatures are recorded shortly after shutdown. Regular monitoring of these outputs is then instituted (hourly if practicable). The temperatures are examined for indications which are not consistent with expectations based on previous experience. During start up fuel element and channel gas outlet (CGO) temperatures are measured to identify whether any individual channel is operating at anomalous higher temperatures.

Any difficulties that arise during refuelling operations are reviewed for potential channel damage/distortion issues. A burst fuel can detection system is installed which continually monitors all channels in the core. Damage to fuel could occur either by core distortion and mechanical damage to the fuel, or by overheating of fuel caused by channel blockage or channel bypass flow. If this were to occur, it would be readily detected by the system, prompting the appropriate response from the operator.

Structural Integrity Assessment Leg

The structural integrity assessment leg of the safety case addresses the increase in weight loss of the graphite during service and the maintenance of its functional capability in the unlikely event that crack-like defects are produced. A review of the existing safety case for the functional capability of the core with respect to structural integrity issues is undertaken. This takes into account the output from the plant monitoring programme. This work is aimed primarily at demonstrating that the likelihood of brick cracking is small. It also provides a consideration of the possibility that graphite shrinkage might lead to keyway pinching and thereby prevent the as-designed core movement provided by the key/keyway system. This leg comprises three main elements:

- Assessment of the changes either predicted or measured in physical and mechanical properties of the graphite, including weight loss.
- Calculation of the various components of loads and stresses arising in the graphite as a result of individual internal and external mechanisms.
- Assessment of the integrity against a suitable failure criterion.

The failure criterion adopted is a simple ratio of applied combined stress to material failure stress. This ratio is termed a *fractional strength value*, or *utilisation*. For an analysis at the best estimate level, a fractional strength value of 1.0 is considered to correspond to a 50% probability of localised cracking. However, due to a number of approximations the calculated utilisations are conservative and regarded as indicative of the risk of failure rather than definitive.

Consequences Leg– Consideration of Postulated Brick Cracking and Keyway Pinching

An understanding of likely modes of brick failure leads to a view on the probability of significant cracking. It is necessary to assess the potential consequences in the hypothetical event that cracked bricks are present in the cores. Consideration is also given to the potential effects on fuel can temperatures, due to coolant flow bypassing, of more onerous but hypothetical cracks. The potential for impairment of control rod movement is considered since should there be more than one crack in a brick, there is the potential for a detached fragment of graphite to either impede control rod entry or impair fuel cooling. The potential for keyway pinching to impede the entry of control rods also forms part of the argument.

Reactor Shutdown Reactivity Margin Safety Case

When a reactor trips, automatically due to a fault signal (unplanned) or manually by the operator (planned), all available control rods fall into the core, under gravity, to shut the reactor down. The reactor safety case demonstrates that the associated reactivity loss is always sufficient to shut-down the reactor, with a significant excess known as the shutdown margin (SDM). It is also necessary to demonstrate a hold-down margin, ensuring that the reactor will not unintentionally become re-critical following shutdown. The following factors will affect the reactivity of a shut-down Magnox reactor viz, core temperature, xenon and other reactor poisons, removal of control rods, removal of flux flattening absorbers, refuelling, presence of air and presence of moisture. Following a shutdown, the reactivity of the core initially falls due to increases in the concentration of xenon-135 which continues to be produced by radioactive decay because it is no longer removed by neutron absorption. After about 12 hours, however, the rate of production of xenon-135 drops and the core free reactivity (the reactivity neglecting control rods and absorbers) will eventually rise above its at-power value. Clearly, the removal of control rods and absorbers from a shutdown core will result in reactivity increase, and because of this, Operating Rules place restrictions on their removal. Similarly, Identified Operating Instructions are used to control refuelling activities, including the movement of fuel and absorbers into and out of the reactor cores (Western, D. J. and Corkerton, P.A. 1998)

As stated above, fuel and core temperatures generally fall during a period of shutdown, reducing the reactivity of the core. There are no explicit limits placed on the temperature of a shutdown reactor containing CO_2 gas. Due to the presence of nitrogen, admission of air increases neutron absorption and hence reduces reactivity. However, due to the risk of oxide formation on any potentially exposed uranium, there is an upper limit on the surface temperature of the fuel cans before it is permissible to admit air into the reactor. Oxidation of uranium would release fission products into the gas circuit to unacceptable levels. Water is an effective neutron moderator (as well as an absorber) and, in a Magnox reactor with high levels of graphite weight-loss, the presence of water within the core can increase reactivity.

Shutdown and hold-down studies are undertaken using endorsed methods and modern assessment criteria. Shutdown margins (SDM) are calculated for each of the assessed core states, and are defined as the reactivity difference between a pessimistic initial critical core state and an assessed core state, minus allowances for operational changes, systematic uncertainties and random uncertainties. Appropriate SDM acceptance criteria are defined consistent with the likelihood of the assessed core state arising. The methods are based upon the core physics code PANTHER which is used to assess normal operation and a range of bounding, post-shutdown loss of core cooling faults. In addition to loss of core cooling faults, it has been necessary in some cases to consider the potential for core reactivity increases arising as consequence of water ingress faults. This has required the development of a revised PANTHER methodology to represent the effects of hydrogen in the interstitial channels of the reactor core.

Long Term Graphite Transient Safety Case

The Long Term Graphite Transient (LTGT) is an assumed fault scenario involving a breach of the pressure circuit leading to depressurisation of the reactor pressure vessel and subsequent air ingress to the core. The reactor is tripped and there is a reduction in forced cooling of the core. Core temperatures are initially driven up by decay heat, and as the transient progresses, oxidation of the graphite and release of stored energy occurs, increasing temperatures further. Hence the reactor has to be sufficiently cooled to avoid significant oxidation of the bulk graphite, spontaneous ignition of the Magnox fuel clad, and subsequent exposure of uranium and release of fission products.

The severity of the LTGT is dependent on moderator graphite properties that may deteriorate with cumulative core irradiation. In particular, stored energy (which may be released as the core heats up) and other properties that affect the graphite oxidation are important. The maximum temperature reached in a LTGT is dependent upon both the power of the channel and the channel temperatures at the start of the transient. The relevant graphite properties are monitored routinely according to an agreed programme of sampling and testing. On the basis of this updated sets of data, extrapolated forward in mean core irradiation are recommended for use in LTGT fault studies.

CONCLUDING COMMENTS

As the reactors approach the end of their operating lives there are several challenges associated with the graphite within their cores.

In the structural integrity safety case area, graphite weight loss is leading to deterioration in material properties. Since the graphite is a complex mixture of filler and binder, with an initial pore density of ~20%, there is variability in mechanical property data. This is enhanced when undertaking measurements on service exposed, high weight loss material. As a consequence there is a need for a conservative approach in the safety case. This in turn presents difficulties in defining the approach to monitoring and inspection. The key challenges are furthering knowledge and understanding of graphite degradation to improve integrity predictions throughout a reactor operational period, improving data measurement capabilities to reduce uncertainties and developing appropriate monitoring and inspection strategies. These challenges are being addressed through sensitivity studies on material properties data, together with statistical analysis of these data. Moreover there is a substantial project to improve the inspection, monitoring and consequences legs of the safety case through research, development and implementation of a large number of projects.

The principle challenge to the reactor SDM safety case is associated with water ingress into high graphite weight loss cores reducing the SDM. However, revisions to the modelling techniques (PANTHER) have led to an improvement in our assessment capability and hence a reduction in the pessimisms in the argument. Currently the position to the end of generation at each site is considered satisfactory.

Recent data measurements for the LTGT safety cases have determined that the deterioration in chemical properties with mean core irradiation is modest. Hence the LTGT cases are considered secure to the end of generation at each site.

Acknowledgements: The authors would like to acknowledge Magnox Electric Ltd permission to publish this work. They would also like to thank several members of Magnox Electric Ltd staff for helpful contributions and comments in preparing this paper.

References

Best, J. V., Stephen, W. J. and Wickham, A. J. (1985). Radiolytic graphite oxidation. *Progress in Nuclear Energy, 16, [2], 127-178.*

Western, D. J. and Corkerton, P. A. (1998). Management of the nuclear safety case. *IMechE Conference Transactions, 1998-6, 219-226.*

AGR Core Design, Operation and Safety Functions

Alan G Steer
Engineering Division, British Energy Generation Ltd.,
Barnett Way, Barnwood, Gloucester, UK.
Email: alan.steer@british-energy.com

Abstract
An overview is presented of the design and operation of the UK's 2nd generation of graphite moderated, CO_2 cooled, Advanced Gas-cooled Reactors (AGR) of which 14 are in operation at 6 sites in England and Scotland. Special emphasis is placed on the roles played by the graphite components during normal operations and under fault conditions. The potential adverse consequences of graphite ageing under irradiation for the key functions of maintaining fuel temperatures within safe limits and of enabling the safe shutdown and hold down of the cores are discussed.

Keywords
AGR, Design, Operation

INTRODUCTION

The Advanced Gas-cooled Reactor (AGR) is the second-generation UK commercial reactor system. Like the first generation Magnox reactors, the AGR uses graphite for moderation with carbon dioxide as coolant. A number of significant changes were introduced as the result of operational experience with the Magnox reactors. The system was designed for high thermal efficiency by providing high-temperature and high-pressure steam conditions equivalent to those from coal and oil fired boilers in more conventional power stations. Each reactor was designed to produce around 1600 MW of heat, which is sufficient to generate about 660 MW of electrical power.

The original AGR stations of Dungeness B, Hinkley Point B, Hunterston B, Heysham stage 1 and Hartlepool were designed during the late 1960s, and constructed and commissioned during the 1970s and early 1980s. Two additional AGR stations, Heysham stage 2 and Torness, were ordered in the late 1970s, and constructed and commissioned during the 1980s. Their design was based on that of the most successful of the original AGR stations at Hinkley Point B and Hunterston B, but incorporating a large number of changes mostly to meet more stringent, modern design requirements, such as seismic qualification. All fourteen reactors are currently operated by British Energy at the six previously mentioned sites, of which four are in England and two are in Scotland.

THE REACTOR SYSTEM

All the primary coolant circuit (Figure 1) is contained within a Pre-stressed Concrete Pressure Vessel (PCPV). The coolant is carbon dioxide gas, CO_2, which is pressurised to 40 bar to give sufficient heat capacity. CO_2 was chosen because it has a low neutron capture cross-section, it does not become radioactive on its passage through the core and it is relatively cheap and readily available. At the heart of the reactor is the core, which consists of the enriched uranium oxide fuel sitting within channels formed by columns of hollow, cylindrical graphite blocks. The fuel is the source of both the heat and fissile neutrons. The heat is converted within the boilers to steam, which will in turn pass through the turbines to drive the electricity generators. The fissile neutrons, once moderated by the graphite, sustain the fission chain reaction.

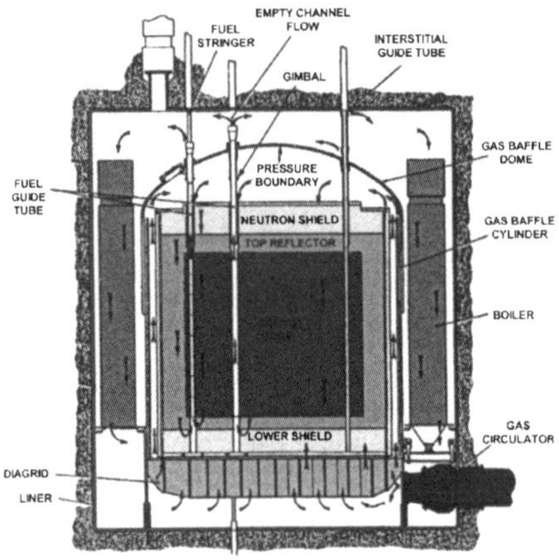

FIGURE 1: AGR core internals and coolant flows.

The heat output from the reactor is varied using the control rods to regulate the chain reaction. The control rods are steel cylinders with neutron absorbing boron inserts that move within their own channels formed by the graphite blocks. The control rods also provide the primary means of shutting down and holding down the reactor. Nitrogen injection systems provide a secondary capability for short-term shut down and hold down, while systems for injecting water and boron beads provide a tertiary capability for long-term hold down. There are significant variations between stations in the designs and capability of the secondary and tertiary systems.

THE FUEL

The reactor is fuelled with uranium dioxide in the form of cylindrical pellets, which are hollow. Around 60 pellets are sealed in a thin-walled stainless steel tube ('can') slightly less than 1 m long to form a fuel pin. The purpose of the can or clad is to contain the fission products, many of which are highly radioactive. Thirty-six pins are arranged in three concentric rings within a graphite sleeve to form a fuel element (Figure 2). Location of the pin cluster together with the central tie bar guide tube within the fuel element is by means of stainless steel braces at the top and middle positions and a support grid at the bottom. The pins are fixed solidly to the support grid, but are able to slide in the axial direction through the braces to accommodate thermal expansion and sleeve shrinkage. The grid and braces are supported by circumferential slots on the inside of the graphite sleeve, which forms the barrier between the downward external re-entrant flows through the core and the upward internal flows past the fuel pins. The sleeve graphite experiences high levels of irradiation and high temperatures leading to high corrosion rates. As the graphite sleeves are replaced each time the channel is refuelled, this is not a long-term issue.

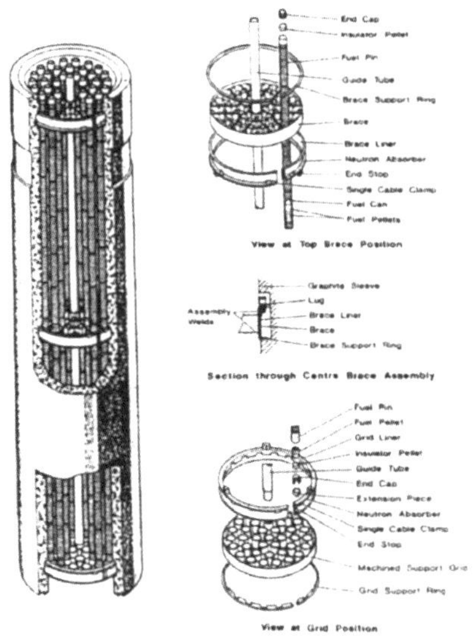

FIGURE 2: Fuel element with graphite sleeve and pins.

Eight fuel elements are sandwiched between a bottom and top reflector unit and assembled on a central tie bar to form a fuel stringer. The reflectors consist of short graphite sleeves, while the tie bar is a 1 cm diameter, nimonic alloy bar that carries the weight of the stringer during loading and unloading of the assembly. Once in the reactor, the weight of the stringer is taken by the core support structure through the bottom reflector standing on a steel support unit. The stringer is then a free-standing structure within the fuel channel with the only lateral restraint being provided above the top reflector by either a steel brush or a graphite seal.

Above the stringer is a plug unit and together these form a fuel assembly, some 26 m long (Figure 3). Each fuel assembly is handled as a single unit by the fuelling machine. Gimbal joints are incorporated to accommodate any small misalignments in the path onto the reactor. The fuel plug unit comprises a closure unit to seal the pressure vessel boundary at the top of the fuel standpipe, a biological shield to reduce radiation levels on pile cap, and a heat shield to reduce heat flow into the standpipe and surrounding concrete. Below, there is a gag unit and actuator to adjust gas flow through the channel so that it matches fuel heat production, and a scatter plug to reduce irradiation of the plug unit components. In addition, a gas sampling line is provided to detect fuel pin clad failure. Thermocouples are provided to measure the temperature of channel gas outlet for control, indication and fault detection purposes.

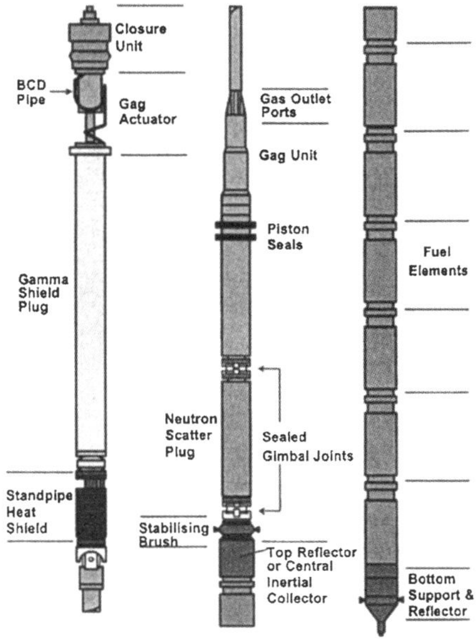

FIGURE 3: AGR Fuel assembly.

Each reactor contains around 300 fuel assemblies. The original intention was that old fuel assemblies should be exchanged for new ones while the reactor remained at-power, but flow-induced impacts between fuel sleeves and the channel walls at some stations means that at-power refuelling is restricted to the reactors at Hinkley Point B, Hunterston B, Heysham stage 2 and Torness. At the other stations, fuel exchanges are made with the reactors shut down. At all stations, refuelling is done in small batches of channels, with each fuel assembly remaining in the reactor for around 5-7 years. As fuel burn-up is most efficient in the centre of the core and least around the outside, shuffling fuel from peripheral to central channels provides better utilisation of the fuel. Not all stations have adopted this fuel management strategy and fuel shuffling has only been implemented at Hinkley Point B, Hunterston B, Hartlepool and Heysham stage 1.

THE CORE

The permanent structure of each core (Figure 4) is a cylindrical array, approximately circular in shape, of columns of graphite blocks, or bricks, of which there are two basic types. The large bricks are approximately circular in plan with a diameter of about 460 mm and a height of about 850-900 mm. They are arranged in columns on a square lattice to provide straight vertical channels for the fuel stringers. The small bricks are about 190 mm square and arranged in columns that occupy the interstitial positions. These columns provide straight vertical channels for the control rods and other ancillary devices. Each core consists of around 11-13 rings of columns of around 12 large bricks, with the inner 9-10 rings having hollow bricks to contain the fuel and the outer 2-3 rings having solid bricks to act as a neutron reflector.

FIGURE 4: Partially complete AGR Graphite Core.

Radial keys and keyways stabilise the columns of large and small bricks (Figure 5). The arrangement of radial keys and keyways allows differential radial expansion between the graphite components and the supporting and restraining steel components to be accommodated by free sliding between the keys and keyways. Non-radial relative movements are always resisted by shear loads between the keys and keyways, with the lateral loads being transferred through the array to the steel restraints around the periphery. The columns of bricks stand on core support plates that are carried by a core support structure that is different for each design of AGR. Apart from their own weight, the columns also carry the weight of the graphite and steel blocks of the Upper Neutron Shield (UNS).

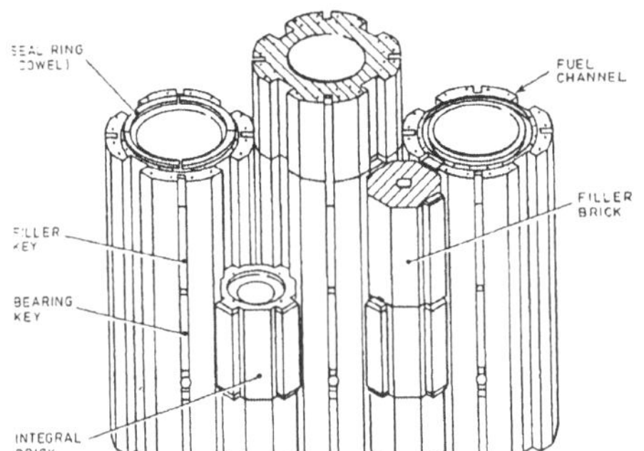

FIGURE 5: AGR Brick Keying Arrangement.

Above and around the graphite core, and attached to the floor of the reactor vault within the PCPV, is a gas baffle consisting of a steel cylinder surmounted by a steel dome. Its purpose is to separate the hot coolant discharged from the tops of the fuel stringers from the cold coolant pumped by the gas circulators into the core from the bottoms of the boilers. There is a modest pressure differential of about 4 bar across the dome producing a significant up thrust on the gas baffle. This is resisted by the weight of the cores at Hinkley Point B, Hunterston B, Heysham stage 2 and Torness, and by fixings within the PCPV at Dungeness B, Hartlepool and Heysham stage 1.

Above each column of bricks that forms a channel for which access is required for fuel assemblies, control rods and ancillary devices, is a steel guide tube supported from the gas baffle. The bottoms of each guide tube are located within the upper bricks of channel in order to provide a clear and uninterrupted path from the bottom of the core through the gas baffle and the standpipe in the PCPV to the pile cap.

THE COOLANT

Gas circulators pump cold CO_2 from the bottoms of the boilers at about 285 °C through the gas baffle into the bottom of the core. Here, the coolant separates into two substantial fractions with one part flowing beneath the core support structure to enter directly the inlet of the fuel assemblies (Figure 1). This direct flow is designed to keep the temperatures of the steel core support structures at around 300 °C. The other fraction of coolant flows upward past the side of the core into the space above the core and below the gas baffle dome before re-entering the core downward from the top. This re-entrant flow is designed to keep the steel core restraint structures and the gas baffle temperatures at around 300 °C. Maintaining the steel components at these temperatures avoids the possibility of breakaway corrosion and the need to use special high temperature steels.

Some 6% of the total heat input from the fuel is generated in the graphite of the core bricks by neutron and γ-ray heating, and the downward flow through the core is designed to remove this heat while maintaining the graphite brick operating temperatures within the range of about 320 °C to 450 °C. These temperatures are above 300 °C, below which the storage of Wigner energy is a concern, and below about 550 °C, above which excessive thermal oxidation might occur. At the bottom of the core, the re-entrant flow at a temperature of about 390 °C mixes with the cooler direct flow before the combined flow enters the bottom of the fuel assemblies. On its way past the fuel pins, the coolant is heated to a temperature of around 630-650 °C prior to discharge above the gas baffle dome and entry to the tops of the boilers.

Radiolytic oxidation of the graphite core components is a concern at all brick temperatures. It is caused by the increased chemical activity of the CO_2 coolant as a result of γ-ray irradiation. It can be inhibited by adding suitable hydrocarbon gases, such as methane, to the coolant and maintaining a regular supply of the inhibitor to the interior of the graphite components. In the early AGR cores, the coolant flows downwards in both the passages between bricks and the annular spaces between the fuel sleeves and the channel walls. At each fuel brick interface, there are transverse passages that allow cross-flows from the outsides of the bricks to the fuel channels. Consequently, coolant and inhibitor can only penetrate bricks from their outer surfaces slowly by diffusion. To encourage this process, large numbers of vertical diffusion holes, also known as methane holes, were bored through the bricks. In the latest AGR cores, the fuel brick interfaces were sealed to impress a flow of coolant through the graphite bricks from their outsides to their bores, reducing the need for so many diffusion holes.

THE CONTROL RODS

There are around 80 control rods (Figure 6) in each reactor and these are split into two main groups according to their function. About half, known as black, bulk or coarse rods, absorb all the incident neutrons. They are fully withdrawn during normal operation. Their purpose is to shut down the reactor and to maintain long-term hold down. The rest, known as grey, trimming or auto rods, absorb only a portion of the incident neutrons. They are partially inserted during normal operation and are moved continuously by the reactor control systems to regulate the chain reaction and, thus, the power output.

All the rods are articulated to enable them to negotiate any small misalignments of the channel and are suspended on chains from the actuator plug unit in the PCPV. Conically shaped noses are provided to prevent snagging of the rods during entry by any ledges or lips between channel bricks. Control is achieved by regulating the electrical supplies to the motor operated winding gear. For a reactor trip, an electromagnetic clutch is de-energised to allow rod insertion under gravity. The rate of insertion is reduced towards the end of travel by a disc brake incorporated in the actuator plug unit.

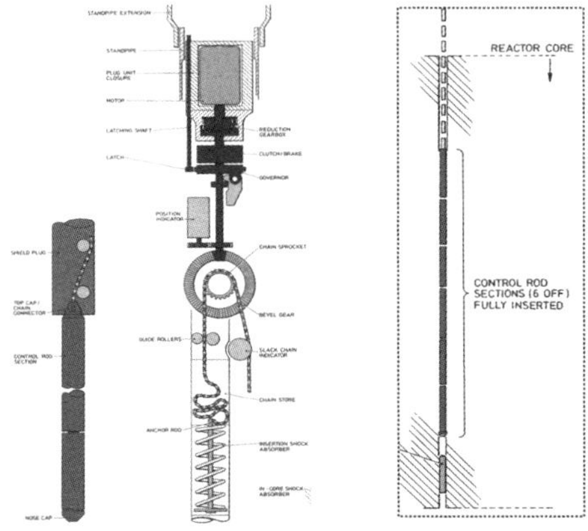

FIGURE 6: Control rod and actuator.

GRAPHITE

Two types of graphite are used in AGR cores. The permanent components are made from Gilsocarbon graphite, which used natural Gilsocarbon cokes as filler particles and flour impregnated with pitch to produce near-isotropic properties. The moderator bricks use graphite that is double impregnated, while the reflector and shielding bricks use graphite that is single impregnated. The fuel sleeves are made using needle coke particles double impregnated with pitch to produce slightly denser, near-isotropic graphite.

The main purpose of graphite is to provide moderation for the nuclear reaction by slowing down the fissile neutrons to increase the probability of capture by other Uranium-235 atoms in the enriched uranium fuel to initiate other fission events and so sustain the chain reaction. The moderation process produces a cascade of carbon atom displacements that amount to about 25 dpa during a reactor lifetime. This causes changes to the physical properties of graphite, including elastic modulus, strength, and Coefficient of Thermal Expansion (CTE). When not subjected to neutron irradiation and at temperatures well below the graphitisation temperature of around 2800 °C, graphite is dimensionally stable in both its stressed and unstressed states. This behaviour changes during neutron irradiation, when graphite undergoes dimensional change in its unstressed state, and readily creeps under stress.

In terms of component integrity, the most important effect of neutron irradiation is dimensional change, which makes the near-isotropic types of graphite used in the permanent AGR core components shrink at low dose before it makes them expand again at high dose. Because of the way that the dose attenuates through the brick thickness, this has the effect of making the fuel brick bores shrink faster than their outsides early in life. This process generates tensile stresses at the bore and compressive stresses at the outside of the brick, with the potential to initiate cracks at the bore. Once the graphite starts to expand again, the process is reversed to generate compressive stresses at the bore and tensile stresses at the outside of the brick late in life, with the potential to initiate cracks at the radial keyway roots.

The long-term behaviour of the graphite and its consequential effects on the behaviour of core components is further complicated by radiolytic oxidation, which causes exponential decays in modulus and strength, and causes the long-term changes in graphite properties to be delayed. The loss of graphite will affect the reactivity of the reactors through loss of moderation and might increase the risk of debris production.

SAFETY FUNCTIONS

The management of safety at UK nuclear sites is based on the risk of exposure of the public to dose. Since the vast majority of the inventory of radioactive materials in a power reactor resides in the fuel, the safe operation of the reactor is managed by ensuring the integrity of the fuel by maintaining its temperature within acceptable limits under all possible reactor conditions. It is necessary to maintain these conditions under fault and hazard conditions, during fuel exchanges and while shut down, as well as when the reactor is at power and producing heat for electricity generation. In practice, this means ensuring that the reactor systems maintain adequate coolant flows to the fuel pins, and unimpeded movements of the control rods and fuel under all these conditions. The core graphite components play important roles in maintaining these functions.

The requirement to ensure adequate flows to the pins means that sufficient coolant flows must be supplied to the inlets of the fuel stringers and upwards past the fuel pins. This requires that the core must retain its shape sufficiently for there to be enough open channels through and past the graphite core bricks, and for there to be clear passages at the bottoms of the fuel stringers. It also means that the flow separation between the downward flow of coolant through the annuli between the fuel sleeves and the channel walls, and upward flow within the fuel sleeves must be maintained. Breaches of this boundary could potentially occur if the fuel elements were distorted by a lack of straightness in the fuel channel wall forcing adjacent fuel sleeves to gap at their interfaces. They might also be caused if fuel sleeves were to be crushed by shearing of doubly axially cracked bricks, or following gross impact damage as the result of impacting between the fuel stringers and channel walls during either on-load fuel exchanges or seismic events.

The requirement to maintain unimpeded movements of the control rods means that the channels within which they move must have sufficient bore size and straightness. The control rods might be impeded if the relative tilts and ledges at brick interfaces caused their rigid sections to come into three points of contact with the channel walls, or if they exceeded the maximum allowable articulation angles between sections of control rod. Similarly, the requirement to maintain unimpeded movements of the fuel means that the channels within which they move must have sufficient bore size and straightness to avoid the fuel sleeves coming into three points of contact with the channel walls.

Common to all these requirements is the necessity for the channels to be continuous, and for them to be straight and open enough to enable the above functions to operate successfully when required. In addition, during fuel exchanges and when the core is under seismic excitation, the relative motion of the fuel sleeves and the fuel channels must be low enough to prevent gross impact damage. The primary means of maintaining core shape and limiting dynamic motion is through the combined actions of the radial keys and keyways (Figure 7). The functionality that they provide would be threatened by damage to the keys, the keyways and the bricks as the result of internal stresses, external loads or some combination of both, and by excessive distortions of the bricks. The extent to which such undesirable behaviour might occur and the time scales involved ultimately depend on how the graphite properties in the cores change with irradiation.

FIGURE 7: AGR System of Radial Keys and Keyways.

Consequently, assessments of changes to graphite properties and graphite component integrity, and the effects of potential damage to graphite components on the functionality of the safety systems, combined with regular in-service core inspections and continuous monitoring of core systems, are integral to the lifetime management of British Energy's fleet of AGR stations.

CONCLUSIONS

An outline has been presented of the main design and operating parameters of the Advanced Gas-cooled Reactors. The requirements placed on the core for their safe operation have been described and ways in which these might be affected by changes in the conditions of the graphite components through the reactor lifetimes has been discussed.

Acknowledgements
The author would like to thank British Energy Generation Ltd. for permission to present this paper.

An Overview of British Energy's Graphite Core Assessment Methodology

Mark R Bradford
Engineering Division, British Energy, Barnett Way, Barnwood, Gloucester, UK.
Email: mark.r.bradford@british-energy.com

Abstract
This paper presents an overview of the methodology employed by British Energy for the assessment of core functionality. It sets the context for the technical papers presented throughout this book which can be considered to be divided into three main areas:

- Material behaviour - The development of models of material behaviour, built upon irradiation data, which are then embedded within various assessment codes.
- Component behaviour - Stress analyses of components determine the predicted likelihood of degradation at a given point in reactor life.
- Whole core behaviour - The use of predicted, measured, or assumed component configurations within models that investigate the whole core response to imposed loading condition.

This paper briefly describes how these components are built up into the core assessment methodology, and how rig experiments and reactor inspection and monitoring data are utilised.

Keywords
Assessment methods, AGR, safety case

INTRODUCTION

The fundamental nuclear safety requirements of the graphite core in the Advanced Gas-cooled Reactor (AGR) are:

- To allow the unimpeded movement of control rods and fuel; and
- To direct the flow of coolant gas so as to ensure adequate cooling of the fuel and the core structure, both in normal and fault conditions.

Both of these requirements require the fuel and control rod channels to be adequately straight and free from significant discontinuities and obstructions. There are also:

- Limits on the allowable amount of core weight loss by radiolytic oxidation; and
- A requirement to ensure safe handling of fuel.

In order to demonstrate compliance with the above requirements, inspection and monitoring are carried out. However, any safety case for operation also requires an assessment of the likely future core state and behaviour under normal operation and fault conditions. To this end, a comprehensive assessment methodology has been developed. The purpose of this paper is to give a "top-down" overview of that methodology and to place into context the following technical papers presented throughout this book.

OVERVIEW OF THE AGR ASSESSMENT METHODOLOGY

The Aspects of the Assessment Route

A flowchart of the assessment route is shown in Figure 1. The central portion of the Figure is concerned with the assessment methods themselves, whilst the divisions on the left demonstrate how the individual components fall under various broad headings, *i.e.*: Consequences; Core Behaviour; Component Behaviour and Material Behaviour. The section on the right of the figure relates to sources of supporting data, which will be considered later in this paper. The consequences of the core behaviour are not covered in significant detail at in this book, but the remaining aspects are aligned with the three main sections in this book.

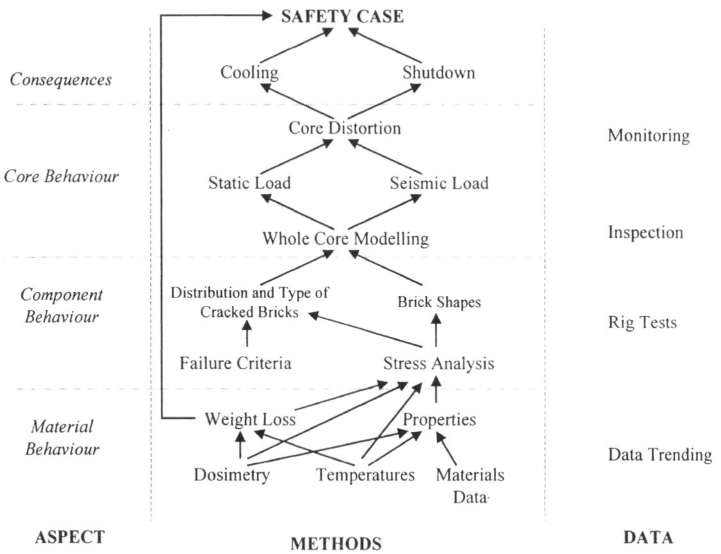

FIGURE 1: The overview of the AGR core assessment methodology.

If the methodology is considered from the top down, it can be seen that any challenge to the safety case fundamentals of adequate shutdown and fuel cooling capability arises as a consequence of core distortion. Consequently, it is necessary to understand the *Core Behaviour* both on a local and a global level. The core behaviour in turn depends on the individual *Component Behaviour* and the component behaviour depends on the *Material Behaviour* (*i.e.* the response of the graphite to the irradiation environment).

An Overview of the Individual Components of the Assessment Route

Core Behaviour

Currently, the assessment route is biased towards the calculation of bounding core distortions under static or quasi-static load (*i.e.* the loads that arise during normal operation, driven by thermal transients, changes in brick shape, *etc*) or as a result of the dynamic loads imposed by a seismic hazard. Both assessment routes make use of computational models of part, or the whole, of the reactor cores. These models are tested and validated against results from a number of rigs with representative components of various scales (from 1/8th scale to full size bricks) and sizes (from 3x3x2 to the whole core).

Component Behaviour

The whole core models require, as input, the number and distribution of components of varying state. The states can be intact or cracked and, if cracked, singly or multiply cracked with the cracks having a number of defined morphologies. For all states it is necessary to have an estimate of the associated brick shapes, particularly the clearances within the axial and radial keying systems that provide the core with its inherent stability.

Estimates of the temporal and spatial distribution of cracked components, as well as their morphology, are derived from 3-dimensional finite element-based stress analysis, combined with suitable failure criteria. The distribution calculations tend to be statistically based and make use of quantified uncertainties in the input data to the stress analysis calculations. The brick shapes are a standard output from the finite element analysis packages used. Validation of the models is then by comparison with reactor measurements.

Material Behaviour

Despite all its complexity, finite element stress analysis is simply a numerical toolbox - the strength or weakness of which lies in the quality of the input information. Within the AGR assessment methodology, this input information is the models of the material behaviour under the imposed loading conditions - in this case fast neutron irradiation and radiolytic oxidation within the constrained geometry of a reactor moderator brick. The calculations therefore require material properties models, together with information on the irradiation conditions (temperature, dose and oxidation). Each of these input variables is itself the output of an associated calculational method, with its own requirement for validation against rig and plant data and support from fundamental research.

The Use of Bounding Assessments within the Calculational Route

Before moving on to consider the sources and use of supporting data, a few words need to be said regarding the use of bounding assessments and assumptions within the calculational route. It has already been noted that some assessments derive bounding states. This not only applies to whole core behaviour, but also to stress analysis aimed at defining the earliest times to onset of certain modes of moderator brick cracking. These approaches have been used for many years where data is sparse or uncertain and can, under the right circumstances, provide confidence that safety assessments are conservative. However, they are not suitable for defining true core safety margins and, consequently, there is now a move towards the use of complementary best estimate analysis, with defined uncertainty margins.

Sources and Use of Supporting Data

Four sources of data to support the assessment route will be considered here: Monitoring, Inspection, Rig Tests and Trending.

Monitoring

Monitoring data is defined as that which is gathered during the normal day-to-day operation of the plant. It includes information such as the recording of the movements of control rods, the loads experienced by the fuel stringer tie-bar during refuelling, and comparison of thermal and neutronic power. The practicability of on-line methods of determining core distortion is also being investigated.

From this list it can be seen that Monitoring data naturally feeds into the assessments towards the top (safety case) end of the route (Figure 2).

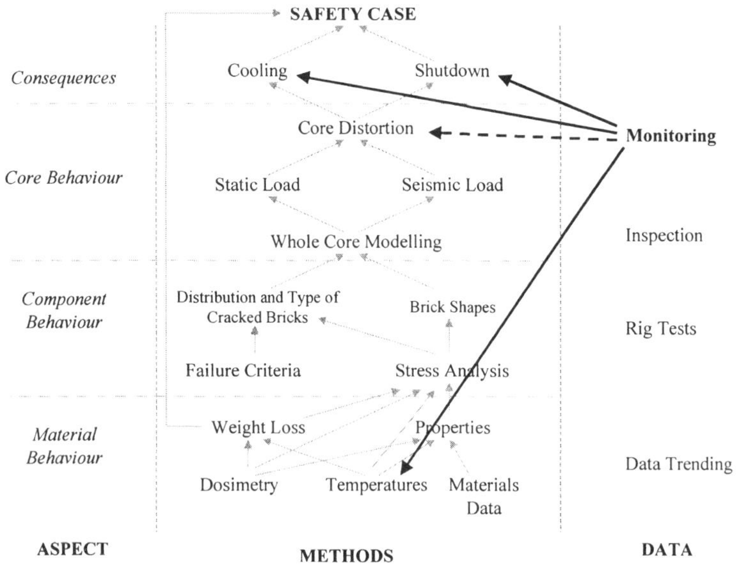

FIGURE 2: The use of monitoring data within the AGR core assessment methodology

Inspection

Inspection data is defined as that which is gathered during periodic shutdowns. It includes:

- Measurement of shrinkage of bricks and distortion of fuel channels with a Channel Bore Measurement Unit;
- Visual inspection using a video camera; and
- Sampling of irradiated graphite, either trepanning of small cores of graphite from the moderator bricks or removal of specifically designed sample carriers from interstitial channels.

Inspection data is therefore of use throughout the whole of the assessment route (Figure 3), providing information on core distortion, brick shapes, the distribution and type of cracked bricks, weight loss and graphite properties.

Rig Tests

Rig tests tend to be used to provide information targeted at a specific aspect of the assessment methodology (Figure 4). For example:

- In the early stages of the gas-cooled reactor programme, Materials Test Reactor (MTR) experiments were used to provide information on material properties for design basis calculations. With the resurgence of interest in graphite moderated cores for future reactor designs, new programmes are being commissioned and they may also have a role to play in determining plant lifetimes.

- Tests on the strength of unirradiated moderator bricks were conducted to define the failure criterion currently used.
- A number of rigs of varying size have been used to provide data to validate whole core response models.

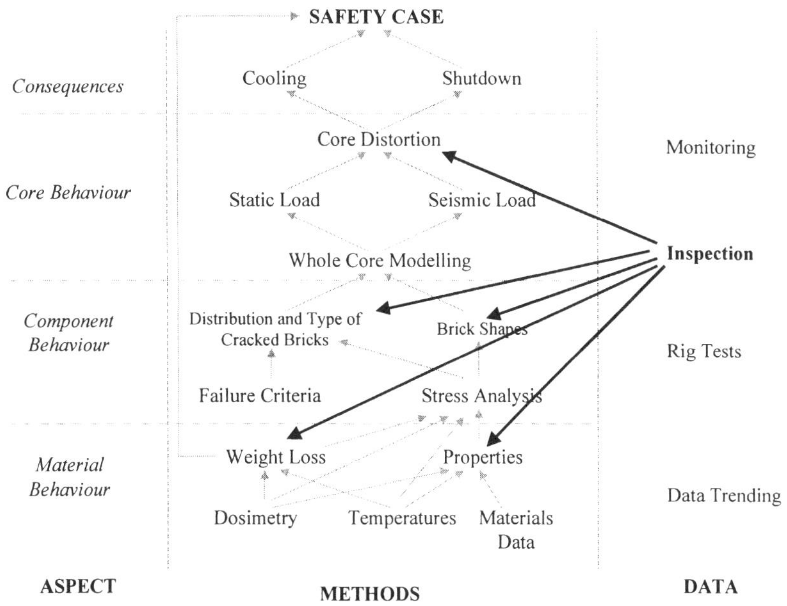

FIGURE 3: The use of inspection data within the AGR core assessment methodology.

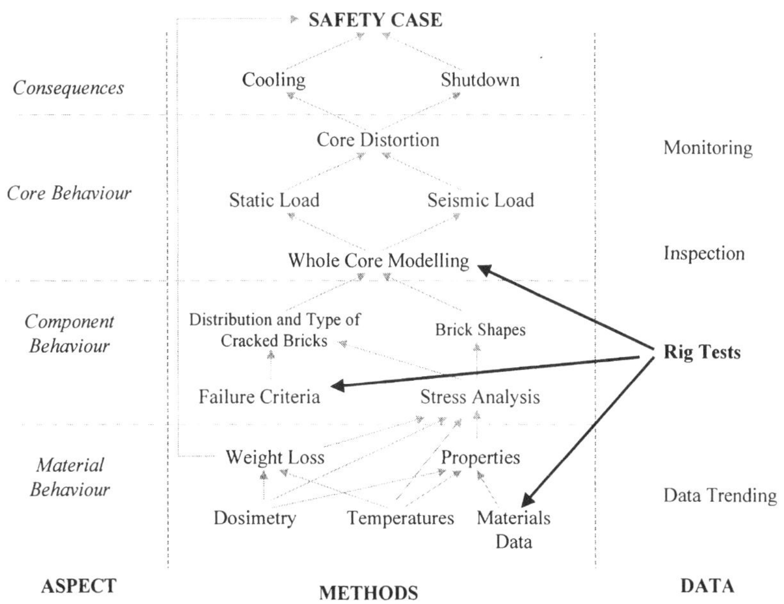

FIGURE 4: The use of rig data within the AGR core assessment methodology.

Data Trending

In addition to the prediction of future core state by means of detailed calculational methods, it is also possible to gain an insight into likely future behaviour by the trending of operational and inspection data (Figure 5).

The trending of operational data is an intrinsic component of core monitoring, the trending of inspection data is an aspect of the assessment route that is in its infancy. Currently, it is used to estimate future levels of moderator brick bore cracking at some stations. However, the possibility of extending the techniques to other data such as weight loss and material properties is currently being investigated.

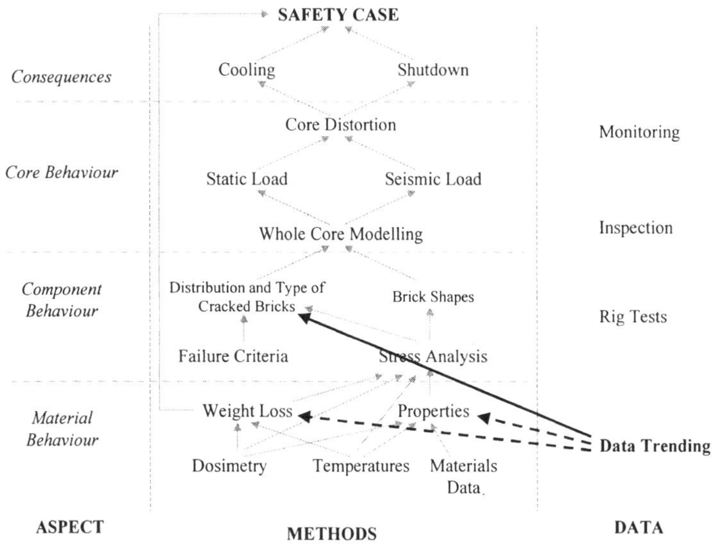

FIGURE 5: The use, and potential use, of data trending within the AGR core assessment methodology.

CONCLUSIONS

This paper has presented an overview of the methodology employed by British Energy for the assessment of core functionality. It sets the context for the technical papers presented throughout this book dealing with the "Ageing Management of Graphite Reactor Cores".

Acknowledgements

The author would like to thank British Energy Generation Ltd for permission to present this paper.

UK Regulatory Strategy for Management of Graphite Ageing in Gas-Cooled, Graphite-Moderated Reactors

Graham B Heys
HM Principal Inspector (Nuclear Installations),
Health & Safety Executive, HM Nuclear Installations Inspectorate, St Peter's House, Balliol Road, Bootle, Merseyside, UK.
Email: graham.heys@hse.gsi.gov.uk

Abstract
The UK Nuclear Installations Inspectorate's regulatory strategy for management of ageing of nuclear graphite core components in the operating Magnox and Advanced Gas-cooled Reactors will be described. The approach adopted to secure continuing adequate and ALARP safety cases for the graphite cores will be considered in the context of the UK's goal setting, non-prescriptive, regulatory regime. The role of independent advice and research will be explained.

Keywords
Regulation, strategy, graphite-ageing

INTRODUCTION

The Health and Safety Executive's (HSE) HM Nuclear Installations Inspectorate (NII) is the body in the UK tasked with regulation of activities at nuclear licensed sites. Graphite components and structures in gas-cooled reactors are increasingly of greater significance as a result of predictions and observation of cracked components. In response a regulatory strategy was developed to ensure the continued safety of these reactors. The objectives of the regulatory strategy are to: provide more information, increase knowledge and understanding of ageing-processes in nuclear graphite; to monitor ageing-processes by enhanced inspection, surveillance and monitoring programmes; to mitigate ageing by reasonably practicable plant modifications supported by a clear understanding of the limit of tolerability for component and core defectiveness.

HEALTH AND SAFETY LEGISLATION AND REGULATORY FRAMEWORK

The main legislation governing the safety of employees and the general public at nuclear installations is the Health and Safety at Work, etc. Act 1974 (HM Government, 1974) and associated relevant statutory provisions. The Nuclear Installations Act 1965, as amended, (HM Government, 1965) is a relevant statutory provision and gives HSE the power to attach to the nuclear site licence such conditions that may appear necessary or desirable in the interests of safety. There are 36 such licence conditions that enable the NII to exercise regulatory control of operations on a nuclear licensed site. These licence conditions require safety cases to be produced for all activities that may affect safety on the site. The legislation is non-prescriptive; NII does not prescribe detailed standards or codes of practice for nuclear plant. Responsibility for the safe design, construction, manufacture, modification, commissioning and operation of nuclear installations rests with the licensee; that is with the station operator. The licensee is required to formulate the design safety criteria and standards, and the construction, commissioning and operating arrangements and procedures that will be used.

The duty of the NII is to see that appropriate standards are developed, achieved and maintained by the licensee, to ensure that the necessary safety precautions are taken and to regulate and monitor the safety of the plant by means of its powers under the licence. NII sets safety goals and the aim is for licensees to set out how they will meet these goals. NII examines how these are met by the licensees and takes enforcement action where necessary. NII seeks to satisfy itself that licensees have the organisation for this and that they are carrying out their functions effectively.

REGULATORY GUIDANCE ON MANAGEMENT OF AGEING OF NUCLEAR POWER PLANT COMPONENTS

HSE publication "Reducing Risks, Protecting People" (HSE, 2001), also known as R2P2, sets out the overall framework for decision making by HSE to ensure consistency and coherence across the full range of risks falling within the scope of the Health and Safety at Work Act. This framework is based upon the method that HSE applies to control the risk at nuclear power stations, originally published in 1988 as "The Tolerability of Risks from Nuclear Power Stations" (HSE, 1992), also known as TOR. These publications present HSE's position on the tolerable levels of individual and societal risk to workers and the public from nuclear power stations. R2P2 and TOR explain the ALARP (As Low As Reasonably Practicable) principle that overarches regulation of the industry to ensure that licensees comply with the Health and Safety at Work Act and relevant statutory provisions.

The NII has developed a set of Safety Assessment Principles (SAPs) (HSE, 1992) as guidance to inspectors for assessment of safety cases. These were developed primarily for NII staff, but they indicate to the public what standards HSE expects to see in designs for new nuclear plant. Whilst the SAPs were developed for new nuclear plant the SAPs are also applied to 'old plant', or those that exist today. The ALARP principle is of particular importance to such assessments; and the age of the plant and its projected life are important factors to be taken into account when judging the reasonable practicability of safety improvements. In addition to the SAPs, Technical Assessment Guides (TAGs) are available that provide guidance to NII inspectors on the interpretation and application of the SAPs.

The International Atomic Energy Agency (IAEA) has produced guidance for the development of strategies to manage ageing of nuclear power plant components (IAEA, 1991, 1992, 1999). Generic methodologies are proposed for management of ageing of structures, systems and components in nuclear facilities and recommend that an effective "ageing management" programme should address three essential elements:

- Understanding of the ageing degradation processes.
- Monitoring of ageing that is capable of detecting component degradation before failure.
- Timely mitigation of ageing and its effects (*e.g.* through maintenance, replacement, or changes in operating conditions) to ensure that required safety margins are maintained.

The IAEA (IAEA, 1999) highlights the importance of having a "systematic ageing management programme" and presents a process for its management. In the UK The Management of Health and Safety At Work Regulations (HM Government, 1999) and associated Approved Code of Practice and Guidance (HSC, 2000) are relevant to management of health and safety risks. As graphite components and core structures age the health and safety risks may increase if not properly managed and so an effective programme for management of ageing in accordance with Regulation 5 of The Management of Health and Safety At Work Regulations that includes effective planning, organisation, control, monitoring and review of preventative and protective measures is necessary. HSE (HSE, 1996) has also provided non-statutory guidance on the management of health and safety at nuclear installations. Whilst it is not compulsory to follow all aspects of this guidance it represents the NII's view of what constitutes good practice in the management of health and safety at nuclear installations. In considering the elements of effective planning, examples of standards used on nuclear plant include those developed by the IAEA.

NII REGULATORY STRATEGY FOR MANAGEMENT OF GRAPHITE AGEING

Consideration of Current Position on Nuclear Graphite Technology

Graphite components and structures in gas-cooled reactors are increasingly of greater significance as a result of predictions and observation of cracked components. The only structural applications of graphite tend to be in the core, and associated components, of gas-cooled reactors. Such applications tend to be onerous in terms of material performance due to the very high radiation doses that some graphite components may experience in particular where replacement is not practicable. There is an absence of adequate reliability data for nuclear graphite structural components and as a result assessment is based on established engineering practice. In addition, graphite structures in gas-cooled reactors often tend to have many components and are of relatively loose construction. Due to differences in design and safety function these complex structures in some instances may be quite defect tolerant although in others safety functions may exhibit low defect[2] tolerance.

Established engineering practice, used to predict the performance and behaviour of graphite components and structures, tends by necessity to be highly complex with many inputs and interactions. Unlike, for example, structural integrity methodologies for steel pressure vessels these methodologies are licensee bespoke, empirically based, and often lack in-depth understanding of graphite behaviour exacerbating the need for bounding data. In-service volumetric inspection has not yet been developed for graphite components and so inspection must rely on visual techniques and mensuration.

Underlying data scatter is often significant and, when combined with the complexity of the methodologies, and uncertainty, results in many possible conjectured predictions of the conditions and behaviour of graphite components and structures. This may lead to uncertainty in the likelihood and consequences of challenge to safety functions of graphite components and structures.

Regulatory Guidance on Handling Uncertainty

HSE publication, R2P2, (HSE, 2001) discusses ways of handling uncertainty in line with the Precautionary Principle. The Precautionary Principle describes the philosophy that should be adopted for addressing hazards subject to high scientific uncertainty, and rules out lack of

---

[2] The definition of "defect" for a graphite component or structure is taken to mean any deviation from nominal and covers dimensional tolerance due to shrinkage, crack-like defects etc.

scientific certainty as a reason for not taking preventative action. The Precautionary Principle is normally invoked in assessment of graphite core safety cases.

Strategy Development
Given these significant age-related phenomena NII need to decide whether graphite core safety cases are adequate and ALARP. To enable NII to address these issues and to ensure that regulatory decisions are soundly based, proportionate to the risk, consistent, and transparent, NII undertook a strategic review of the technology supporting graphite core safety cases. The first stage of this undertaking was to perform a review of nuclear graphite technology to identify its limitations (Marsden, 2003), referred to later as the Technology Review. The second stage was to develop a regulatory strategy to ensure effective regulation and influence on licensees' strategies for management of ageing of graphite reactor cores.

The output from the Technology Review and experience of recent assessments of graphite core safety cases were reviewed and following a debate with NII's graphite Nuclear Topic Group (NTG) four strategic objectives were agreed:

1. Develop a revised Technical Assessment Guide (TAG) to provide a clear framework for judging the adequacy of graphite core safety cases. The objective of the revised guide was to ensure consistency in the assessment of safety cases and clarity of NII's expectations.
2. Secure the maintenance and improvement of graphite core safety cases for operating reactors to ensure that risks remain As Low As Reasonably Practicable (ALARP).
3. Develop and maintain a credible and cost effective graphite research programme for operating reactors to reduce uncertainty by improving knowledge and understanding.
4. Ensure that NII is well informed on key technical issues through independent advice and independent peer review. The objective was to enable NII to make sound regulatory decisions and improve understanding of the strengths and weaknesses of graphite core safety cases.

A regulatory strategy was developed to achieve these four objectives and is discussed further below.

Progress with Achievement of Strategic Objectives

Revised Assessment Guidance, Strategic Objective 1
A revised Technical Assessment Guide (TAG) (HSE/NSD, 2005) was recently approved to enable an inspector to judge the adequacy of a graphite-core safety case where component defectiveness is predicted or observed. This TAG has a section on management of ageing in graphite reactor cores. The TAG recognizes that the graphite core safety case may be a life-limiting feature for some of the graphite-moderated gas-cooled reactors and there is therefore a need to consider whether it is appropriate for the licensee to have a core ageing management strategy and programme in place to support continued operation. This is particularly important in the context of Periodic Safety Reviews (PSRs) that require identification of ageing processes that may limit the life of the plant.

The TAG considers the effect of uncertainty and the contribution of each leg of the safety case to its overall adequacy and identifies likely conditions to define the limit of tolerability. British Energy and British Nuclear Group were consulted to seek their views on the graphite

TAG both at conception and final draft stage. Peer review was undertaken both by NII specialists, in particular the NII graphite NTG, and the NII Graphite Technical Advisory Committee (GTAC). During the development of the revised guidance the approach was presented to the licensees' and was effective in securing significant improvements in safety cases, their structure and content.

Development of the TAG

A fundamental review of the approach adopted for assessment of graphite core safety cases was undertaken. To mitigate uncertainty and complexity, a multi-legged approach to safety cases for graphite components and structures, based upon independent and diverse arguments, is expected. The rigour of application and robustness of the supporting data and information should be based upon: the safety categorisation of the graphite components and structures; the potential radiological consequences of their failure; and the perceived or actual proximity of a challenge to graphite component and structure safety function. The multi-legged arguments when taken together with the various elements of established engineering practice provide defence-in-depth.

The graphite TAG presents the following possible legs of a graphite core safety case as:

- Design.
- Manufacture, construction and commissioning (MCC).
- Component and core condition assessment (CCCA).
- Defect tolerance assessment (DTA).
- Analysis of consequences of core ageing on fuel and core cooling, shutdown, hold-down and fuel integrity.
- Monitoring.
- Examination, inspection, surveillance, sampling and testing (EISST).

Whilst it is possible to establish other legs for a graphite core safety case this structure was selected as it presents each of the separate contributors to the safety case in a form such that it is straightforward to assess their contribution to ALARP. Each leg potentially makes an independent contribution to achievement of safety.

Essentially the design, which in many respects mirrors the CCCA leg, defines the as-designed function of the core and its components. This should enable the assessor to understand the detailed function of the core design and therefore more easily able to establish if changes in operation or functionality of components may challenge core safety function. The MCC leg of the safety case defines the as-constructed condition of the core in terms of component as-manufactured properties, component defectiveness, manufacturing and design concessions, component clearances, *etc.* The CCCA leg of the safety case is the predictor of future component and core behaviour. The DTA should assess the tolerance of core safety functions to component defectiveness whereas the consequences leg assesses the radiological consequences of any assumed or predicted defects. Essentially the difference is that if the DTA leg demonstrates tolerance then any consequences leg is supportive and a redundant and diverse argument. Monitoring, defined in the TAG as "... the gathering of information on plant condition or operating parameters by means of installed or other equipment or systems during plant operation", provides means by which core safety functions and any challenges may be detected and responded to during operation rather than having to wait until a shutdown to examine component and core condition. EISST, defined in the TAG as "...the gathering of information on plant condition or operating parameters by means of activities undertaken during a reactor shutdown such as a periodic shutdown", provides information on how the actual components and core are behaving.

A key step in the assessment process for an apparent multi-legged safety case is to establish if each leg is independent, for example any assumptions are not derived from other legs of the safety case. The most robust case would be one where all legs are independent and there is diversity in the arguments presented. If more than one leg is underpinned by common data, assumptions or methodologies then the strength of a multi-legged case may be undermined as there is no longer independence in the arguments presented.

Improvements in Graphite Core Safety Cases, Strategic Objective 2
Strategic Objective 2 was to secure the maintenance and improvement of graphite core safety cases for operating reactors to ensure that risks remain ALARP. A number of actions were identified to achieve this:

- Seek further enhancement of examination, inspection, surveillance, sampling and testing programmes.
- Seek enhanced monitoring and trending of core behaviour.
- Investigate the scope, when and where appropriate, for ALARP plant modifications and methods of operation.
- Seek validation of all key inputs and outputs of the methods adopted in the CCCA leg of the safety case if reasonably practicable.
- Establish the sensitivity of predictions of component condition to variation in assumptions, data and methods.
- Obtain materials data that bounds future operation.
- Establish the limit of tolerability in terms of numbers and morphology of cracked bricks their distribution, frequency and location (or configuration).

The licensees have already made good progress in achieving these goals.

Research, Strategic Objective 3
The Health and Safety at Work Act places a duty on the Health and Safety Commission to make arrangements for carrying out research and also to encourage research by others. The Nuclear Installations Act enables HSE to recover from the licensees' expenses incurred in connection with the carrying out of nuclear safety research. Under the current arrangements NII produces annually a Nuclear Research Index (NRI) that identifies nuclear safety research requirements, and expected outcomes, for each technical area (graphite being one of them). The reactor licensees subsequently, in advance of each financial year, produce a Nuclear Safety Research Schedule that details the research projects that they propose to fund and how these will enable the NRI requirements to be addressed and enable subsequent closure. NII agrees this programme and monitors progress throughout the year. NII may also fund research under the HSE Levy programme for maintenance of independent capability or where it is judged necessary to complement or undertake additional research to the licensees'.

In 2005 major amendments were made to the graphite section of the NRI to work towards achieving Strategic Objective 3. Research is specifically needed to:

- Improve the fundamental understanding of ageing processes and cracking in graphite components to help in prediction of component and core condition.
- Obtain materials data that bounds future operating conditions.
- Establish limiting component, channel, and core condition or limit of tolerability and associated safety margins.
- Underpin existing safety cases by improved understanding.
- Improve the capability of existing inspection and monitoring techniques and develop improved techniques.

NII is working towards establishing collaborative graphite research programmes with those developing new gas-cooled reactors. The first meeting to discuss collaboration took place in January 2006.

Independent Advice, Strategic Objective 4
Arising from the Technology Review a decision was taken to establish the Graphite Technical Advisory Committee (GTAC) to provide advice to NII. The key objectives of GTAC are:

- To provide NII with timely technical advice on issues associated with graphite core integrity.
- To provide answers to complex problems where there is currently judged to be significant knowledge uncertainty and limited predictability.
- To ensure that NII is well informed on key technical issues and therefore able to make sound regulatory decisions.
- To mitigate uncertainty.
- To improve NII's understanding of the strengths and weaknesses in analytical aspects of graphite core safety cases.

The first meeting of GTAC was held in January 2004. NII asked GTAC to provide answers to a programme of 23 questions on all aspects of graphite component and core data and methodologies, their uses and interpretation. This programme of 23 questions is scheduled was completed in April 2006.

CONCLUSIONS

The NII's regulatory strategy to manage ageing of graphite components and structures in gas-cooled graphite-moderated reactors is described. The key elements are: improved guidance to ensure consistency and proportionality in assessment of safety cases; to obtain more information, knowledge and understanding of graphite component and core ageing through enhanced inspection and monitoring and research; to mitigate ageing by reasonably practicable plant modifications supported by a clear understanding of the limit of tolerability for component and core defectiveness.

As a result of this regulatory intervention strategy significant ALARP improvements in safety cases have already been achieved. Further improvements are anticipated through enhanced inspection, monitoring, and research thereby ensuring that the risks continue to be ALARP as graphite component ageing progresses. Independent advice and research is ensuring that NII is well informed on key technical issues and therefore able to make sound regulatory decisions.

Acknowledgements
The views expressed in this paper are those of the author and do not necessarily represent the views of HSE or HM Nuclear Installations Inspectorate.

References

HM Government (1965). The Nuclear Installations Act 1965. Her Majesty's Stationery Office.

HM Government (1974). Health and Safety At Work etc Act, ISBN 0 11 141439 X. Her Majesty's Stationery Office.

HM Government (1999). The Management of Health and Safety At Work Regulations 1999, ISBN 0 11 085625 2, Her Majesty's Stationery Office.

Health & Safety Commission (HSC) (2000). Management of Health and Safety At Work, Management of Health and Safety At Work Regulations 1999, Approved Code of Practice and Guidance, ISBN 0 7176 2488 9, Her Majesty's Stationery Office.

Health & Safety Executive (1992). The Tolerability of Risk From Nuclear Power Stations. Her Majesty's Stationery Office. ISBN 0 11 886368 1.

Health & Safety Executive (1992). Safety Assessment Principles For Nuclear Plants. Her Majesty's Stationery Office. ISBN 0 11 882043 5.

Health & Safety Executive (1996). Managing For Safety At Nuclear installations, ISBN 0 7176 1185 X, HSE Books.

Health & Safety Executive (2001). Reducing Risks, Protecting People, HSE's Decision Making Process. HSE Books. ISBN 0 7176 2151 0.

Health & Safety Executive, Nuclear Safety Directorate (2005). NSD Business Management System, Technical Assessment Guide, Graphite Reactor Cores, T/AST/029 Issue 002.

International Atomic Energy Agency (1991). Data Collection and Record Keeping for the Management of Nuclear Power Plant Ageing, Safety Series No. 50-P-3, IAEA, Vienna.

International Atomic Energy Agency (1992). Methodology for the Management of Ageing of Nuclear Power Plant Components Important to Safety, Technical Reports Series No. 338, IAEA, Vienna.

International Atomic Energy Agency (1999). Implementation and Review of a Nuclear Power Plant Ageing Management Programme, Safety Report Series No. 15, IAEA, Vienna.

Marsden, B J. (2003). AGR and Magnox, Nuclear Graphite Technology, Strategic Review, Report reference NGRG R117 Issue 1. Manchester School of Engineering, The University of Manchester.

PART B

Advances in the Fundamental Knowledge of Graphite Behaviour under Irradiation

The Effects of Thermal Annealing on the Mechanical Properties of PGA Graphite

Martin P Metcalfe and John F B Payne
Nexia Solutions Ltd, Berkeley Centre, Berkeley, Gloucestershire, UK.
Email: martin.p.metcalfe@nexiasolutions.com, john.b.payne@nexiasolutions.com

Abstract
The two principal graphite ageing processes in a Magnox reactor core are fast neutron damage and radiolytic graphite oxidation. Graphite strength, Young's modulus and thermal conductivity are affected by both these processes. Relationships describing the separate effects of these two processes on PGA graphite have been derived from Materials Testing Reactor data. The combined effect of the two processes is assumed to be multiplicative in accordance with the so-called "Product Rule". At sufficiently low doses, property changes due to fast neutron irradiation are caused principally by dislocations in the graphite crystal lattice produced by the displacement of carbon atoms. These dislocations can be removed by thermal annealing, giving rise to recovery of graphite properties. The success of the Product Rule in describing the combined effects of irradiation damage and oxidation on strength and Young's modulus has been tested on PGA graphite by the use of thermal annealing. Results from early Materials Testing Reactors are briefly reviewed, and new data for material irradiated recently in the Idaho National Laboratories Advanced Test Reactor are presented together with baseline measurements made on virgin PGA graphite.

Keywords
Thermal annealing, PGA graphite, mechanical properties

INTRODUCTION

Relationships describing the individual ageing processes of fast neutron irradiation and radiolytic graphite oxidation for Pile Grade 'A' (PGA) graphite used in Magnox reactors have been derived from Materials Testing Reactor (MTR) data. The MTR data for radiolytic oxidation were derived by removal (by thermal annealing) of effects due to fast neutron irradiation. The combined effect of these two ageing processes is assumed to be multiplicative in accordance with the so-called "Product Rule". The Product Rule assumption is central to the compilation of predicted material properties for core structural integrity assessments. Recent studies on behalf of British Nuclear Group of the behaviour of PGA graphite strength with weight loss have employed a method of enhancement of radiolytic weight loss with thermal oxidation to achieve target weight losses. The temperatures necessary for thermal oxidation are such that some thermal annealing takes place. Samples were subjected to controlled thermal annealing to complete this process and to permit a direct comparison of strength test results with MTR data for radiolytic oxidation. The work presented here investigates the effectiveness of thermal annealing to remove radiation damage and the success of the Product Rule in describing the combined effects of fast neutron irradiation and radiolytic oxidation.

IRRADIATION DAMAGE

Irradiation damage in graphite has been reviewed extensively (see for example Kelly, 1981). Neutron collision damage in graphite results in the formation of structural defects, such as displaced atoms, vacancies and higher-order aggregates that raise the internal energy. The subsequent interaction of these defects with each other and the surrounding hexagonal lattice also causes dimensional distortion and physical property changes (Ewels *et al.*, 2003). A stacking fault, enclosed by vacancy or interstitial loops, is formed when atoms dislodged from the basal planes form a small new section of basal plane intercalated between the original basal planes. The new basal planes grow as irradiation proceeds. For a single crystal, the process of intercalation of additional basal planes could continue indefinitely, causing crystal growth perpendicular to the basal planes and shrinkage parallel to the basal planes. In polycrystalline graphite, these changes in crystal dimensions can be accommodated by voids in the material, until the voids are closed.

The elastic properties of polycrystalline material are mainly determined by the shear modulus of the crystals which is most affected by basal dislocations. The gliding of these dislocations produces the elastic shear in the crystal in response to an applied shear stress. Irradiation-induced lattice defects tend to "pin" the basal dislocations, reducing the elastic strain produced by a given stress. This contribution to the change in polycrystalline graphite properties with irradiation is referred to as the "pinning term". The population of lattice defects will saturate before the average atom has been displaced once (corresponding to Equivalent DIDO Nickel Dose (EDND) of ~7×10^{20} neutrons cm^{-2}) hence the "pinning term" saturates also. When the lattice defects are removed by thermal annealing, the basal dislocations are unpinned, and the crystal shear modulus returns to its lower, unirradiated, value.

The elastic properties of polycrystalline material also depend on the relation between the large scale stresses and the shear stresses on the component crystals. This relation can be changed by irradiation, since the microstructure of the material will change as the crystal dimensions change. This contribution to the change in polycrystalline graphite properties is referred to as the "structure term". The overall change in Young's modulus of polycrystalline graphite is thus the product of the "pinning" and "structure" terms. Once large new sections of basal plane have grown, they are unlikely to be removed by annealing, so the "structure term" contribution to Young's modulus is unlikely to be removed by thermal annealing.

The fluence and temperature at which irradiation damage begins to change the structure term will determine whether thermal annealing can successfully be used on graphite from Magnox reactor cores to remove the effects of fast neutron irradiation on mechanical properties. Radiolytic oxidation of graphite may be regarded as changing the "structure term" and of course its effect is not removed by annealing. Provided the contribution of fast neutron irradiation to the change in the structure term is much less than the effect of radiolytic oxidation, the oxidation and fast neutron irradiation induced effects will be the product of the "pinning" and oxidation controlled "structure" terms.

In Gilsocarbon graphite used in Advanced Gas-cooled Reactors, turn-round in dimensional change rate reflecting the increasing effect of irradiation on the structure term is delayed by graphite oxidation and appears to take effect at EDND between ~60 and ~80×10^{20} neutrons cm^{-2} for 13% and 23% oxidised material respectively. Such a transition point is not likely to be dissimilar for oxidised PGA graphite and, noting that lifetime fast neutron fluence in Magnox cores will be $<60 \times 10^{20}$ neutrons cm^{-2} (EDND), any contribution from the structure term is not expected to be significant.

MTR INVESTIGATIONS OF THERMAL ANNEALING

Simmons (1965) reviews the effects of thermal annealing on Young's modulus. The UKAEA conducted a series of studies on the effects of thermal annealing of graphite, including an investigation of the behaviour of highly crystalline well-oriented pyrolytic graphite and PGA graphite at temperatures between 150 °C and 350 °C and fluence in the range 7.7-18.5×10^{20} neutrons cm^{-2} (EDND) (Kelly *et al*, 1966). The observed behaviour of Young's modulus of PGA graphite is illustrated in Figure 1. The changes in modulus produced by annealing are attributed to pinning of basal dislocations. For these fluences and irradiation temperatures, recovery in Young's modulus was largely complete at thermal annealing temperatures of ~1500 °C. A more complicated and unexplained behaviour was observed at higher annealing temperatures with an apparent rise in Young's modulus when annealed at higher temperatures.

Later studies on extruded reactor grade graphite (Matsuo and Honda, 1973) confirmed the modulus recovery on annealing observed by Kelly *et al.* (but not the unexplained rise at higher annealing temperatures). As illustrated in Figure 2, thermal annealing at temperatures >2000 °C for one hour resulted in >90% recovery of Young's modulus. Although thermal neutron fluences are quoted, assuming a 2:1 ratio of fast to thermal for the neutron spectrum of a light water reactor, the two higher fluences are likely to be sufficient to saturate pinning.

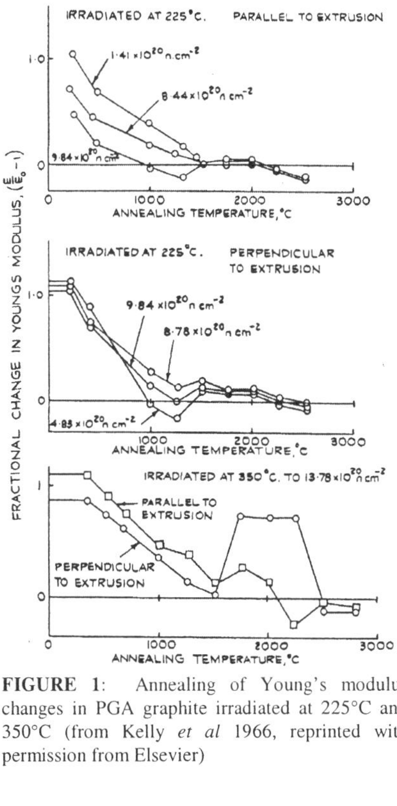

FIGURE 1: Annealing of Young's modulus changes in PGA graphite irradiated at 225°C and 350°C (from Kelly *et al* 1966, reprinted with permission from Elsevier)

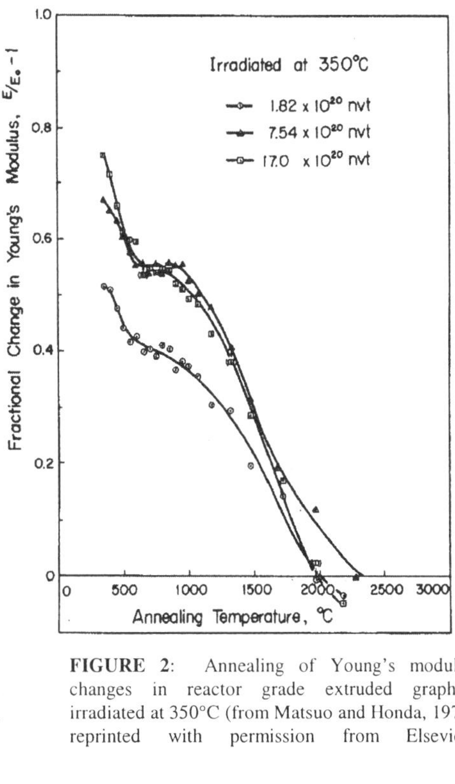

FIGURE 2: Annealing of Young's modulus changes in reactor grade extruded graphite irradiated at 350°C (from Matsuo and Honda, 1973, reprinted with permission from Elsevier)

Despite the unexplained behaviour illustrated in Figure 1, the work by Kelly *et al* (1966) provided the basis for defining and justifying the annealing conditions adopted in later MTR experiments which generated the current PGA database for radiolytic oxidation. The thermal annealing conditions adopted by the UKAEA to obtain strength and modulus data for PGA graphite with fast neutron damage removed were heating in a tube furnace for approximately one hour at temperatures in the range 1800-2000 °C. For the current work, samples were heated in a furnace with an inert atmosphere for one hour at approximately 2100 °C. The irradiation conditions of these treated samples are detailed as appropriate in the discussion below.

CURRENT THERMAL ANNEALING STUDIES

The studies presented comprise three parts: the effect of thermal annealing on the Young's modulus and compressive strength of virgin graphite and the effects of thermal annealing on the Young's modulus only of irradiated graphite with and without oxidation. The absence of thermal oxidation of samples during the annealing process was confirmed by weighing samples before and after treatment.

Effect of Thermal Annealing on Virgin Graphite

The purpose of this first stage of the investigation was to assess what effect if any thermal annealing under the chosen conditions had on virgin PGA graphite. 46 control specimens (ø19 × 19 mm) from three Bradwell graphite heats were tested. All 46 samples underwent DYM measurements (ultrasonic time-of-flight method). 22 out of the 46 samples were then annealed at 2100 °C for 1 hour in an inert environment and subjected to DYM and compressive strength tests post-annealing. The 24 non-annealed samples were subjected to compressive strength tests for comparison with the corresponding annealed values. The results are shown in Figures 3 and 4.

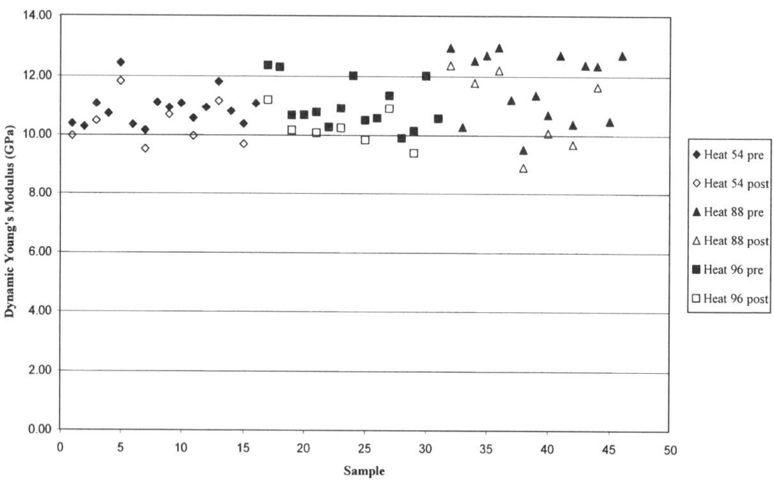

FIGURE 3: Effect of thermal annealing on the DYM of virgin PGA Graphite.

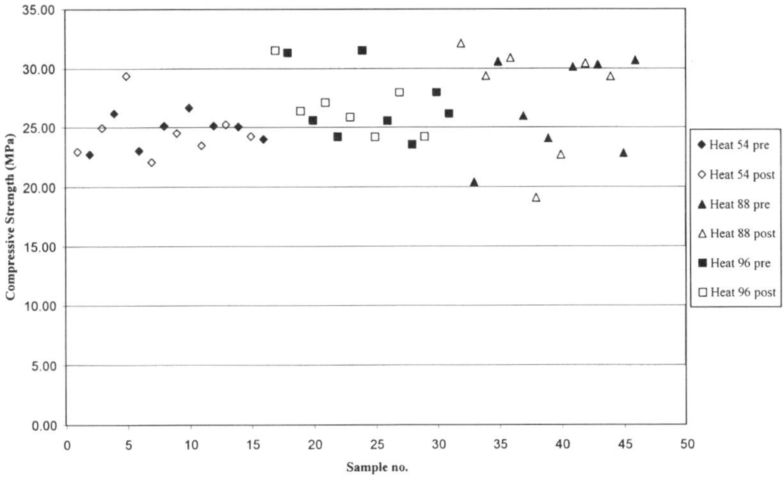

FIGURE 4: Effect of thermal annealing on the compressive strength of virgin PGA graphite.

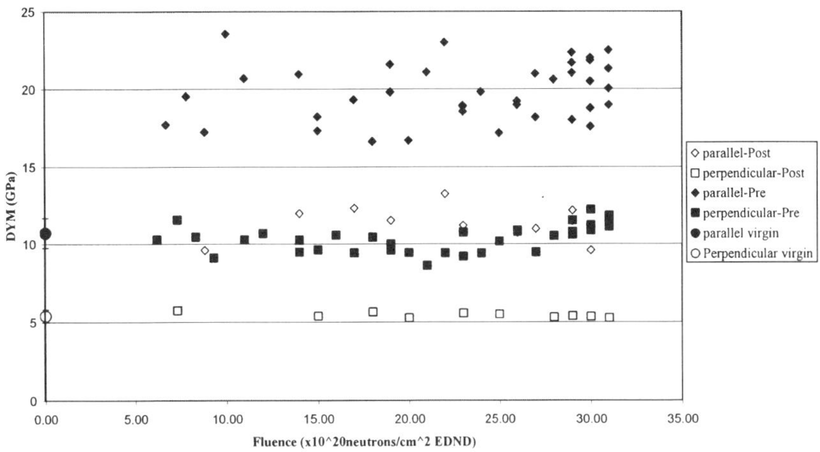

FIGURE 5: Effect of thermal annealing on the DYM of PGA graphite irradiated in an inert environment.

This investigation has shown that the annealing process has a small, but noticeable effect on the measured DYM values. Specifically, the DYM values appear to have reduced by 0.5-0.7 GPa compared to the virgin un-annealed values. Compressive strength does not show a similar trend. However, this could be due to the larger scatter of the strength measurements, as the means of the non-annealed samples were 24.71 ± 1.40 MPa, 26.95 ± 3.04 MPa and 26.84 ± 4.09 MPa for heats 54, 96 and 88, respectively, with the corresponding mean strengths of the annealed samples being 24.58 ± 2.20 MPa, 26.72 ± 2.52 MPa and 27.68 ± 4.85 MPa. Overall, these results show that there is no significant change in virgin strength on annealing. If a small reduction in strength had been observed (based upon the observed small reduction in DYM) then strengths measured post-annealing would be conservative.

Effect of Thermal Annealing on Irradiated Un-Oxidised Graphite

PGA graphite samples have been irradiated in the Idaho National Laboratory Advanced Test Reactor (ATR). Approximately 70 pairs of samples were irradiated, one of each pair in an inert atmosphere, and one in carbon dioxide with typical Magnox reactor levels of hydrogen and CO. The samples were irradiated to fluences in the range ~7-35×10^{20} neutrons cm^{-2} EDND at a temperature of ~400 °C. A sub-set of the samples from the inert capsule have been thermally annealed with DYM being measured before and after treatment. The results are shown in Figure 5. It should also be noted that Figure 5 includes the virgin DYM values used in the MTR database, since the sample moduli before irradiation were measured using an early version of the DYM measurement equipment and the values are not directly comparable. Measurements show that thermal annealing causes the DYM to recover to its unirradiated value.

Effect of Thermal Annealing on Irradiated Oxidised Graphite

In addition to the annealing of a sub-set of the ATR-irradiated inert capsule samples, a further smaller subset of samples from the oxidising capsule were tested. The DYM of six samples was re-measured after annealing. Due to the reduced modulus of the annealed graphite and the highly porous structure of the most oxidised samples, the ultrasonic pulses used to measure DYM were strongly attenuated. Consequently, the measured DYM of the highest weight loss sample (54.5%) is considered unreliable. The densities and weight losses of the samples are preliminary estimates based on averages of the heights and diameters from laser mensuration data. The weight losses of the samples irradiated in CO_2 ranged from 23.1% to 54.5%.

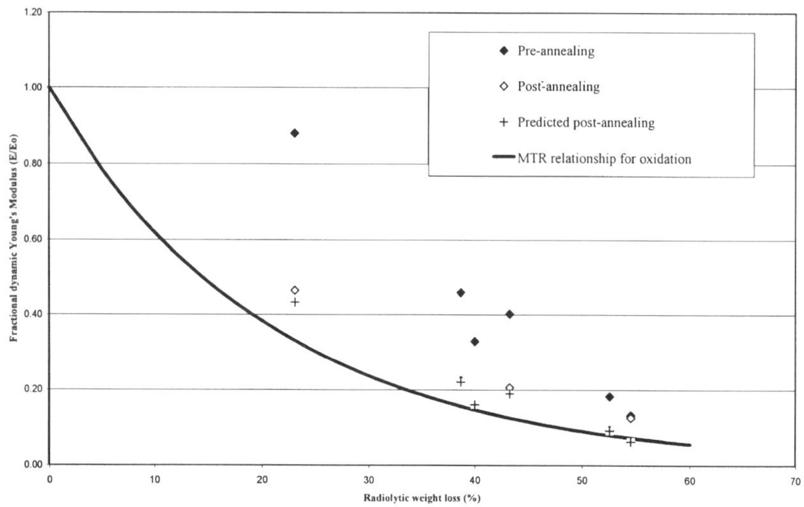

FIGURE 6: Effect of thermal annealing on the DYM of PGA graphite irradiated in an oxidising environment.

Figure 6 shows the pre- and post-annealed DYM measurements for each of the samples. Post-anneal values predicted on the assumption that the oxidation effect multiplies the fast neutron effect, which is removed by annealing, are in good agreement with measured post-annealed DYM values (the fast neutron effects are calculated from MTR data for un-oxidised graphite). Excluding the unreliable 54.5% oxidized sample measurement, the average ratio of measured to predicted post-annealed DYM is 1.04. Figure 6 also shows the relationship $E(x) = E(0)e^{-4.8x}$ between Young's modulus, E, and fractional radiolytic weight loss, x, that was derived by fitting to MTR data for thermally annealed samples. It should be noted that the ultrasonic time-of-flight method for measuring DYM produces values typically 15% greater than predicted for irradiated graphite. This is believed to be an artefact of the measurement technique and is allowed for in current safety cases.

Effect of Thermal Annealing on Monitoring Samples from Magnox Reactors

Graphite samples have been trepanned from Magnox reactor graphite cores on a regular basis since 1982. The DYM of a sub-set of these samples has been measured before and after thermal annealing. Figure 7 summarises the data, which covers fluences approximately in the range 19–50× 10^{20} neutrons cm^{-2} EDND, irradiation temperatures approximately in the range 220–380 °C and weight losses approximately in the range 6–32%.

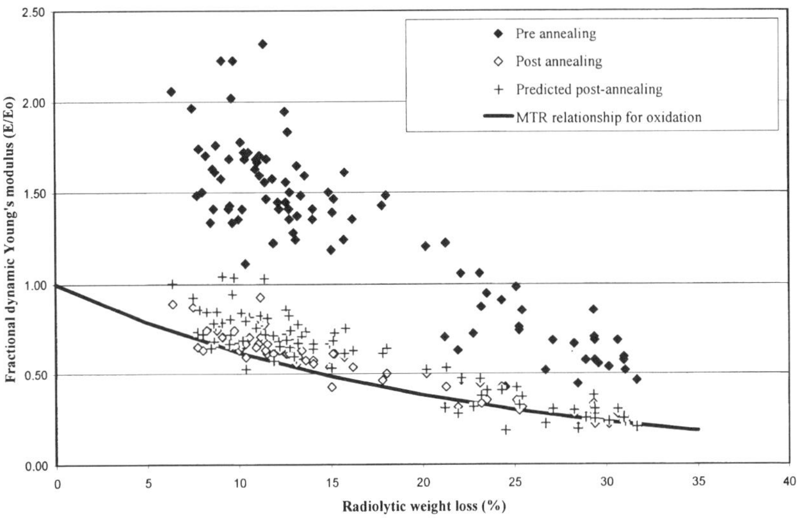

FIGURE 7: Effect of thermal annealing on the DYM of PGA graphite samples trepanned from Magnox reactor graphite cores.

The pre-annealed DYM data are presented together with the post-annealed values predicted from them, the post-annealed DYM data, and the $E_x/E_0 = e^{-4.8x}$ relationship for radiolytic oxidation (where E_0 is the virgin modulus, E_x is the modulus at fractional weight loss x). The observed ~15% discrepancy between measured and predicted DYM values for irradiated graphite should be noted. The average ratio of measured to "predicted" (the measured pre-annealed value adjusted to remove the irradiation damage based upon MTR predictions) post-annealed DYM is 0.91. It is interesting to note that this observed ratio is consistent with the reported behaviour in Figures 1 and 2, possibly indicating the presence of some structural damage.

DISCUSSION

The data presented here show that thermal annealing has no significant effect on the strength and DYM of virgin PGA graphite. The DYM measurements on graphite irradiated by up to 50×10^{20} neutrons cm^{-2}, in both inert and oxidising environments, show that fractional changes in DYM induced by irradiation, and removed by thermal annealing, are independent of radiolytic oxidation in the range 0-50% weight loss (approximately). This validates the Product Rule for DYM for this range of oxidation and irradiation within the limitations of the accuracy and reproducibility of the measurement techniques employed. Failure in irradiated graphite has been observed to occur at a fixed ultimate strain energy per unit volume (Simmons, 1965 and Losty and Orchard, 1962) which implies a square root relationship between irradiation induced changes in strength and Young's modulus. Whilst the precise relationship between strength and Young's modulus changes is not apparent from Magnox reactor graphite data, it is believed that the effect of thermal annealing on the strength of graphite can be inferred from the DYM results. There are currently no data for the effects of thermal annealing on material that has been subjected to a combination of radiolytic and thermal oxidation. The Product Rule successfully describes the combined effects of fast neutron irradiation and radiolytic weight loss on DYM of PGA graphite for the quoted fluences, irradiation temperatures and weight losses.

CONCLUSIONS

The thermal annealing of PGA graphite for approximately one hour in an inert atmosphere at 2100 °C has been observed to remove the effects of irradiation damage on DYM, and by inference strength, at fluences of up to 50×10^{20} neutrons cm^{-2} EDND and radiolytic weight losses of up to approximately 50%. A precise measure of the efficiency of the annealing process is limited by the accuracy and reproducibility of measurement techniques, but for these conditions removal is judged to be effectively complete. The Product Rule assumption that the separate effects of fast neutron damage and radiolytic weight loss on DYM, and by inference strength, can be combined multiplicatively is supported by PGA graphite data for this same fluence and weight loss range.

Acknowledgements

The authors would like to acknowledge British Nuclear Group Ltd permission for publishing this work.

References

Ewels, C. P., Telling, R. H., El-Barbary, A. A., Heggie, M. I. and Briddon, P. R. (2003). Metastable Frenkel pair defect in graphite: Source of Wigner energy? *Physical Review Letters*, [91], 2.

Kelly, B. T., Martin, W. H., Price, A. M., Dolby, P. and Smith, K. (1966). The annealing of irradiation damage in graphite. *J Nuc Mat*, 20, 195-209.

Kelly, B. T. (1981). Physics of graphite. Applied Science Publishers, London and New Jersey.

Losty, H. H. W. and Orchard, J. S. (1962). *Proceedings of the Fifth Conference on Carbon*, Pergamon Press, Vol 1, 519.

Matsuo, H. and Honda, T. (1973). Changes in the Young's modulus of neutron-irradiated reactor grade graphite by thermal annealing. *J Nuc Mat*, 48, 207-209.

Simmons, J. H. W. (1965). Radiation damage in graphite, Pergamon Press, Oxford.

Development of a Model of Dimensional Change in AGR Graphites Irradiated in Inert Environments

Ernest D Eason
Modeling & Computing Services, Boulder, Colorado, USA.
Email: eeason@ix.netcom.com

Graham Hall and Barry J Marsden
Nuclear Graphite Research Group, School of Mechanical, Aerospace & Civil Engineering, University of Manchester, PO Box 88, Manchester, UK.
Email: graham.n.hall@manchester.ac.uk; barry.marsden@manchester.ac.uk

Abstract
Dimensional and material property changes have been measured over many years on Gilsocarbon graphite subjected to fast neutron damage, radiolytic oxidation or a combination of both. The data are being gathered together now in a comprehensive database and used to develop dose-damage relationships of irradiated graphite material properties for use in irradiated graphite component stress analyses. This paper covers analysis of the dimensional change characteristics of Gilsocarbon graphite when irradiated in an inert environment. The analysis included pattern recognition, development of an empirical model of dimensional change, and validation of the model. The new model provides a good fit over the range of the four primary modelling variables: dose, irradiation temperature, graphite type, and measurement direction. It is convenient to use, featuring an analytic function of temperature and dose that eliminates the earlier need for interpolation on those variables. The new model also provides analytic expressions for dimensional change rate, which are needed for stress analyses, and it explicitly accounts for the small amount of anisotropy in the Gilsocarbon materials.

Keywords
Gilsocarbon graphite, AGR, dimensional change, irradiation, inert environment, modelling

INTRODUCTION

Advanced Gas-cooled Reactors (AGRs) use a graphite core made from Gilsonite coke, which is usually referred to as Gilsocarbon. The effects of fast neutron irradiation in an AGR lead to changes in the structural, mechanical and thermal properties of Gilsocarbon. These properties are further affected by ionising radiation (gamma and neutron) in combination with the carbon dioxide coolant, leading to radiolytic oxidation. Property changes also depend on irradiation temperature, coolant composition and direction of loading.

The methodology currently employed in modelling graphite behaviour in AGRs is an empirically based approach, which relies on a variety of assumptions and scaling factors to relate data from Materials Test Reactors (MTRs) to the operational behaviour. The Nuclear Safety Directorate (NSD) of the Health and Safety Executive (HSE) has initiated a project to obtain independent advice on the dose-damage relationships.

The effort includes collecting and reviewing all of the available data on Gilsocarbon materials to produce a comprehensive database including all variables (numeric) and factors (non-numeric) such as graphite material code or type, dose, temperature, grain direction, irradiation atmosphere, *etc.* Then all of the available data can be analysed to determine which variables and factors are statistically significant and to identify the trends in the data in order to produce improved irradiation models. Such an approach also gives guidance to aid understanding and point to areas where further data or researches are required. This paper summarises the modelling and analysis work on the dimensional change characteristics of AGR graphite when irradiated in an inert environment.

DATASET DEVELOPMENT

The collection and review of data and the development of the datasets were conducted at the University of Manchester by the Nuclear Graphite Research Group (Hall, 2005). The process undertaken is not examined in this paper, but the end result is approximately 22,000 data points that encompass 17 material properties of 88 types of graphite (*i.e.*, different identifying codes), all based on Gilsocarbon coke, in 6 environment/loading conditions (virgin, irradiated, pre-radiolytically oxidised, radiolytically oxidised, combined irradiation and radiolytically oxidation, and trepanned) over a prototypic AGR temperature range of 350-600°C.

The database in the AGR temperature range includes about 1,900 data points with measured dimensional change in an inert environment, with a maximum Equivalent DIDO Nickel Dose (EDND) of up to 367 × 10^{20} n/cm^2. Of these, about 1500 points are for relevant materials and are sufficiently complete to be usable for model calibration or validation. The dataset for combined irradiation and radiolytic oxidation environments is considerably smaller, containing fewer than 400 data points. Figure 1 visually compares the two datasets. The maximum dose in the radiolytic oxidation environments is much lower, at 72 × 10^{20} n/cm^2 EDND. The reduced number of points and dose range complicate the development of a model of the oxidation effects, so modelling the inert environment data is the logical first step.

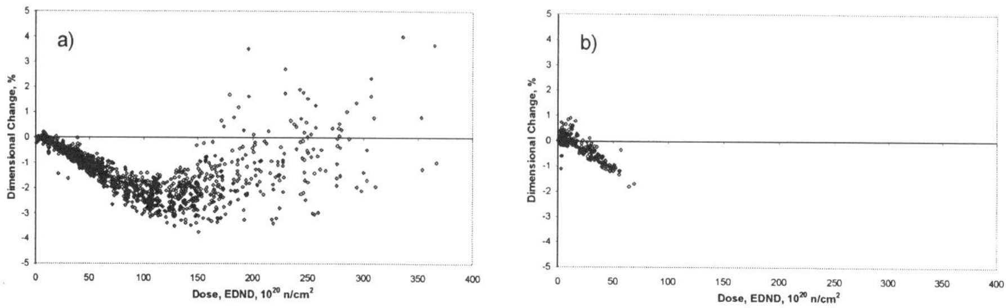

FIGURE 1: MTR data for Gilsocarbon graphite (350-600°C): a) irradiation in an inert environment; and b) combined irradiation and radiolytic oxidation.

Modelling Variables

Several variables were considered in the modelling process, but the final choices are dimensional change (dependant variable) versus dose, irradiation temperature, graphite group, and measurement direction. These variables were originally chosen based on pattern recognition and exploratory modelling, but final residual analysis also confirmed that only these independent variables are needed. The four chosen predictor variables are reasonably independent of each other, which is important for numerical stability in fitting the data.

Grouping of Graphites for Analysis

Many different Gilsocarbon graphite codes were used historically to identify specific materials, some of which correspond to the same or very similar material. Data with various graphite codes have been combined into nine graphite groups for analysis, such that each group contains graphites used at specific nuclear plants (or not used in any plant), made by the same forming techniques (moulded, extruded, and pressed), with similar initial density and similar amplitude of dimensional change. None of the groups has fewer than 50 data points. Additional graphite groups contain data with unknown measurement direction and other missing information, and in some cases, special graphite grades made for research purposes, which were not used in any nuclear plant. All of these data were considered in the development of the model form, but only the relevant data were used in the calibration and validation procedures. Thus, dimensional change curves that are specific to a particular AGR plant are not biased by data on materials which may be used in a different plant or for research purposes.

Separation of Data into Calibration and Validation Sets

After defining the graphite groups, the dataset was randomly split prior to the modelling process into a 90% calibration subset and a 10% validation subset. The purpose was to reserve a validation sample of data that were not used in model development to provide proof that the fitted model has predictive capability.

DIMENSIONAL CHANGE MODEL

The analysis tasks included pattern recognition, exploratory modelling, model calibration, and validation. A pattern recognition tool, TAC, was applied to the dimensional change dataset to identify key variables and analyse their degree of independence. Additional TAC runs with multiple predictor variables were made in various subsets, such as narrow temperature ranges and specific graphite groups, for insight on the optimal functional form for fitting the data. The ability to study the effect of each variable with the other variables simultaneously accounted for is the main advantage of TAC over more conventional analysis approaches. From the TAC runs, the strongest modelling variables were identified and the dependence of turnaround dose on temperature was characterised.

Model of Turnaround Dose

The minimum of the dose curves over a range of temperatures was extracted from TAC analyses and a simple model was developed for the dose at minimum dimensional change, $EDND_m$, which is also referred to as turnaround dose. The turnaround dose depends on measurement direction as well as temperature of irradiation, as shown in Equation [1].

$$EDND_m = f(direction)(2211 - T_{irr})^{3.17} \qquad [1]$$

This form gives a reasonable fit within the range of temperatures of interest to AGRs, approximately 350 – 600 °C, and credible results up to about 1100 °C. The shape of the function is shown in Figure 2.

Definition of Dose Ratio, DR

A dimensionless dose ratio (DR) is defined as the actual dose (EDND) divided by the turnaround dose $EDND_m$ given by Equation [1]. The reason for defining the dose ratio is the observation during exploratory modelling that the different shapes of dimensional change vs. dose curves at various temperatures nearly coincide in a "master curve" if expressed in dose ratio units. Because data at many temperatures can be modelled together, the subsequent modelling steps are simplified.

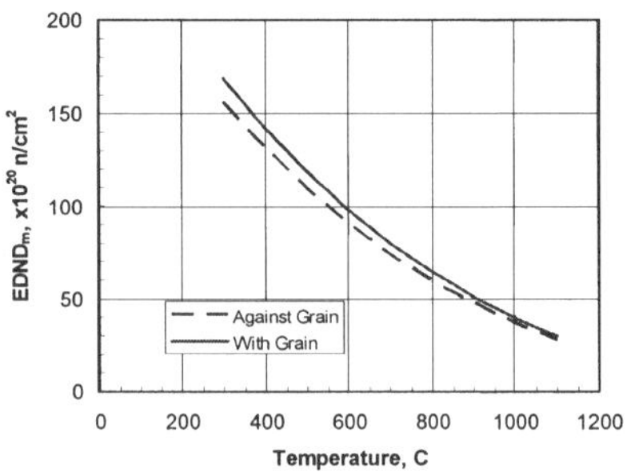

FIGURE 2: Model of turnaround dose, EDND$_m$.

General Dimensional Change Model Form

A convenient fitting form with sufficient flexibility to fit dimensional change data as a function of many different temperatures and graphite types, in both measurement directions, is given by Equation [2].

$$DimChg = DR^{B3}[B1(DR - A4)^2 - B2] \qquad [2]$$

There are three fitting functions in this model form, A4, B2, and B3. A4 is a value near 1.0 that varies for each graphite type and measurement direction. B2 is a function of T_{irr}, graphite type, and measurement direction, and B3 depends on T_{irr}. The function B1 depends on B2, A4, and an initial crossover dose ratio, DR_0. Figure 2 illustrates the general model form and shows the amplitude of the shrinkage B2, and the initial crossover dose ratio DR_0. The variation of A4, B1, B2, and B3 with irradiation temperature, graphite group, and measurement direction is explicitly modelled in the new model.

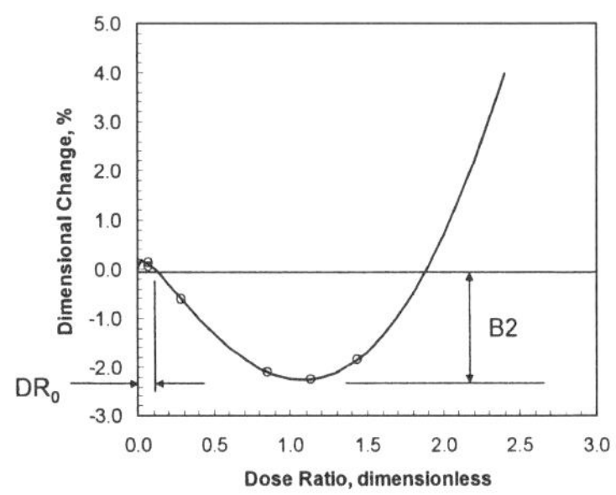

FIGURE 3: Typical curve at 400 °C, measured against grain.

COMPARISON OF DIMENSIONAL CHANGE MODEL AND DATA

For each of the nine graphite groups, the calibrated dimensional change model has been compared with the data in the two measurement directions at various irradiation temperatures, examples of which are given below. Figures 4a and 4b show that the model corresponds well to data for graphite group 4 over two temperature ranges (360-390 °C and 425-475 °C), even with the significant difference in directionality for this graphite group. The directionality effect was not as great in graphite group 5, but the model still compares well to the data over the 425-475 °C and 570-600 °C temperature ranges (Figures 4c and 4d respectively). Additional comparisons with two other graphite groups are given below in Figure 6.

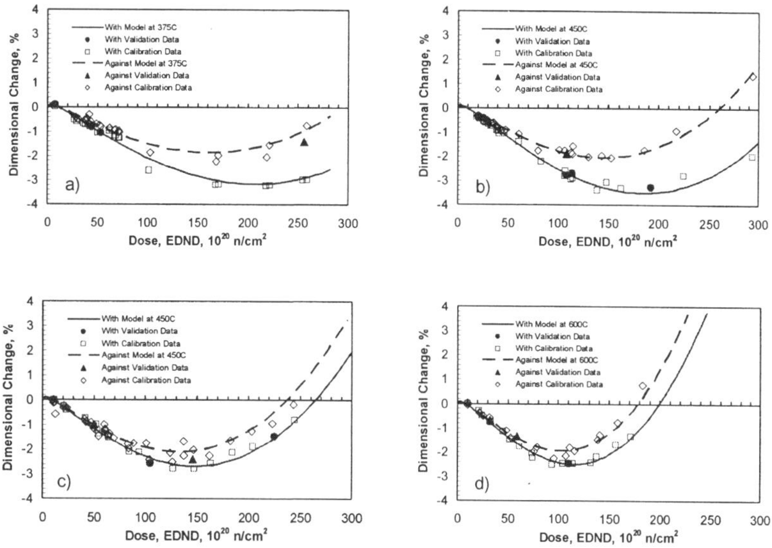

FIGURE 4: Dimensional change curve compared with data: a) graphite group 4, 360-390 °C; b) graphite group 4, 425-475 °C; c) graphite group 5, 425-475 °C; and, d) graphite group 5, 570-600 °C.

Based on Figure 4 and many similar comparisons for each graphite group and temperature range with sufficient data, the model is clearly able to change the curve shape as a function of irradiation temperature, graphite group, and measurement direction to match the data. In most cases, the agreement between the data and the relevant curve is quite good. The agreement is often even better than indicated by the plots, as some of the apparent "scatter" about the model curves can be attributed to the fact that the model is plotted at a single temperature and the data usually include a range of different temperatures.

Quality of Fit

There are several ways to assess the quality of the least squares fit for the model. A standard approach is to compare the model graphically with the data in appropriate groups, as shown by Figure 4. In 44 such comparisons over the range of temperature, graphite groups, and measurement direction, there is generally good agreement between the data and the new model.

A more quantitative comparison can be made using the standard error of the model. Over the entire calibration set, the standard error is Se = 0.19 in percent Dimensional Change units, based on 1269 points. Thus the model on average estimates the typical dimensional change of 1-3% with an uncertainty of a few tenths of a percent.

Another approach for evaluating goodness of fit, which takes all model variables into account on a single plot, is to compare model estimates of dimensional change and actual dimensional change. A perfect fit with no scatter would have all data on the solid 1:1 line shown in Figure 5. In reality, there is always some scatter about the 1:1 line, so dashed lines are plotted in Figure 5 representing 5th and 95th percentiles of a normal distribution about the 1:1 line with Se = 0.19%. From the fact that the calibration points in Figure 5a scatter close to the 1:1 line, mainly within the dashed bounds and without obvious under or over prediction, it is apparent that the model provides a good estimate of actual dimensional change over the range of the data.

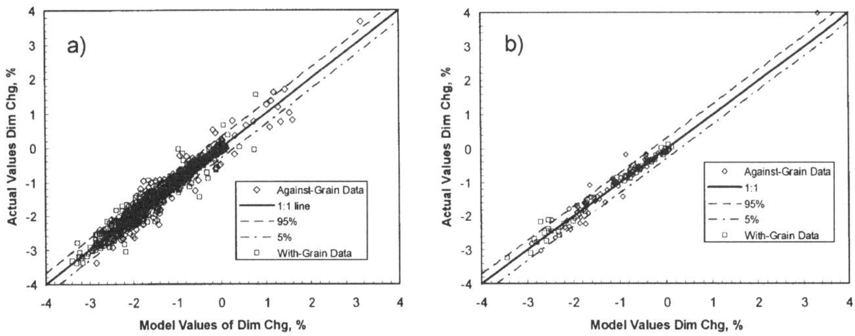

FIGURE 5: Actual versus model values of dimensional change; a) calibration dataset; b) validation dataset.

Residual plots were also generated, plotting the residuals (difference between each data point and the model) against each variable as another indication of quality of fit. The residual plots for variables that are in the model did not give any indication of inaccurately modelled trends, and residual plots for independent variables not in the model did not identify any un-modelled trends that would suggest that additional variables should be added to the model.

Model Validation

The validation points that were not used in model development show excellent agreement with the model as shown in Figure 5b. The standard deviation of residuals about the model for the validation set is 0.18, essentially the same as Se = 0.19 for the calibration set, indicating an equally good fit. This result proves that the model is capable of predicting data that were not used in model development.

Additional validation, though not quite as persuasive as the 10% random validation sample, comes from datasets which were not used in calibration because they had unknown measurement direction. These data are not a randomly selected, thus representative sample, but if the model has predictive capability, they should agree reasonably with either the Against or With direction model curves, or with both curves, since data tested in both directions could be mixed in the unknown-direction datasets. Reasonable agreement with one or both curves was found, giving further indication of predictive capability.

Comparison with Industry-Accepted (Paper 28) Model of Dimensional Change

The existing industry-accepted models for dimensional change are contained in an "in-house" document commonly referred to as Paper 28. The Paper 28 model of dimensional change is designed to interpolate between a curve for maximum weight loss (Fit-A) and one for no weight loss (Fit-B) to estimate the response of AGR core materials in various oxidising environments. The curves in Paper 28 are given in tabular form at several Equivalent DIDO Temperatures (EDTs) and for various doses up to 200 × 10^{20} n/cm^2 EDND. The relationship between actual irradiation temperature, T_{irr}, and EDT depends upon the test reactor flux, so that different test reactors have different actual temperature at the same EDT. For comparison with MTR data from inert environments and the new model, the Paper 28 curve with no weight loss (Fit-B) at 450°C EDT is used.

Figures 6a and 6b show comparisons of the Paper 28 model and the new model for two graphite groups that have substantial amounts of data at 450 °C EDT. All of the data in Figure 6 are from irradiation in the PLUTO test reactor, for which T_{irr} = 450 °C at 450 °C EDT. It is apparent that the new model provides a better fit, both at low dose in either measurement direction and near turnaround in the Against direction for graphite group 9. The new model also is a substantially better fit for doses at and beyond the limit of the Paper 28 curve (*i.e.*, ≥ 200 × 10^{20} n/cm^2 EDND) as shown for graphite group 2 in Figure 6a.

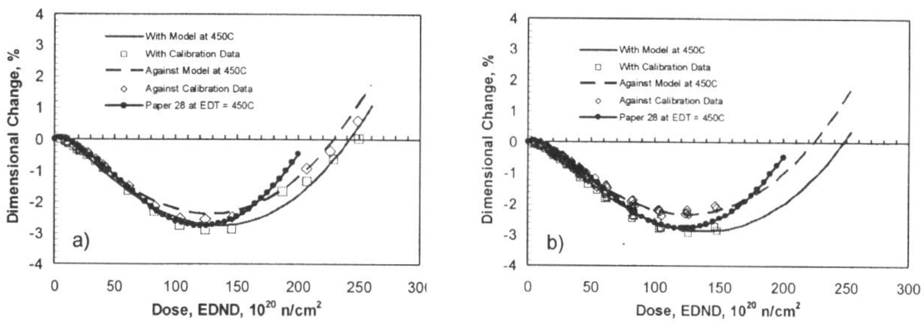

FIGURE 6: Comparison of models and data; a) graphite group 2, 450 °C EDT; and, b) graphite group 9, 450°C EDT.

All available inert-environment data points at 450 °C EDT, for any graphite group and any test reactor, are compared with the Paper 28 model and the new model in Figure 7. The data in Figure 7a show considerable scatter about the Paper 28 model (the 1:1 line) for model values of Dim Chg < -1% and a consistent bias for model values of Dim Chg > -1%. The new model in Figure 7b shows neither excessive scatter nor overall bias relative to the same data. The statistical analysis finds that the standard deviation and mean of the residuals about the Paper 28 model are both larger than the corresponding measures for the new model, by amounts that are statistically significant. Thus the new model is a significantly better fit to the data.

Planned Future Model Revision

Though the Paper 28 Fit-B model is an inferior fit to the relevant data in an inert environment, it should be noted that the Paper 28 model is able to treat the effects of weight loss from an oxidising environment, an effect that is not yet incorporated in the new model.

An analysis is planned for the near future to extend the inert-environment model to treat oxidising conditions.

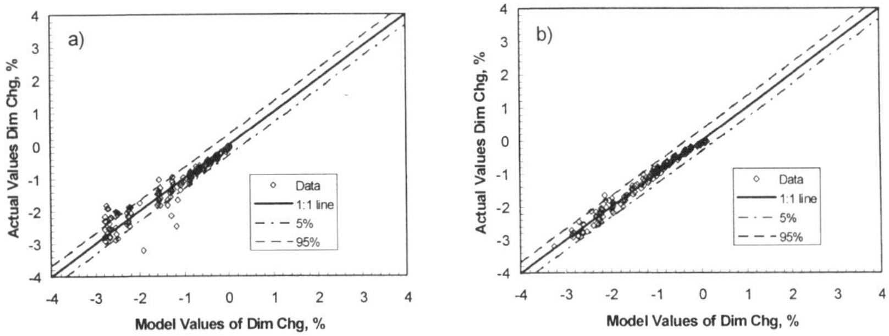

FIGURE 7: Actual versus model values of dimensional change at 450 °C EDT: a) Paper 28 model; and b) new model.

CONCLUSIONS

- An improved model was developed for estimating the dimensional change due to irradiation in inert environments. The new model provides a good fit to the available data over the range of the four primary modelling variables: dose, irradiation temperature, graphite type, and measurement direction.

- The new model has demonstrated predictive capability for data not used in fitting, including a randomly-selected validation dataset that was not used during calibration and other data with unknown measurement direction, which could not be used in calibration but could be compared to the final model.

- The new model is a significantly better fit to the data than the Paper 28 Fit-B model in comparisons on inert environment data where both models are applicable.

- The new model is analytic in irradiation temperature and dose, eliminating the need for interpolation as is required with the Paper 28 dimensional change model. It also provides analytic expressions for dimensional change rate, which are more convenient for stress analysis, and it explicitly accounts for the small amount of anisotropy in Gilsocarbon materials.

- The new model does not yet incorporate any effects of radiolytic oxidation, although model revisions to account for such effects are planned.

Acknowledgements

The authors would like to thank the Health and Safety Executive (Nuclear Safety Directorate) for financial support. The views expressed in this paper are those of the authors and do not necessarily represent those of the Health and Safety Executive.

References

Hall, G. (2005). Data and references for the analysis of models defining the irradiation behaviour of graphite material properties in advanced gas cooled reactor cores, NGRG R140, University of Manchester.

Relating Measurements of Mechanical Properties of Nuclear Graphites to Reactor Conditions: A Review of the Effects of Temperature and Pressure

Brian McEnaney
Department of Mechanical Engineering, University of Bath, Bath, UK.
Email: B.McEnaney@bath.ac.uk

Abstract
The measurement of properties of trepanned specimens of core graphite during post irradiation examination, PIE, usually takes place at ambient temperatures and pressures. In making such measurements it is either implicitly or explicitly assumed that the differences in temperature and pressure between PIE conditions and reactor operating conditions are either negligible or are conservative. This paper presents a review of evidence for the effect of pressure and temperature upon the Young's modulus and strength of reactor graphites and considers the possible relevance to the operation of Magnox reactors and AGR.

Keywords: Mechanical properties; temperature effects; pressure effects

INTRODUCTION

Post irradiation examination (PIE) of trepanned core graphite specimens is a key component of the strategy for managing the ageing of Magnox reactors and Advanced Gas Cooled reactors (AGR). The properties measured as part of the PIE of trepanned AGR graphite samples include: bulk density, open and closed pore volumes, differential thermal oxidation, DTO[3], gas diffusivity, dynamic Young's modulus, coefficient of thermal expansion and flexural strength. PIE enables changes in the properties of the core graphites to be compared with changes in the properties that are predicted from models for the behaviour of the graphite in the reactor environment.

PIE usually takes place at ambient pressure and temperature and there is a need to know the relationships between the measurements made under PIE conditions and the properties of the graphite under reactor conditions. Aspects of the reactor environment that may be expected to have an influence upon the properties of the graphite include: irradiation and the temperature, pressure and oxidation potential of the coolant gas. For AGR and Magnox reactors the influence of irradiation and oxidation are addressed in the equations that are used to predict the behaviour of the graphite. For example, the effect of irradiation and oxidation on the dynamic Young's modulus of irradiated AGR moderator graphites, E_d, is given by

$$E_d = {}_0E_d . P . S . W, \quad [1]$$

where ${}_0E_d$ is the dynamic Young's modulus of the unirradiated graphite, P is saturated pinning term to account for the pinning of mobile dislocations, S is a structure term to allow for structural changes in the graphite due to irradiation damage and W is a weight loss term

---

[3] DTO estimates the amount of carbonaceous deposit on a graphite sample by exploiting the difference in air reactivity between the deposit and the graphite substrate.

due to oxidation. In using such equations it is either implicitly or explicitly assumed that the effects of differences in temperature and pressure between reactor operating conditions and PIE conditions are either negligible or conservative, for example, that strength measured during PIE is lower than the strength of the graphite under reactor conditions. Some evidence for the effect of temperature and pressure upon the Young's modulus and strength of nuclear graphites is described in this paper and its possible relevance to reactor operation is considered.

EFFECT OF TEMPERATURE UPON YOUNG'S MODULUS

The effect of measurement temperature upon the dynamic Young's modulus of a nuclear grade graphite was reported by Mason and Knibbs in the 1960s (Mason and Knibbs, 1967). The material was described as '*a British nuclear grade carbon produced from a petroleum coke filler and Ford's coal tar pitch binder*'. Series of specimens heat-treated in the range 1200-3000 °C were prepared from the starting material, cut both parallel and perpendicular to the extrusion direction. This suggests that, the starting material was a baked product that had been heat-treated to less than 1200 °C. Measurements of Young's modulus were made by the flexural vibration method generally on slender cylindrical specimens (101.6 mm long x 3.2 mm diameter) over the temperature range 20-1000 °C. Measurements at sub-ambient temperatures were also made while slowly warming the specimens from liquid nitrogen temperatures.

The major part of the paper (Mason and Knibbs, 1967) is devoted to the effects of heat-treatment temperature on Young's modulus, which were related to changes in shear stiffness and dislocation pinning/unpinning. For the purposes of this paper the most useful results are for the material that was heat-treated to 3000 °C since this material is expected to be the most graphitic. The results are in Figure 1. At sub-ambient temperatures there is a reduction in Young's modulus with increasing temperature. The fall in the value of Young's modulus continues with increasing measurement temperatures above ambient to reach a minimum at ~300 °C. At higher temperatures there is an increase in Young's modulus that continues to 1000 °C. There is hysteresis when Young's modulus is then measured at progressively decreasing temperatures.

The Young's modulus of a crystalline material is expected to decrease with increasing temperature on thermodynamic grounds and Mason and Knibbs attributed the decrease at low temperatures to this factor, Figure 1. The increase in Young's modulus at high temperatures was attributed to the progressive closure of Mrozowski cracks as temperature increased. These cracks result from relief of thermal stresses produced by anisotropic shrinkage of graphite crystallites on cooling from graphitisation temperatures. Mason and Knibbs offer a tentative explanation for the presence of hysteresis. It is proposed that thermal stresses are required to open Mrozowski cracks on the cooling cycle, whereas no such stress, or possibly reverse stresses are required to close them during the heating cycle. Thus, at a given temperature, there are more closed cracks (and hence a higher modulus) on the cooling cycle than on the heating cycle. Figure 1 shows that the Young's modulus of the graphite at room temperature is ~6.0 GPa. This is much lower than the values for PGA and IM1-24 graphite, which are 14.14 ± 1.42 GPa and 10.90 ± 0.25 GPa, respectively (Neighbour and McEnaney, 1996). This suggests that the baked carbon block from which the series of heat-treated carbons were obtained by Mason and Knibbs was unimpregnated.

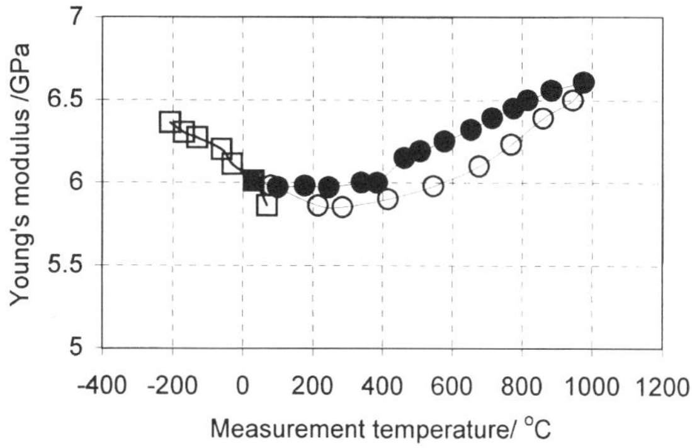

FIGURE 1: Effect of measurement temperature on Young's modulus parallel to the extrusion direction for a unirradiated nuclear grade carbon heat-treated to 3000 °C; □ low temperature measurements with temperature increasing; ○ high temperature measurements with temperature increasing; ● high temperature measurements with temperature decreasing. Redrawn from (Mason and Knibbs, 1967). Reprinted with permission from Elsevier.

Qualitatively similar results to those in Figure 1 for the effect of temperature upon Young's modulus were found for IG-11 graphite (Maruyama *et al*, 1987) and more recently for UCAR PCEA graphite (Miller *et al*, 2005). These results will be considered in more detail below. The precise variation of Young's modulus with temperature will depend upon the nature of the graphite, since the extent and distribution of Mrozowski cracks will depend upon the thermal history of the sample as well as the microstructure of the precursors materials.

EFFECT OF PRESSURE UPON YOUNG'S MODULUS

The effect of gas pressure on the 4-point bend strength of several nuclear graphites was studied by Boey and Bacon (1986b). The graphites were a Gilsocarbon material designated A and two pitch coke graphites designated as C1 and C2; the materials were machined into beams 500 mm long and 5 mm square cross section. The test equipment was enclosed in a stainless steel pressure vessel, which was pressurised to 4.14 MPa (600 psi) using nitrogen.

Some typical results for Gilsocarbon graphite A are in Figure 2. The slopes of the stress-strain plots in Figure 2 show that flexural stiffness of the graphites is higher at the elevated pressure. Boey and Bacon attribute the increase in stiffness to the shrinkage of some closed pores as a result of the strains ($\varepsilon \sim 10^{-4}$ at 4.14 MPa) induced by the pressurising gas in open pores. Another factor leading to an increase in stiffness at higher pressure is the restricted shear deformation of crystallites following the shrinkage of the closed porosity (the flexural strength results are also considered below).

In another paper by the same authors a similar explanation was advanced for the effect of rubber coatings on the bulk modulus, *K*, of some nuclear graphites (Boey and Bacon, 1986a). In this work hydrostatic measurements of strain in the circumferential and longitudinal directions were made on cylindrical specimens, 25 mm long x 20 mm diameter. The samples were pressurised using nitrogen gas to 4.14 MPa (600 psi) and strain measurements made in steps of 0.69 MPa (100 psi). The specimens were then de-pressurised and measurement made at similar pressure intervals. Measurements were made on uncoated graphites and on graphites specimens that had been covered with a rubber coating.

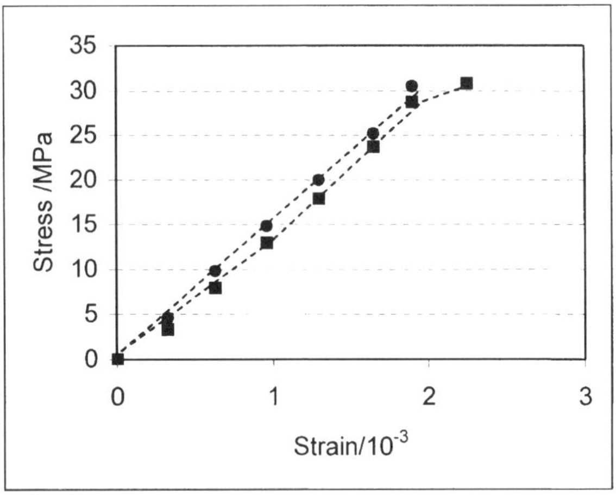

FIGURE 2: Effect of gas pressure upon the bend strength of a Gilsocarbon graphite A; ■ 0.10 MPa; ● 4.14 MPa. Redrawn from Boey and Bacon (1986b). Reprinted with permission from Elsevier.

This work showed that application of a rubber coating increases the strain at a given hydrostatic pressure leading to a lower bulk modulus. For example, for a Gilsocarbon graphite A the bulk moduli for the uncoated and coated specimens were 10.7 and 6.1 GPa, respectively. For uncoated graphites the pressurising gas can penetrate open pore volume, which results in some shrinkage of closed pores and restricts the shear deformation of the graphite. Access of the pressurising gas to open pores is not possible in the case of coated specimens and so shrinkage of the total pore volume contributes to the bulk modulus for coated specimens.

EFFECT OF TEMPERATURE ON STRENGTH OF NUCLEAR GRAPHITES

The effect of AGR reactor environment upon the fracture of four nuclear graphites was studied by Jones and Reed (1991). The graphites are a Gilsocarbon graphite IM1-24, VFT pitchcoke graphite used as a sleeve graphite, and two candidate sleeve graphites EO4 and P2 Nittetsu graphites. Jones and Reed measured three-point bend strengths in a specially designed controlled atmosphere test rig that could operate from ambient conditions up to AGR operating temperatures and pressures. Tests were performed using combinations of the following experimental variables:

- Temperature, 20 °C or 400 °C.
- Pressure, 0.10 or 4.14 MPa (1 bar or 40 bar).
- Atmosphere, dry CO_2 (0-35 vpm H_2O) or wet CO_2 (300-370 vpm H_2O).
- Open laboratory conditions, 20 °C, 0.10 MPa air, ~10^4 vpm H_2O.

A comparison of the strengths of the four graphites under reactor conditions and under open laboratory conditions is shown in Figure 3. For each graphite, the strength measured under open laboratory conditions is significantly less than that measured under reactor conditions. This is attributed to the effect of adsorbed water vapour on the fracture surface energy at room temperature.

The strength, σ_f, of a brittle, homogeneous, isotropic material containing a crack of length c is given by the Griffith equation

$$\sigma_f = \left(\frac{2E\gamma}{\pi c}\right)^{1/2} \quad [2]$$

where E is Young's modulus and γ is the fracture surface energy. For a crystalline solid, E decreases with increasing temperature, as noted above. The temperature coefficient of γ is given by

$$d\gamma/dT = -S^s \quad [3]$$

where S^s is the surface entropy, which is positive, thus, $d\gamma/dT$ is negative, *i.e.*, γ decreases with increasing temperature.

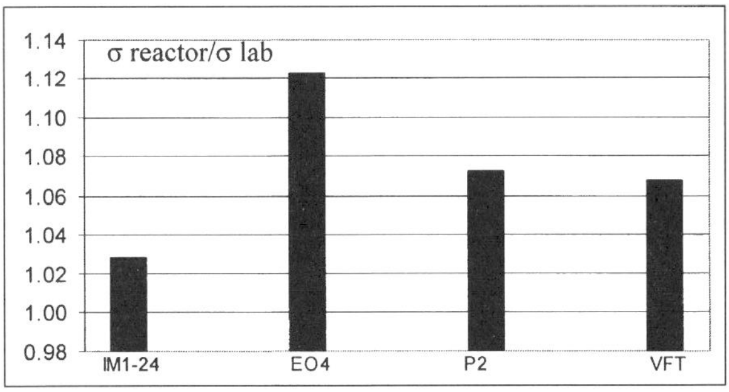

FIGURE 3: A comparison of the bend strengths of four nuclear graphites under reactor conditions (400 °C, 4.14 MPa CO_2, 300-400 vpm H_2O) and open laboratory conditions (20 °C, 0.10 MPa CO_2, ~10^4 vpm H_2O). Data from Jones and Reed (1991); used with permission.

Thus, from the Griffith equation, the fracture strength should decrease with increasing temperature, assuming that c is independent of temperature, whereas an increase in strength was observed by Jones and Reed, Figure 3. This was attributed to a lowering of the surface energy of the graphite by adsorption of water vapour onto the surface of the graphite under ambient conditions, which presumably outweighs the temperature effects described above. Similar results were obtained for 4-point bend strengths of a fine-grained graphite, IG-11 (Maruyama *et al*, 1987) measured in vacuum, 10^{-3} Pa, and in air at room temperature and pressure. The strength measured in air was ~20% less than the strength measured in vacuum. In the case of the graphites studied by Jones and Reed the differences ranged from ~3% for IM1-24 graphite to ~12% for EO4 graphite, indicating that the extent of the effect is dependent upon the microstructure of the graphite. From the Equation [2], $\gamma \sim \sigma_f^2$, assuming constant E and c. Thus, from the Griffith equation, the 3% and 12% reductions in strength for IM1-24 and EO4 graphites correspond to 6% and 20% reductions in γ respectively.

It is now recognised that for graphites the term 2γ in Equation [2] includes the energy dissipated by plastic deformation and sub-critical cracking in addition to the surface energy created by propagation of the critical crack that leads to failure (Kelly 1981). The critical value $2\gamma_c$ may be equated to the fracture toughness, G_c. Thus adsorption of water vapour may lower the surface energy of sub-critical cracks as well as that of the critical crack. It is also possible that water vapour may affect other energy dissipation mechanisms for example by reducing frictional effects during crack bridging (Ouagne *et al.*, 2002).

There is other evidence that the strength of graphite measured in vacuum is higher than in atmospheric air (Rowe, 1962) and that the presence of water vapour under atmospheric conditions has a deleterious effect upon the strength of graphites (Logsdail, 1968, Vitovec and Stachurski 1972).

THE EFFECT OF PRESSURE UPON THE STRENGTH OF NUCLEAR GRAPHITES

Figure 2 and Table 1 show that the increase in flexural stiffness on increasing pressure of nitrogen at ambient temperatures is accompanied by a slight reduction in strength and a larger decrease in failure strain. For example the reduction in mean strength and failure strain of Gilsocarbon A graphite on pressurisation was ~2% and 13%, respectively. The mean strengths for the Gilsocarbon graphites are within one standard deviation of each other.

TABLE 1: Four point bend strengths, σ_f, and failure strains, ε_f, of three nuclear graphites measured at P = 0.10 MPa and P = 4.14 MPa. Results are mean ± one standard deviation. Data from Boey and Bacon (1986b).

Graphite	σ_f/MPa (P= 0.10MPa)	$10^6\varepsilon_f$ (P= 0.10 MPa)	σ_f/MPa (P= 4.14 MPa)	$10^6\varepsilon_f$ (P= 4.14 MPa)
Gilsocarbon A	31.1 ± 2.8	2190 ± 230	30.4 ± 3.2	1910 ± 150
Pitchcoke C1	32.2 ± 1.7	2710 ± 170	29.9 ± 1.7	2230 ± 150
Pitchcoke C2	27.2 ± 1.6	2510 ± 180	26.2 ± 1.4	2060 ± 185

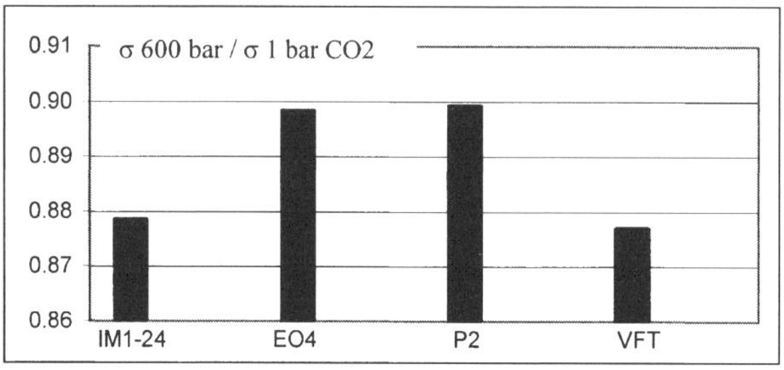

FIGURE 4: A comparison of the flexural strengths of four nuclear graphites at reactor pressure 4.14 MPa (600 bar) CO_2 and ambient pressure 0.10 MPa (1 bar) CO_2. For both tests the temperature was 400 °C and the water vapour content was 0.35 vpm. Source data from Jones and Reed (1991); used with permission.

Similar strength results were found by Jones and Reed for measurements made at reactor temperatures and low water vapour concentrations (Jones and Reed, 1991). A comparison of flexural strengths at 4.14 MPa CO_2 and 0.10 MPa CO_2 at 400 °C and the same water vapour content (0.35 vpm H_2O) also shows lower strengths at the higher pressure, Figure 4. The ratio $\sigma_{4.14\ MPa}/\sigma_{0.10\ MPa}$ ranges from ~0.88 for IM1-24 and VFT graphites to 0.90 for EO4 and P2 graphites. The equivalent ratios for the measurements made at room temperature by Boey and Bacon on graphites A1, C1 and C2 range from 0.93-0.96 (Boey and Bacon, 1986b).

Boey and Bacon point out that their results are not consistent with the finding of Losty and Orchard that graphites fail at a constant strain energy density (Losty and Orchard 1962). However, in an extensive survey of the fracture of polygranular graphites Tucker *et al* conclude that fracture models based upon constant strain energy density (and critical strain and critical stress) are unsatisfactory in most loading conditions (Tucker *et al*, 1986).

DISCUSSION

Figure 5 shows the difference between Young's modulus at the measurement temperature, E_T, and at 20 °C, E_{20}, expressed as $100(E_T - E_{20})/E_{20}$, as a function of the measurement temperature for two nuclear graphites IG-11 (Maruyama *et al*, 1987) and PCEA axial direction (Miller *et al*, 2005). For PCEA graphite there is a shallow minimum at ~100 °C to 300 °C where E_T is less than E_{20} by ~1-2%. For IG-11 graphite Young's modulus does not change with temperature up to ~300 °C. Thus, for both graphites, E_T is rather insensitive to measurement temperature in this temperature range; this is probably because this temperature range is the transition zone between decreasing moduli at low temperatures and increasing moduli at high temperatures for this graphite, as in Figure 1. At higher temperatures the two graphites show very similar behaviour. Above ~400 °C, E_T is greater than E_{20} for both curves rising to a maximum difference of 5.5% at 800 °C. This increase results from Mrozowski crack closure as discussed above. Thus, these results indicate that the effect of temperature on Young's modulus is small in the operating temperature ranges of Magnox reactors (typical coolant temperatures: 230-360 °C) and AGR (typical core temperatures: 325-450 °C). The larger discrepancies that are found at the upper end of the temperature range in Figure 5 suggest that measurements of Young's modulus at ambient temperature will be less than those at the operating temperature.

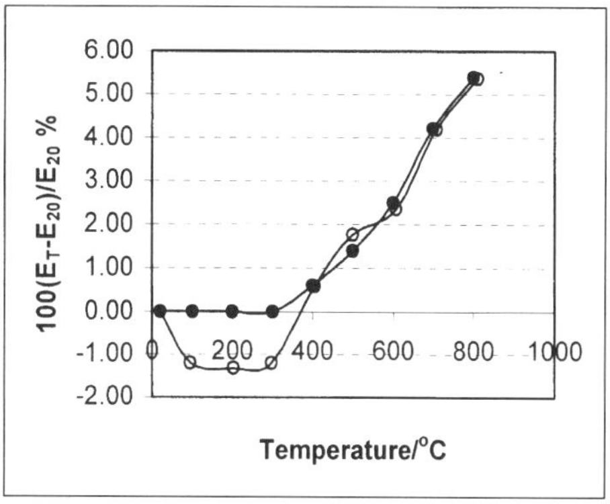

FIGURE 5: A comparison of Young's modulus of two nuclear graphites measured at 20 °C, E_{20}, and at higher temperatures, E_T. ● IG-11 graphite (data from Maruyama *et al*, 1987); ○ PCEA axial direction (data from Miller *et al*, 2005).

There is clear evidence that the flexural modulus of nuclear graphites increases with increasing gas pressure, Figure 2. This is attributed to the inhibition of shear deformation and partial shrinkage of closed pores brought about by the strains induced in the graphite by the gas pressure in open pores. Thus, as far as the effect of pressure is concerned, measurements of Young's modulus at ambient pressure are likely to be less than the value at reactor operating pressures. Under conditions where the effects of variations in water vapour content are eliminated, the strength of graphites at high pressure is slightly less than that measured at low pressures, both at ambient temperatures (Boey and Bacon, 1986b) and at 400 °C (Jones and Reed, 1991). There is clear evidence from several studies that the strength of graphites measured under ambient laboratory conditions is less than under reactor conditions. This effect is usually ascribed to the lowering of the surface energy of the graphite by adsorption of water vapour, a factor that appears to dominate over the expected effects of temperature and pressure upon the strength of graphites.

CONCLUSION

In conclusion, a number of factors need to be borne in mind when considering the relevance of the work described in the previous sections to reactor operation. First, very few of the experiments reported above were designed specifically to compare the properties of the graphites under reactor conditions to those measured during PIE, the work of Jones and Reed being an exception (Jones and Reed, 1991). Second, the methods used to measure properties, especially Young's modulus, are usually not the same as the method developed for PIE of small samples trepanned from the reactors and the effects of specimen size and test methodologies are important issues in this field. Finally, properties of graphites are strongly related to their microstructures and so results obtained for one graphite type may not apply to another. More specifically, the results for an unirradiated graphite may be different from those for the irradiated and radiolytically oxidised graphite. On the other hand, if a variety of graphites show the same result or trend, then it is reasonable to expect that other graphites may behave in a similar way. For example, this is probably the case for the effects of temperature upon Young's modulus and of water vapour upon the strength of graphites.

References

Boey, S. Y. and Bacon, D. J. (1986a). Deformation of polycrystalline graphite under pressure. *Carbon*, 24, 557-564.

Boey, S. Y. and Bacon, D. J. (1986b). Bend strength of graphite under pressure. *Carbon*, 24, 571-574.

Jones, D. D. and Reed, J. (1991). The effect of reactor environment on the fracture strength of IM1-24, VFT and Nittetsu graphite. Nuclear Electric Report TD/FCD/REP/0086/B, CSDMC/P46/B/1.

Kelly, B. T., (1981). The Physics of Graphite. Applied Science Publishers, London, pp. 128-129.

Losty, H. H. W. and Orchard, J. S. (1962). The strength of graphite. *Proceedings of the 5th Conference on Carbon*, Vol. 1, Pergamon Press, New York, pp. 519-532.

Logsdail, D. H. (1968). The effect of gaseous environment on flexural strength of graphite. Abstract. *Carbon*, 6, 212.

Maruyama, T., Eto, M. and Oku, T. (1987). Elastic modulus and bend strength of a nuclear graphite at high temperature. *Carbon* 25, 723-726.

Mason, I. B. and Knibbs, R. H. (1967). The Young's modulus of carbon and graphite artefacts. *Carbon*, 5, 493-506.

Miller, D. J., Homerin, P. and Ball D. R. (2005). Properties of UCAR nuclear graphites grades PCEA and PCIB SFG at elevated temperatures, *Proceedings of the 6th International nuclear graphite specialist meeting, INGSM-6*, Chamonix, France, in production.

Neighbour, G. B. and McEnaney, B. (1996). A systematic study of acoustic emission from nuclear graphites. *Graphite moderator lifecycle behaviour*, IAEA-TECDOC-901, IAEA Vienna, Austria, pp. 299-312.

Ouagne, P., Neighbour, G. B. and McEnaney, B. (2002). Crack growth resistance in nuclear graphites, *J. Phys D: Appl. Phys.*, 35, 927-934.

Rowe, G. W. (1962). High temperature strength of clean graphite. *Nuclear Engineering*, 7, 102-103.

Tucker, M. O., Rose, A. P. G. and Burchell, T. D. (1986). The fracture of polygranular graphites. *Carbon*, 24, 581-602.

Vitovec, F. H. and Stachurski, Z. H. (1972). Effect of environment on the mechanical anisotropy of polycrystalline graphite. *Carbon*, 10, 417-428.

Flexural Strength of Porous Graphite

Milorad Novovic, Graham Heys* and Paul Bowen
Department of Metallurgy and Materials, University of Birmingham, Birmingham, UK.
*Nuclear Installations Inspectorate, St Peter's House, Balliol Road, Bootle, UK.
Email: m.novovic@bham.ac.uk

Abstract
This paper presents the results on the four-point bend strength of porous graphite. The potential influence of pore size, specimen size, test temperature and notches on the flexural strength have been investigated for the two types of highly porous graphite with the same porosity level of 52% but very different pore sizes. It has been found that the fine-pore graphite exhibited superior strength compared to the coarse-pore graphite grade, at both room temperature and the temperature of 300 °C, and in both un-notched and notched specimen condition. The test temperature, specimen size and the presence of a notch have been found not to have a marked influence on the strength distribution in either of the two graphite grades investigated.

Keywords
Flexural strength, porous graphite

INTRODUCTION

The integrity of the nuclear reactor fuel bricks is assessed in terms of the stress over strength ratios. The stresses in the bricks arise as a combination of internal and external loading. Externally generated stresses are small and are usually not a cause of concern. The internal stresses arise as a consequence of a dimensional change of graphite due to the neutron irradiation and thermal expansion or contraction (as a result of the through-wall temperature gradients) and variability of the coefficient of thermal expansion (which is also a consequence of the effects of neutron irradiation on the structure of graphite).

The prediction of the strength of graphite bricks after radiolytic oxidation and neutron irradiation poses a significant challenge. As no reliable data exist for the strength of the full size bricks under representative loading conditions, the strength values for the full size bricks in the irradiated condition are calculated by extrapolation of the strength values obtained from a range of small bend specimens – virgin (heat-certificate) test pieces, installed specimens and trepanned samples. The use of flexural strength values rather than the uniaxial tensile strength values is rationalised partly by the finite element analyses of stress patterns in graphite fuel bricks which show that the stress patterns of fuel bricks are approximately related to such bend configuration. The most significant stress component is axial (uniaxial bend) which is the same stress field applied in four-point bend (pure bending) test to measure flexural strength. While the statistical distributions of graphite strength can be routinely evaluated for start-of-life unirradiated material, there are severe restrictions on test-piece sizes and test-piece numbers available in irradiated material from surveillance or test-reactor schemes. Because of the large differences between the sizes of these specimens and the actual graphite brick the stress-volume relationships need to be established with care. Effects of stress concentrations (*i.e.* notches) introduce additional complexity into consideration of these stress-volume relationships. There is therefore a need to find suitable surrogate materials on which strength distributions for high weight loss (highly porous) graphites can be established.

EXPERIMENTAL

Material

The material investigated in the present programme comprised two types of porous graphite, designated PG 25 and PG 60, characterised by the same porosity level but different pore sizes. The graphites were produced by the UCAR Carbon Company Inc., and are commercially available as graphite filters. The material was delivered as two blocks of approximately 250×300×350 mm^3 in size. Pore sizes and their distributions are expected to be more uniform in these graphites than those present in the nuclear graphite after radiolytic oxidation and neutron irradiation. The manufacturer's specification for the typical properties for these graphites is summarised in Table 1.

TABLE 1: Typical properties of porous graphite at room temperature.

Property		PG 25	PG 60
Bulk Density	g / cm^3	1.05	1.03
Porosity	%	52	53
Compressive Strength	MPa	4.1	2.8
Average Pore Diameter	μm	33	119
C.T.E.	10^{-6} / °C	2	2

Specimen Machining

Four-point bend test pieces of two sizes were machined from the two graphite blocks for testing in the *unnotched (plain)* specimen condition to coincide with specimen sizes used for differing measurements of flexural strength of reactor graphite by the nuclear industry: ¼×¼×6" (6.35×6.35×152.40 mm^3) and ¾×¾×6" (19.05×19.05×152.40 mm^3). For testing in the *notched* condition, ½×½×3.15" specimens with relatively severe central notch of 4.33 mm depth, 45° included angle and 0.25 mm root radius were employed. For this specific notch profile the elastic stress concentration factor is 4.15.

The machining involved slicing the layers of graphite material from the top surface of the blocks to an appropriate thickness, then cutting these layers into specimen blanks which were finally ground to exact specified dimensions. The surface condition of the finished specimens produced by grinding was very good with the dimensional accuracy of ± 25 μm. For the current testing programme, all specimens were tested in the same orientation. All test-pieces were numbered with respect to their location and orientation relative to the original graphite block.

Four-Point Bend Testing

The flexural strength measurements were carried out at room temperature and a test temperature of 300 °C in four-point bend (pure bend) testing arrangement. Larger *unnotched* specimens (*i.e.* ¾×¾×6" specimens) and all *notched* test-pieces were tested on a digital, screw-driven Denison-Mayes machine fitted with a calibrated load cell of 1 kN capacity. Smaller unnotched specimens (*i.e.* ¼×¼×6" specimens) were tested on an ESH servo-controlled electro-hydraulic testing machine fitted with a 0.1 kN load cell. Tests carried out at 300 °C were performed in air and at atmospheric pressure, using an Instron environmental chamber with temperature control of ±1 °C. All tests were performed in displacement control with the loading rate kept constant at 0.5 mm/min.

The loading configuration for the two *unnotched* specimen sizes was the same with the four-point bend inner span of 40 mm and outer span of 120 mm, while for the *notched* specimens the inner span of 40 mm in combination with the outer span of 63.5 mm was utilised. Load versus displacement records were obtained for all bend tests and the maximum load (load at the point of fracture) was used to calculate the flexural strength using the equation derived in the theory of stresses in bending beams, Figure 1. For comparison purposes, the flexural strength for notched samples was calculated as the maximum nominal 'net-section' (*i.e.* reduced cross-section) bending stress at the notch root.

$$\sigma_{max}^{unnotched} = \frac{3 \cdot P_{max} \cdot S_1}{b \cdot h^2}$$

$$\sigma_{max}^{notched} = \frac{3 \cdot P_{max} \cdot S_1}{b \cdot (h-d)^2}$$

b - specimen thickness;
h - specimen width;
d - notch depth;

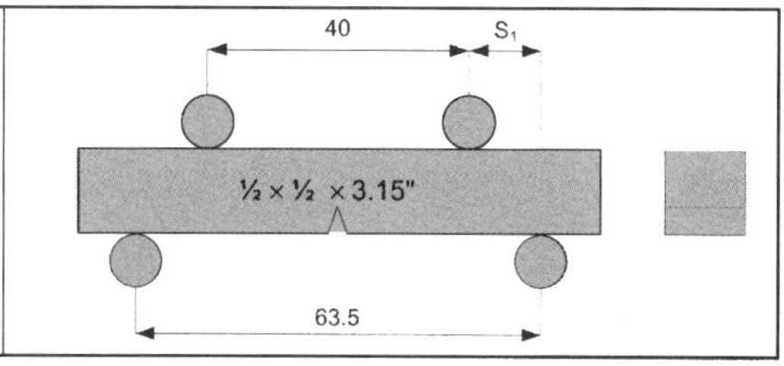

FIGURE 1: Evaluation of flexural strength from four-point bend test.

Selected fracture surfaces of bend specimens were examined using a JEOL 700F field-emission gun scanning electron microscope at an accelerating voltage of 20 kV.

RESULTS AND DISCUSSION

A total of 138 *unnotched* specimens have been tested at room temperature with 35 specimens in each of the following conditions: PG 25 - ¼×¼×6", PG 25 - ¾×¾×6" and PG 60 - ¾×¾×6", while 33 tests have been completed for the PG 60 - ¼×¼×6" condition. Tests on ten PG 25 - ¾×¾×6" and ten PG 60 - ¾×¾×6" specimens were completed at 300 °C. In *notched* specimen conditions, sets of ten test-pieces were tested at room temperature and temperature of 300 °C for both graphite grades, totalling 40 notched specimens to date.

The individual strength values and the corresponding histogram plots for the PG 25 and PG 60 graphite grades obtained from the ¼×¼" and ¾×¾" specimens tested at room temperature are plotted in Figure 2. This figure indicates that the flexural strength values for PG 25 graphite are greater than the strength values for PG 60 for both specimen sizes employed. Based on the datasets of 35 test results, the PG 25 graphite dataset for ¾×¾×6" specimen size is characterised with the mean value of 4.30 MPa (and a standard deviation of 0.46 MPa) compared to the mean strength of 2.39 MPa (and standard deviation of 0.19 MPa) for PG 60 graphite. Similarly, for the ¼×¼×6" specimen size, PG25 graphite exhibited the mean strength of 4.13 MPa with standard deviation of 0.66 MPa, while PG 60 graphite mean strength value was 2.26 MPa with a standard deviation of 0.20 MPa.

While Figure 2 confirms the superior strength of PG 25 compared to PG 60 graphite, it also indicates the much greater variability of PG 25 graphite strength values. For ¾×¾" test-pieces, the strength values for the PG 60 graphite are tightly distributed with a range of 0.78 MPa, while the PG 25 distribution demonstrated a range of 1.81 MPa. For ¼×¼" test-pieces, PG 60 graphite strength values are also characterised with much smaller variability with the minimum value of 1.86 and maximum value of 2.64 MPa compared with 2.93 and 5.31 MPa for the minimum and maximum strength value obtained for the PG 25 graphite (note, however, that in all cases the "signal to noise ratio" (mean value to standard deviation ratio), was approximately 10:1). For this specimen size, Figure 2 also reveals a very different shape of the strength distributions for the two graphite grades. While PG 60 graphite exhibits a nearly symmetrical distribution of strength values characterised by a small variability, the strength of PG 25 graphite is characterised by an irregular, discontinuous distribution with large variability. However, despite the larger scatter in strength of PG 25 graphite there was no overlap in the measured strength values for the two graphite grades in either of the two investigated specimen sizes.

Figure 2 also indicates that there is no effect of specimen size on the strength values for the PG 25 and PG 60 graphites. For both these graphite grades, the mean values of strength do not seem to be affected by the specimen size (an interesting result with respect to the current industry practice of Weibull extrapolation to a full scale brick). The specimen size appears to have a weak influence on the variability of strength for the PG 25 graphite grade, while the scatter in strength measurements for the PG 60 appears to be closely similar for both specimen sizes.

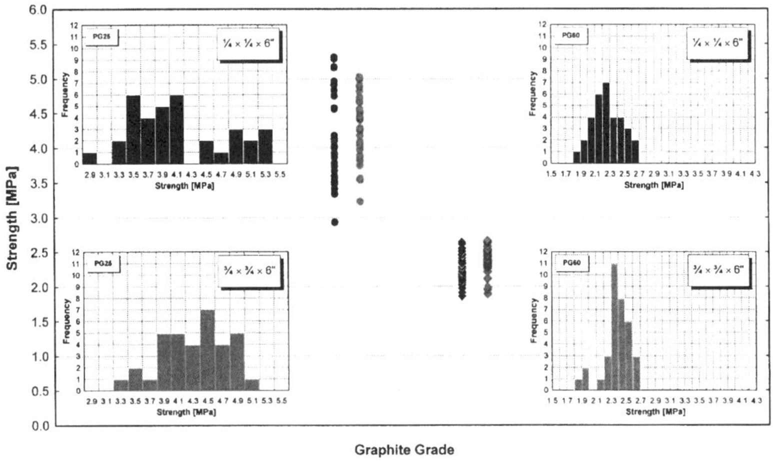

FIGURE 2: Effect of specimen size on the flexural strength of porous graphite.

The individual strength values and the corresponding histogram plots for the PG 25 and PG 60 graphite grades obtained from ¾×¾" specimens tested at 300 °C are plotted (and compared with the equivalent values obtained at room temperature) in Figure 3. It can be seen from this figure that test temperature of 300 °C does not degrade the strength properties of these graphites. In fact, it appears that the strength of PG 60 graphite slightly increases as a result of increase in test temperature (a statistically significant difference between the means according to the *t*-test). Although based on ten measurements only, the variability of strength distributions at 300 °C also seems comparable with that at room temperature. Fractographic examination of specimens tested at 300 °C did not reveal any obvious change in fracture mechanism compared to fracture morphology of specimens tested at room temperature.

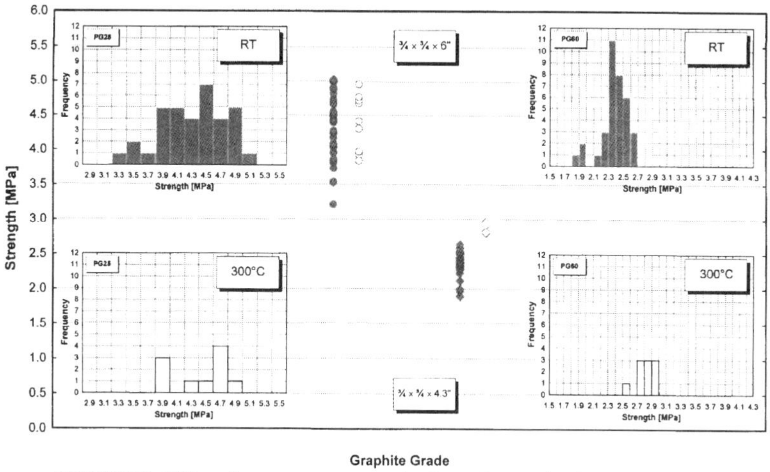

FIGURE 3: Effect of test temperature on the flexural strength of porous graphite.

Figures 4 and 5 show the comparison of flexural strength of *unnotched* (¾×¾") and *notched* (½×½") PG 25 and PG 60 graphites, obtained at room temperature and the temperature of 300 °C, respectively, while Figure 6 shows the actual notch profile and the crack propagation path (on the specimen surface) for the PG 25 graphite. As evident from Figure 4, despite the severity of the notch, the room temperature nominal 'net-section' flexural strength appears to be slightly smaller for PG 25 graphite and slightly greater for PG 60 graphite compared with the unnotched strength values, while the nominal 'net-section' flexural strength appears to be similar to the flexural strength of unnotched samples for both graphite grades at 300 °C (Figure 5). All notched specimens failed at the notched cross-section with the crack propagation path (as observed on both front and back surfaces) following the porosity cavities around solid graphite particles.

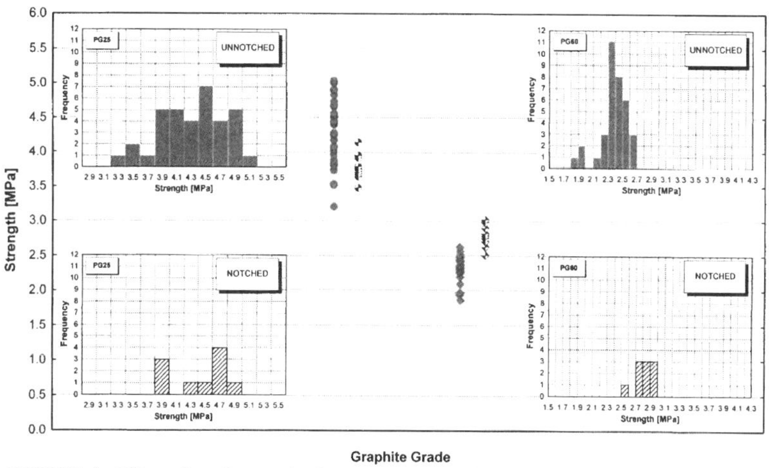

FIGURE 4: Effect of notches on the flexural strength of porous graphite at room temperature.

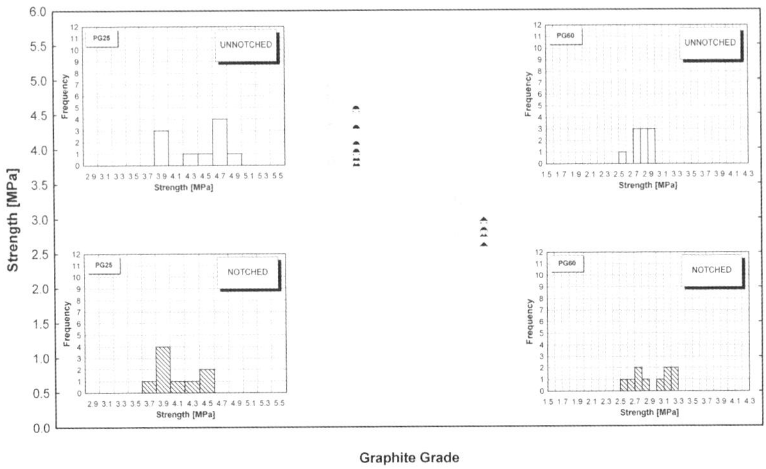

FIGURE 5: Effect of notches on the flexural strength of porous graphite at 300 °C.

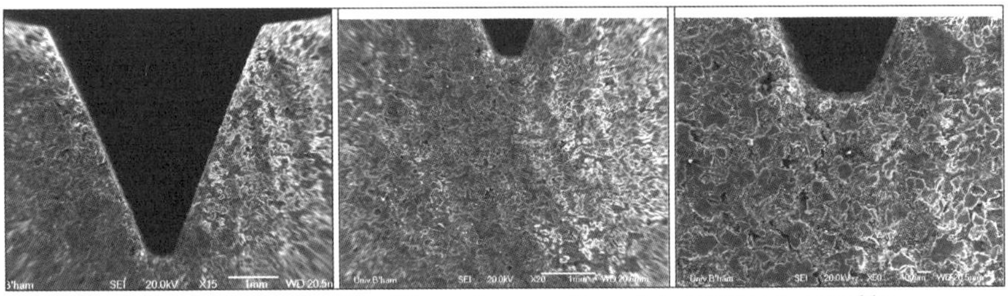

FIGURE 6: Notch profile and the crack propagation path in PG 25 porous graphite.

Independent of specimen size and test temperature, unnotched PG 25 graphite samples fractured in a brittle manner with failure at maximum load. The fracture was characterised by an audible signal, sudden drop in load and complete specimen separation. For the unnotched PG 60 graphite, smaller specimens failed as for PG 25 graphite, while larger specimens were not characterised by a sudden, significant drop in load and by instantaneous separation of the specimen. For this specimen condition, crack growth after maximum load was stable, and almost all bend tests of ¾×¾×6" PG 60 specimens (terminated after a reduction of 50% relative to the maximum load) resulted in some remaining uncracked ligament. Notched PG 25 and PG 60 specimens exhibited behaviour similar to the larger unnotched PG 60 specimens, with no complete specimen separation at the attainment of maximum load. The behaviour observed indicates the inability of the coarse-pore highly porous graphite to store enough elastic energy for fast propagation of a brittle crack.

Figure 7 shows the fracture surfaces of PG 25 (*a, b* and *e*) and PG 60 (*c, d* and *f*) graphite. Images *a* and *c* show the general fracture surface appearance, while images *b, d, e* and *f* give higher magnification details of features on the fracture surfaces, including possible shrinkage cracks and the fracture of the connecting ligament between porosity cavities. The crack initiation sites, however, could not be unambiguously identified.

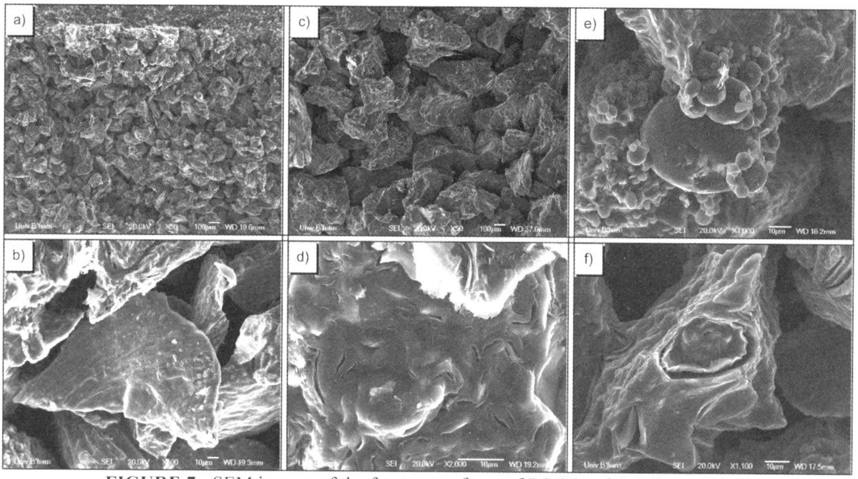

FIGURE 7: SEM images of the fracture surfaces of PG 25 and PG 60 graphites.

It is evident from Figure 2 that there is no effect of specimen size on the strength values for either PG 25 or PG 60 graphite. For both these graphite grades, the mean values of strength are very similar for the two specimen sizes. For the present testing programme, the "stressed-volume" of ¾×¾" specimens was nine times larger compared with the "stressed-volume" in ¼×¼" test-pieces. The insensitivity of the mean strength to an order of magnitude increase in the "stressed-volume" appears to be in contradiction with theoretical assumptions of the weakest-link theory and associated size effect in brittle materials. In addition, as shown in Figures 4 and 5, the apparent 'net-section' flexural strength appears to be similar to the flexural strength of unnotched samples in both graphite grades. These effects may be the result of the microstructural scale of these porous graphites relative to the specimen dimensions, and in particular to the notch root size. The microstructural scale and the large difference in graphite grain size between the two graphite grades are revealed in the low magnification images in Figure 7 (*a* and *c*). Even in finer, PG 25 graphite, the grain size is quite large (greater than 100 μm), while in PG 60 the grain size is approximately 400 μm, much greater than the notch root radius of 250 μm.

CONCLUSIONS

Two graphite grades of the same porosity level (52-53%) but different pore sizes (33 and 119 μm) have been examined in terms of the flexural strength measured in four-point bending. Specimens have been tested in two specimen sizes to match virgin (heat-certificate) sizes and sizes of specimens extracted from reactors. The tests have been carried-out in both *unnotched* and *notched* conditions at room temperature and temperature of 300 °C.

PG 25 graphite (graphite with finer pores) exhibited superior strength compared to PG 60 graphite grade (larger pores) for both specimen sizes tested, at both test temperatures and in both *unnotched* and *notched* specimen conditions.

The room temperature strength values for the PG 60 graphite grade for the two specimen sizes are closely similar. The strength distributions exhibited very small variability, and this variability is not influenced by the specimen size. The average room temperature strength values for the PG 25 graphite in two specimen sizes also appear to be similar. However, the strength values for the PG 25 graphite exhibited larger scatter compared with the PG 60 graphite. The scatter in this graphite appears to be a function of the specimen size with smaller specimens exhibiting larger variability in the measured flexural strength.

The room temperature strength of PG 25 graphite is equivalent to that obtained at 300 °C. In PG 60 graphite, the strength appears to increase slightly as a result of increase in test temperature. Fractographic examinations of specimens tested at 300 °C revealed no obvious change in fracture appearance compared to the fracture morphology observed for specimens tested at room temperature.

Room temperature nominal 'net-section' flexural strength is found to slightly decrease for PG 25 graphite and to slightly increase for PG 60 graphite compared with their respective unnotched strength values. The 'net-section' nominal strengths of notched PG 25 and PG 60 graphites are closely similar to the unnotched strength values at the temperature of 300 °C.

Acknowledgements

The work presented in this paper is sponsored by the Nuclear Installations Inspectorate, Health and Safety Executive, UK. The views expressed in the paper are those of the authors and do not necessarily represent those of the Nuclear Installations Inspectorate (NII).

Observations of Strain Localisation and Failure in Nuclear Graphite

M R Joyce and T J Marrow
School of Materials, Materials Performance Centre, University of Manchester, UK.
Email: mark.joyce@manchester.ac.uk

ABSTRACT

The surface strain fields evolved during disc compression tests on 10 mm radius discs of virgin IM1-24 are obtained via Electronic Speckle Pattern Interferometry. A novel analysis technique is demonstrated, which derives bulk elastic properties with reference to an analytical stress function. Comparison of the predicted and experimentally derived strain distributions allows microstructural scale strain perturbations to be characterised. This provides evidence of strain localisation which develops with increased load. Monitoring the development of persistent regions of localised strain with increasing applied global strain provides a valuable insight into the deformation and fracture processes operating within the material.

KEYWORDS

Mechanical properties, strain mapping, strain localisation

INTRODUCTION

A full understanding of the effects of microstructural degradation on the fracture of nuclear graphite requires models for the nucleation and coalescence of damage under the influence of different strain states, including strain gradients at stress concentrations such as notches. A preliminary step towards this is the characterisation of strain localisation, caused by microstructural features, under applied loading. In-situ strain mapping can provide the necessary observations of this process, albeit only on the sample surface. The aim of this work is to establish a methodology for determining the statistical distribution of microstructural strains in virgin Gilsocarbon graphite (IM1-24), and furthermore to provide an insight into the failure process. It is known that fracture and crack propagation in this material involves the formation of micro-cracks (Hodgkins *et al*, 2005), and it is anticipated that strain localisation is a precursor to crack nucleation.

EXPERIMENTAL METHODOLOGY

Disc compression tests were performed using small, thick, specimens of IM1-24 graphite with a radius and thickness of 10 mm. Reliable measurements of tensile fracture strength in the disc compression test are only possible if the sample fails in tension at its centre (Rudnick *et al*, 1963). In practice this requires that the loading be distributed along a prescribed length of the disc periphery (Awaji and Sato, 1979), which may be achieved using curved anvils. Alternatively, two diametrically opposed flats can be machined into the samples, as suggested by Wang *et al*, (2004). Such an approach eases both sample preparation and experimental set up, and hence was adopted in this work. To investigate the dependence of the measured elastic properties on the sample geometry, specimens with parallel flattened regions with a total width from 1 mm to 6 mm were prepared. The load, P, was applied in displacement control using an Instron 5802 electro-mechanical frame at a rate of 0.05 mm min^{-1} until sample failure occurred at P_{max}. The Q-300 3D ESPI system manufactured by Dantec

Ettemeyer was used to observe in-situ sample deformation. In each test the sample was loaded progressively without interruption, whilst a series of speckle pattern images were recorded at known load levels, in approximately 100 N increments from a small preload until sample failure.

STRAIN FIELD DETERMINATION

The speckle pattern image series collected in each test were analysed using the 'ISTRA' software package (Erne *et al*, 1998) supplied by Dantec Ettemeyer. This uses interferometry to construct separate horizontal and vertical deformation fields between successive images. By accumulating these, the total deformation fields between an image at a given load and a reference image may be obtained. The vertical strain field was then obtained by vertical differentiation of the vertical deformation field, whilst the corresponding operation was performed on the horizontal deformations to calculate the horizontal strain field. These operations and all other image processing were performed using Matlab routines. Due to the relatively noisy nature of ESPI data, spurious deformation values were first rejected using a 3×3 median filter. Numeric differentiation was then carried out in the appropriate direction by considering the average deformation gradient over a length of 13 pixels. The raw results were then smoothed using a mean filter with an averaging box of 9×9 pixels to produce the final strain fields. Typical examples of horizontal and vertical strain fields determined by this method are shown in Figure 1. The disc test geometry ensures that in successful tests the high stress region is restricted to the sample centre, hence all subsequent analysis is restricted to the central 100×100 pixels, which corresponds to an area of approximately 9 mm square.

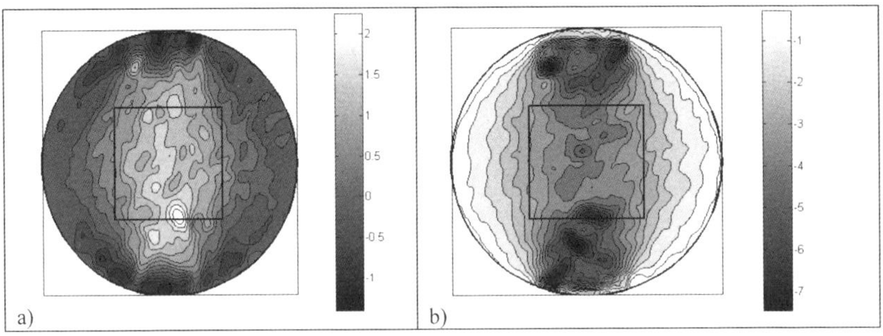

FIGURE 1: Typical measured strain fields from the surface of a disc compression test. This example has a 3mm flattened width and is loaded to 1.4kN ($P/P_{max} \sim 0.3$). The strain magnitude (millistrain) is given by the greyscale; (a) horizontal direction (nominally tensile), (b) vertical direction (nominally compression). In both cases the square box indicates the region to which the modulus and strain magnitude analysis is restricted.

DERIVATION OF ELASTIC MECHANICAL PROPERTIES

If it is assumed that the surface strains are representative of those within the bulk, it is possible to derive material constants by linking the experimentally derived strain fields with an analytical stress function. The centre region of the measured strain field was correlated with a stress solution proposed by Hung and Ma (2003). The direct and shear stresses may be derived at any given distance, r, and angle, θ, within a diametrally loaded thin disc of radius R, with the load P evenly distributed over the angle, α. Their general solutions for radial, σ_r, tangential, σ_θ and shear, $\tau_{r\theta}$, stresses for the range $0 \leq \theta \leq \pi/2$ are given in Equations [1] to [3].

$$\sigma_r(r,\theta) = -\frac{P}{2\pi\alpha Rt}\left\{\begin{aligned}&\left(1-\rho^2\right)\left[\frac{\sin 2(\alpha+\theta)}{\rho^4+1-2\rho^2\cos(\alpha+\theta)}+\frac{\sin 2(\alpha-\theta)}{\rho^4+1-2\rho^2\cos(\alpha-\theta)}\right]\\&+\tan^{-1}\left[\frac{1+\rho^2}{1-\rho^2}\tan(\alpha+\theta)\right]+\tan^{-1}\left[\frac{1+\rho^2}{1-\rho^2}\tan(\alpha-\theta)\right]+\Phi\end{aligned}\right\} \quad [1]$$

$$\sigma_\theta(r,\theta) = -\frac{P}{2\pi\alpha Rt}\left\{\begin{aligned}&-\left(1-\rho^2\right)\left[\frac{\sin 2(\alpha+\theta)}{\rho^4+1-2\rho^2\cos(\alpha+\theta)}+\frac{\sin 2(\alpha-\theta)}{\rho^4+1-2\rho^2\cos(\alpha-\theta)}\right]\\&+\tan^{-1}\left[\frac{1+\rho^2}{1-\rho^2}\tan(\alpha+\theta)\right]+\tan^{-1}\left[\frac{1+\rho^2}{1-\rho^2}\tan(\alpha-\theta)\right]+\Phi\end{aligned}\right\} \quad [2]$$

$$\tau_{r\theta}(r,\theta) = \frac{P}{2\pi\alpha Rt}\left\{\left(1-\rho^2\right)\left[\frac{-\rho^2+\cos 2(\alpha-\theta)}{\rho^4+1-2\rho^2\cos(\alpha-\theta)}-\frac{-\rho^2+\cos 2(\alpha+\theta)}{\rho^4+1-2\rho^2\cos(\alpha+\theta)}\right]\right\} \quad [3]$$

where $\Phi = \begin{cases} 0 \; for \;\; 0 \le \theta \le \frac{\pi}{2} - \alpha \\ \pi \; for \;\; \frac{\pi}{2} - \alpha \le \theta \le \frac{\pi}{2} \end{cases}$, $\rho = \frac{r}{R}$ and t is sample thickness.

The geometry employed in the analytical solution is not identical to the specimens used in the mechanical testing program. This could introduce a systematic error, since the discrepancy becomes larger as flattened width increases. However a finite element study showed the maximum stress error introduced by employing this function is less than 3% for the sample geometries studied (Joyce *et al*, 2005). This error is insignificant when compared to the computational advantage provided by a discrete analytical solution.

It is assumed that IM1-24 may be considered macroscopically homogeneous, isotropic and fully linear elastic. Consequently and by assuming plane stress conditions at the sample surface, analytical strain fields can be derived from the stress solution using values of Young's modulus (E) and Poisson's ratio (υ). This is an approximation for nuclear graphite, which exhibits non-linear elasticity (Taylor *et al*, 1967). The analytical strain fields were produced for each image using the appropriate applied load and arbitrary values of elastic constants (within the range of 1 GPa < E < 15 GPa and $0.1 < \upsilon < 1.0$). In each case the error (*Err*) between the measured strain fields and the corresponding analytical fields within the central region of n pixels (100×100) was calculated according to Equation [4].

$$Err = \frac{\sum_{i=1}^{n}\left[\left(\frac{\varepsilon_{H,\exp}-\varepsilon_{H,an}}{\varepsilon_{H,an}}\right)^2+\left(\frac{\varepsilon_{V,\exp}-\varepsilon_{V,an}}{\varepsilon_{V,an}}\right)^2\right]}{n} \quad [4]$$

where the subscripts 'an' and 'exp' refer to the analytical and experimentally derived strain (ε) values in the horizontal (H) and vertical (V) directions at a given pixel. Bulk elastic properties were derived by employing an optimisation routine to find the values of Young's modulus and Poisson's ratio which, when used to create analytical strain fields, would minimise *Err*. The value of *Err* at which this occurs, Err_{min}, is a measure of how well the analytical function is able to fit the experimentally observed strain fields. High values of Err_{min} suggest that the measured strain fields do not comply with those produced by the uniformly loaded disc test geometry, thereby indicating that the sample is either misaligned or is exhibiting macroscopic cracking. In either case the predicted elastic constant values are likely to be spurious, and were rejected if Err_{min} exceeded 4%. The properties recovered by this method are shown in Figure 2 for a range of sample geometries, as a function of applied load.

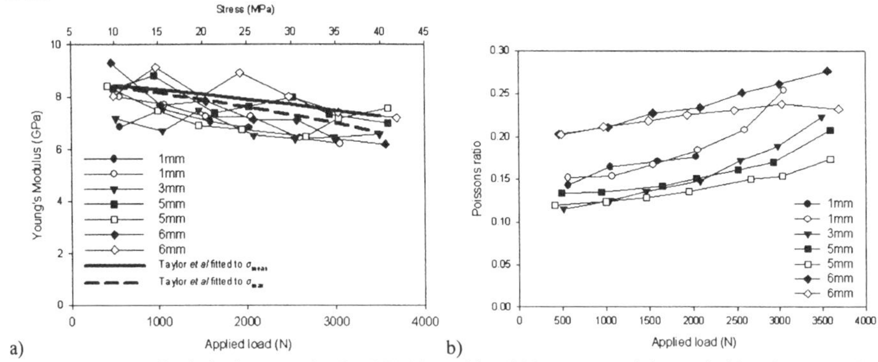

FIGURE 2: Derived elastic properties for IM1-24 graphite: (a) Young's modulus, and, (b) Poissons's ratio. Data for Young's modulus of an isotropic nuclear graphite of similar microstructure (Taylor *et al*, 1967) at stress levels comparable to the mean and maximum compressive stresses evolved in the centre region of the disc samples is also shown.

Results were not obtained for applied loads below ~500 N and above ~3500 N due to the Err_{min} criterion. Within this load range the Young's modulus result is reasonably consistent between the sample geometries, and has an average value of 7.38 GPa ± 0.77 GPa. There is a slight decreasing trend with increasing applied load level. The disc test results may be compared with those from uniaxial compression tests by considering the results at corresponding stress levels in the two sample geometries. The thick lines in Figure 2a show results obtained by Taylor *et al* (1967) from a similar isotropic graphite at stress levels corresponding to the mean and maximum vertical compressive stresses evolved within the central region of the disc compression sample. These lines define a band which correlates well with the disc test results.

The Poisson's ratio shows a generally rising trend with increased load. Strain dependence of Poisson's ratio has not been reported previously. However the average value of 0.18 ± 0.04 is consistent with literature values for this graphite grade (Taylor *et al*, 1967). Both the Young's modulus and Poisson's ratio results appear consistent with expected values, with no apparent influence of sample geometry on the recovered values. The lack of such a trend indicates that the potential systematic error introduced by the chosen analytical function is insignificant. This allows the methodology to be applied to the investigation of strain localisation at the microstructural scale.

MICROSTRUCTURAL SCALE STRAIN LOCALISATION

Microstructural strains should be detectable using the ESPI technique employed to determine bulk properties. As applied the technique has a resolution of ~90 μm within the central region, compared to pore dimensions of up to several hundred μm and filler particles of more than 500 μm. It is evident from the strain maps shown in Figure 1 that significant points of strain localisation are produced on the surface of the disc compression samples during the loading cycle. Characterising the distribution and behaviour of these regions will provide a significant insight into the deformation and fracture behaviour of this material. Since only surface measurements are possible that some of the strain fields observed may be caused by processes operating in the bulk of the material.

In well aligned disc compression tests with appropriately distributed loading, the mode of failure is by vertical splitting (Rudnick 1963), due to tensile stresses evolved at the disc centre. Hence it is assumed that the cracking and strain localisation that is a precursor to failure will be most evident in the horizontal tensile direction. Therefore, the strain localisation has only been characterised in the horizontal direction.

In the disc compression geometry, regions of micro-scale strain localisation cannot be defined by simply considering their deviation from the global mean strain, since unlike uniaxial tension or compression the macroscopic stress distribution is non-uniform. However the analytical solution that was fitted to derive the global elastic constants may be adequately used as a baseline and the micro-strain localisation can then be defined in terms of the magnitude of the discrepancy between measured strains and analytical solution. This is illustrated in Figure 3, which shows the measured horizontal strain field and the corresponding analytical fit for a 3 mm flattened width sample at a load of 2 kN ($P/P_{\max} \sim 0.45$). Several regions of locally raised strain are evident. Clearly at higher strains, when cracking is evident, the elastic properties derived from the fitted analytical solution must be invalid. However the solution may still be adequately used as a baseline from which to characterise the apparent strains, which result from the displacements associated with cracking. Hence analytical strain fields were derived for all load increments irrespective of the associated value of $Err_{\min}$. Data that had previously been rejected due to Err_{min}>4% was thus included.

Assuming homogeneous material properties, the error between the analytical solution and the measured strain fields is a direct measure of the strain localisation at a given point. Hence the influence of microstructure on the distribution of measured strain may be characterised by considering this parameter across the analysed surface area. This is shown in Figure 4. If strain localisation is negligible, the distribution should appear as a delta function, *i.e.* a single line centred on the zero error line. However, as shown in Figure 4a, the microstructure has a significant effect and the distribution adopts a Gaussian shape. Figure 4b shows the error distribution at a load near to failure. The distribution is skewed, with a significant fraction of high strain "outliers" and a shift in the peak position towards lower strains.

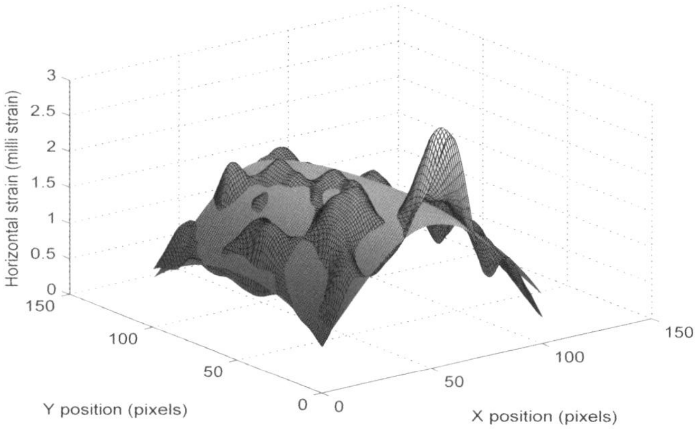

FIGURE 3: Strain localisation in IM1-24, comparison of measured and fitted analytical horizontal strain fields within the central region (100x100 pixels) of a 3 mm flattened width disc compression sample ($P/P_{max} \sim 0.45$).

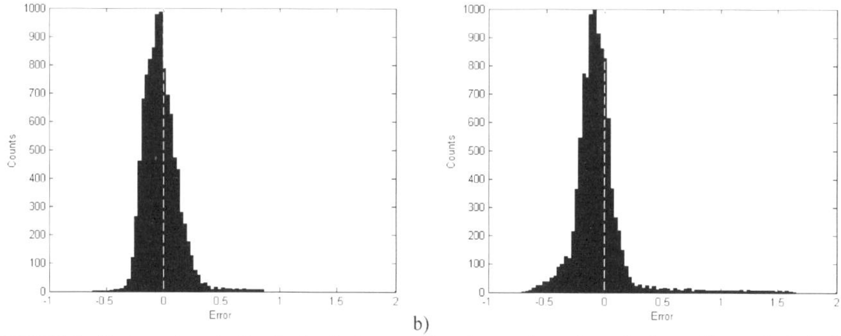

a) b)

FIGURE 4: Horizontal strain distribution in the central region on the surface of a 3 mm flattened width disc compression sample; (a) 1.4 kN ($P/P_{max} \sim 0.3$), (b) 4.5 kN ($P/P_{max} \sim 0.95$).

NON-LINEAR DEFORMATION AND FAILURE INITIATION

The error distribution alone does not give any insight into spatial characteristics of the strain localisation process. Inspection of the tensile strain field with increasing load (*e.g.* Figure 3) shows that the high strains appear to be localised at a finite number of well defined points rather than randomly distributed. The degree of strain accumulation during loading may therefore be characterised for these individual regions of high strain. This procedure has been carried out for the strain field shown in Figure 3 by identifying, and measuring, the local maxima in the error field as they evolve with increasing applied load. In the example shown for a 3 mm flattened width disc, 14 distinct strain localisation outliers were identified within the central 9×9 mm region. The accumulation of maximum strain at these points is shown in Figure 5. The strains increase linearly with load until immediately prior to sample failure. Seven samples have been analysed to date, and in each case nonlinear deformation, significantly prior to failure, was only observed at a single outlier.

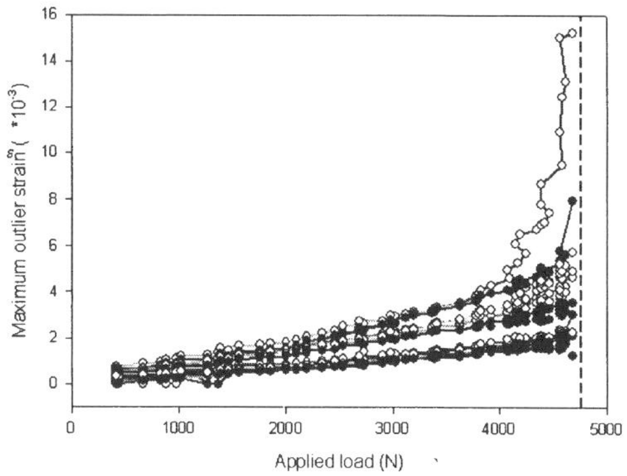

FIGURE 5: Evolution of strain within regions of strain localisation (outliers) with the application of load. The dashed line marks the load at which the sample failed (data for a 3 mm flattened width disc). Similar data was obtained for all other analysed samples.

DISCUSSION

The analysis has shown that elastic properties may be derived by measuring the surface strain fields and then correlating these with an analytical stress solution, assuming homogeneous, isotropic, plane stress behaviour. The elastic moduli were only obtained at moderate to high strains as the errors inherent within the ESPI technique are at their most significant due to the small strain increments relative to the preloaded state at the beginning of the disc compression test. Direct comparison of the Young's modulus results with literature values is not possible due to the non-uniform and biaxial stress state in the disc compression sample. However, plotting the results from uniaxial compression tests at stress levels equivalent to both the mean and maximum compressive stresses evolved in the analysed region of the disc sample defines a band of behaviour which falls within the scatter of the disc test results. This indicates that stress state does not appear to significantly affect Young's modulus. The observed rise in the derived Poisson's ratio with increasing load appears to be significant, and a possible mechanism may be a gradual onset of diffuse micro-cracking. This has been observed, using X-ray micro-tomography, in strained graphite of the same microstructure (Hodgkins *et al*, 2005). The micro-cracks would be expected to be aligned perpendicular to the maximum tensile stress, reducing the elastic modulus in this direction (Talreja *et al*, 1985). The consequent elastic anisotropy would be manifest in the derived Poisson's ratio, since the analysis assumes isotropic behaviour. The observed rise Poisson's ratio is thus understood to be a consequence of the specimen geometry, and is not the intrinsic material property. The implication is that significant micro-cracking occurs at relatively low applied loads in the disc compression test.

Although the analytical function that has been used to derive the elastic moduli is an approximation, it is adequate to use as a baseline for the purpose of defining regions of strain localisation. The error distributions take on a Gaussian form, even at low load conditions (Figure 4). The width of the distribution is significant, compared to that for a homogeneous material of comparable modulus, such as Perspex (Joyce *et al,* 2005). The heterogeneous graphite microstructure therefore has inherent strain localisation, which may be due to porosity, differences in the elastic properties of matrix and coarse filler particles or micro-cracking.

It is suggested that the significant rise in the maximum strain for a single outlier prior to specimen failure, in each of the seven tests analysed to date (Figure 5), is the consequence of significant crack nucleation, at or near to the specimen surface. The increased displacements that are associated with crack opening appear as strain in the ESPI analysis. It is proposed that coalescence of the micro-cracks is the precursor to crack nucleation. Work is in progress, employing X-ray micro-tomography, to determine the critical applied strains or stress condition for nucleation and coalescence of micro-cracks.

CONCLUSIONS

- Young's modulus and Poisson's ratio may be successfully derived from disc compression samples, by optical strain mapping of the surface strain fields during loading.
- The distribution of microstructural strains is generally Gaussian at low loads, with evidence increasing strain localisation at higher loads.
- It is proposed that the failure of the samples is preceded by formation of a critical crack nucleus from the coalescence of micro-cracks.

Acknowledgements
The authors would like to thank British Energy Generation Ltd. for financial and material support under grant GRA/GNSR/6022. The opinions expressed in the paper are those of the authors, and not necessarily those of British Energy plc.

References

Awaji, H. and Sato, S. (1979). Diametral Compressive test method. *J Eng Mat Tech*, 101, 139-147.

Erne, O., Guist, C., Bode, H. and Ettemeyer, A. (1998). Proceedings of SEM conference, USA.

Hodgkins, A., Marrow, T. J., Mummery, P., Marsden, B. and Fok, A. (2005). X-ray tomography observation of crack propagation in nuclear graphite. Proceedings of "Application of x-ray tomography to study fracture and crack growth", Churchill College, Cambridge, 7-8 April 2005.

Hung, K.-M. and Ma, C.-C. (2003). Theoretical analysis and digital photo elastic measurement of circular disks subjected to partially distributed compressions. *J Expt Mech*, 43 [2], 216-224.

Joyce, M. R., Mouterde, B., Jegou. S. and Marrow, T. J. (2005). Mechanical property determination of nuclear graphite by diametral compression testing. To be submitted to *J. of Strain Analysis*.

Rudnick, A., Hunter, A. R. and Holden, F. C. (1963). An analysis of the diametral compression test. *Materials Research and Standards*.

Talreja, R., Yalvac, S., Yats, L. D. and Wetters, D. G. (1992). Transverse cracking and stiffness reduction in cross ply laminates of different matrix toughness. *J Compos Mater*, 26, 1644 –1663.

Taylor, R., Brown, R. G., Gilchrist, K., Hall, E., Hodds, A. T., Kelly, B. T. and Morris, F. (1967). The mechanical properties of reactor graphite. *Carbon*, 5, 519-531.

Wang, Q. Z., Jia, X. M., Kou, S. Q., Zhang, Z. X. and Linqvist, P.-A. (2004). The flattened Brazilian disc specimen used for testing elastic modulus, tensile strength and fracture toughness of brittle rocks: analytical and numerical results. *Int. J. Rock Mech Ming Sci.*, 41, 245-253.

Evaluation of the Brazilian Disc Test in Fracture Strength Measurements of Nuclear Graphite

Haiyan Li, Alex S L Fok[4] and Barry J Marsden
Nuclear Graphite Research Group, School of Mechanical, Aerospace and Civil Engineering, The University of Manchester, PO Box 88, Sackville Street, Manchester, UK.
Email: alex.fok@manchester.ac.uk

Abstract
The Brazilian disc test or disc under diametral compression has been used as a means of measuring the fracture strength of nuclear graphite samples trepanned from the moderator cores as part of the monitoring campaign for ageing graphite-moderated reactors. However, high contact stresses in the vicinity of the loading points may result in failures outside the central region where the main failure is expected, thus giving inaccurate estimates of the fracture strength. Several methods have been suggested to alleviate the problem, *e.g.* the introduction of flats to the top and bottom of the disc or the use of circular anvils. In this paper, the finite element method is used to examine the influence of the above modifications to the test on the stress distributions within the disc. In addition, a continuum damage mechanics failure model is used to simulate the fracture process so as to compare the fracture behaviour of the different configurations. The possible effect of irradiation-induced material degradation on the measured fracture strength is also considered.

Keywords
Brazilian disc test, fracture strength, finite element analysis

INTRODUCTION

Nuclear graphite samples are routinely trepanned from the UK graphite-moderated reactors for testing to monitor irradiation-induced changes in material properties. The trepanning techniques currently used produce cylindrical samples with a circular cross-section. Traditionally, uniaxial tensile tests such as beam in bending have been used to determine the tensile strength of nuclear graphite. These would require the cylindrical trepanned samples to be machined into rectangular beam specimens prior to testing. Given that the samples are expensive to obtain, as only a small number are trepanned during outages, and they have to be handled remotely because of their high doses, it would be advantageous to have alternative tests which require less machining in specimen preparation. To this end, the Brazilian disc test or disc in diametral compression (Awaji and Sato, 1979; Vardar and Finnie, 1975) has been used as an alternative. This is a mechanical test for measuring the tensile strength of brittle materials by compressing a disc along a diameter using anvils (see Figure 1). This sets up tensile stresses in the plane of the disc which are perpendicular to the loaded diameter, the magnitude of the maximum tensile stress at the centre of the disc being about a third of that of the maximum compressive stress. As the disc specimens can easily be machined out from the cylindrical trepanned sample, the machining involved could be minimised. Another advantage of this test is that the tensile stress distribution along the loaded diameter is rather uniform; see later.

---

[4] Author to whom any correspondence should be addressed.

The high contact stresses in the vicinity of the loading points may, however, result in failures outside the central region where the main failure is expected, thus giving inaccurate estimates of the fracture strength. Several methods have been suggested to alleviate the problem by spreading out the load, *e.g.* the introduction of flats to the top and bottom of the disc or the use of circular anvils (Awaji and Sato, 1979; Vardar and Finnie, 1975).

In this paper, the finite element method will be used to examine the influence of the above modifications to the test on the stress distributions within the disc. In addition, a continuum damage mechanics (CDM) failure model (Shi *et al.*, 2004; Zou *et al.* 2004) will be used to simulate the fracture process so as to compare the fracture behaviour of the different configurations. The possible effect of irradiation-induced material degradation on the measured fracture strength will also be considered.

ELASTIC STRESS ANALYSIS

The three different configurations of the disc in diametral compression considered in this paper are shown in Figure 1. These include (a) a full disc with plane anvils; (b) a full disc with circular anvils; and, (c) a disc with flats and plane anvils. The diameter of the discs is 28.8 mm and the thickness is 15 mm. The radius of curvature of the circular anvils (Figure 1b) is 21 mm and the length of the flats in Figure 1c is 6 mm. To study the failure behaviour of each configuration, the elastic stress distributions within the disc are analyzed first using the finite element (FE) method. Note that analytical solutions are available for cases (a) and (b); see *e.g.* Awaji and Sato (1979) and Timoshenko and Goodier (1970).

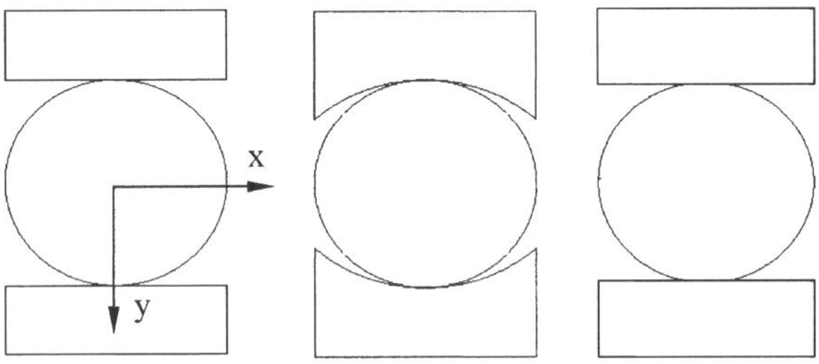

(a) Plane anvils (b) Circular anvils (c) Disc with flats

FIGURE 1: The different configurations of disc in diametral compression.

FE Models

The commercial FE package ABAQUS (HKS, 2004) was used for all the stress analysis. For each configuration, only a quarter of the disc was analysed to exploit symmetry. Eight-node quadrilateral plane-strain elements were used. The bottom anvil was constrained in the y-direction and a uniform pressure was applied on the top anvil. Contacts between the disc and the anvils were considered using the "Contact Surface" facility in ABAQUS (HKS, 2004). The coefficient of friction between the two materials was assumed to be 0.1.

The assumed material properties for the graphite disc and the steel anvils are given in Table 1. Those for graphite are based on the properties of an electrode graphite for the steel-making furnace and are in the direction perpendicular to that of extrusion (Awaji and Sato, 1979).

This graphite was chosen because it allows direct comparison of the results presented in this paper and those reported in Awaji and Sato (1979).

TABLE 1: Material properties for graphite and steel.

	Young's modulus E (GPa)	Poisson's ratio ν	Tensile strength σ_t (MPa)	Fracture toughness G_{IC} (MPa·mm)
Graphite	5.5	0.07	6.3	0.12
Steel	210	0.3	—	—

Elastic Stress Distributions

Figure 2 shows the von Mises stress contours of the disc under the different configurations. Compared with the circular anvil (Figure 2b), the higher stress concentration produced in the full disc by the plane anvil (Figure 2a) at the point of contact can clearly be seen. But away from the point of contact, the stress distributions of the two cases are quite similar. For the disc with flats (Figure 2c), the stress distributions show more differences from the other two cases with a full disc. There also appear to be stress concentrations at the outer edges of the flats, although their magnitudes seem to be lower than those in the full disc loaded by plane anvils.

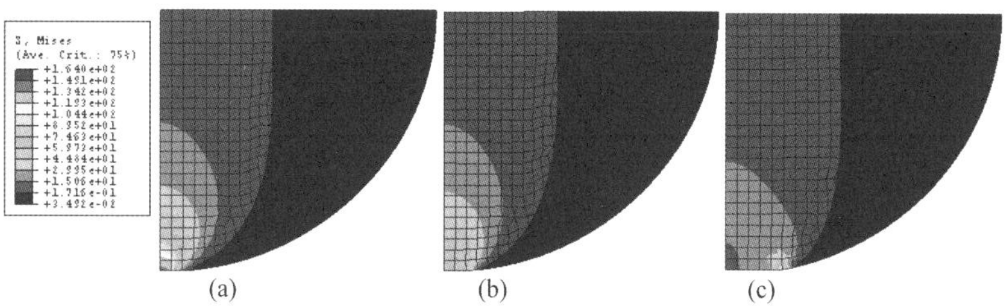

FIGURE 2: Von Mises stress contours in discs under diametral compression: (a) full disc with plane anvils; (b) full disc with circular anvils; and, (c) disc with flats and plane anvils.

Figure 3 shows the distributions of the direct stress perpendicular to the loaded diameter. It can be seen that this stress is tensile in the majority of the disc, but in the vicinity of the point of contact it is compressive. The disc with flats again shows more differences in its stress distribution from the other two cases with a full disc, with the compressive stresses being concentrated around the outer edges of the flats.

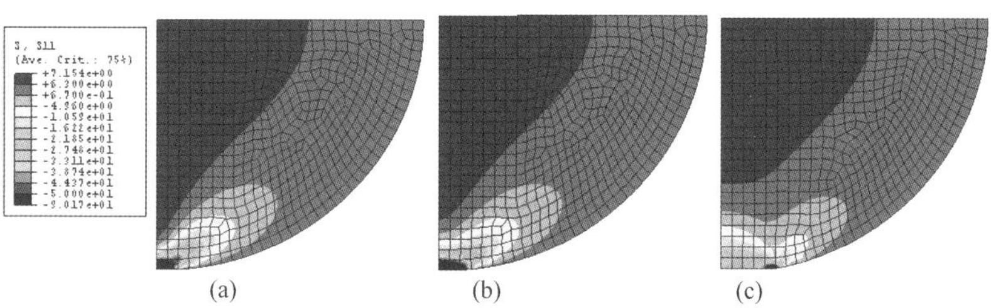

FIGURE 3: σ_{xx} (or σ_{11}) contours in discs under diametral compression: (a) full disc with plane anvils; (b) full disc with circular anvils; and, (c) disc with flats and plane anvils.

Figure 4 shows the distributions of σ_{xx} (or σ_{11}) and σ_{yy} (or σ_{22}) along the x- and y-axes. The stresses are normalized against the theoretical maximum tensile stress at the centre of the disc, *i.e.*

$$\sigma = \frac{P}{\pi R t} \qquad [1]$$

where P is the load, R is the disc's radius and t is its thickness.

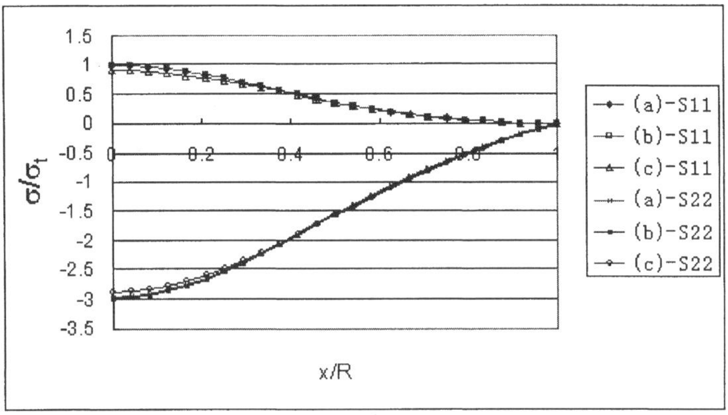

FIGURE 4a: Normalized σ_{xx} (or σ_{11}) and σ_{yy} (or σ_{22}) along x-axis.

It can be seen from Figure 4a that, along the diameter (x-axis) perpendicular to the loaded axis, the stress distributions are essentially unaffected by the shape of the anvils or the presence of the flats, although a small reduction in stress magnitude is seen with the latter. Figure 4a also confirms that the maximum compressive stress in the loading direction at the centre of the disc has a magnitude three times that of the maximum tensile stress in the perpendicular direction. Both of the direct stresses reduce to zero towards the edge of the disc.

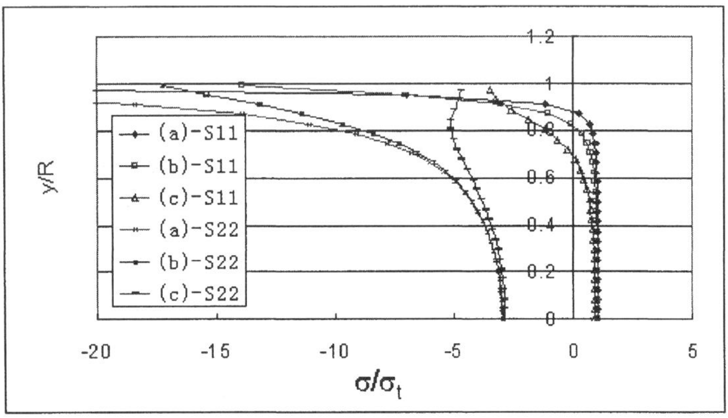

FIGURE 4b: Normalized σ_{xx} (or σ_{11}) and σ_{yy} (or σ_{22}) along y-axis.

As expected, the stress distributions along the loaded diameter (Figure 4b) show more variations between the different configurations, especially close to the loading point. The reduction in the stress concentration using circular anvils or flats on the disc can clearly be seen, with the latter giving more significant reduction. The tensile stress in the x-direction is uniform over 50% of the loaded diameter for all cases.

Contact Area

The contact area between two cylindrical bodies of isotropic materials can be expressed as (Timoshenko and Goodier, 1970):

$$b = \sqrt{\frac{4P}{t}\frac{(k_1 + k_2)R_1 \cdot R_2}{R_1 + R_2}} \qquad [2]$$

where $k_1 = \dfrac{1-\nu_1^2}{\pi E_1}$ and $k_2 = \dfrac{1-\nu_2^2}{\pi E_2}$; [3]

and where b is the half width of contact; P/t is the load per unit length of cylindrical surface; R_1 and R_2 are the radii of curvature of the surfaces (positive for convex surfaces); ν_1 and ν_2 are Possion's ratios and E_1 and E_2 are Young's moduli of the contacting bodies.

Assuming a critical tensile stress of 6.3 MPa (Table 1), Equation [1] gives a critical load of 4275 N. According to Equation [2] and the materials given in Table 1, the analytical contact lengths for configurations (a) and (b) at this critical load are 3.13 mm and 5.43 mm, respectively. Figure 5 gives the FE results of the contact pressure distribution along the contact length for the three configurations at the same critical load. The effectiveness of the circular anvils and the flats in reducing the contact pressure can again be seen, but there are also some high contact pressures at the corners of the flats which result in the high stress concentrations identified in Figures 2c and 3c. The contact lengths for configurations (a) and (b) are roughly 2.32 mm and 4.72 mm, respectively. The differences with the analytical results are probably caused by the coarseness of the FE mesh.

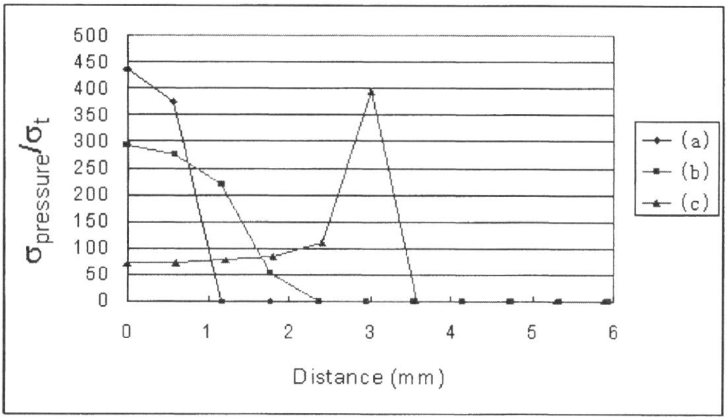

FIGURE 5: Contact pressures along the contact surface of the disc: (a) plane anvils; (b) circular anvils; and, (c) disc with flats and plane anvils.

FAILURE ANALYSIS USING CDM MODEL

The fracture process of the disc under the different configurations was then simulated using a continuum damage mechanics (CDM) failure model for nuclear graphite developed at the University of Manchester (Shi *et al.*, 2004; Zou *et al.* 2004). The CDM model uses both the conventional stress-based and fracture-mechanics-based failure criteria and couples together the effects of mode-I and mode-II loadings. A damage parameter is introduced to the interfacial constitutive law, which relates tractions and displacements across the interface, to take account of the effects of material damage. Therefore, the CDM model can predict both the initiation and propagation of cracks in brittle components with high stress concentration or even stress singularity. The model has been implemented into ABAQUS (HKS, 2004) as interface elements using its User Element Subroutine (UEL). These interface elements have no thickness and can be implanted in between continuum solid elements at positions where potential crack surfaces may form.

The fracture simulation required the whole disc to be modelled. The boundary conditions were similar to those imposed in the elastic stress analysis, but loading was applied through prescribed displacements at the top anvil and reaction forces determined from the restrained end of the bottom anvil. Four-node plane-strain elements were used for the continuum solid and interface elements were placed throughout the disc. The 'Stabilize' facility in ABAQUS (HKS, 2004) was used to improve stability of the numerical analysis during the occurrence of fracture, and ABAQUS would terminate the computation when global instability took place, at which point gross failure of the disc was assumed to occur.

Figure 6 shows the final fracture appearances of the discs at the point of gross failure. In all the cases considered, multiple cracking was predicted, with the main crack being along the direction of the loaded diameter, *i.e.* perpendicular to the tensile stresses, as observed in experiments. Considerable damage due to crushing was also predicted at the points of contact prior to the main failure, with the full disc with circular anvils (case (b)) suffering the least damage through crushing.

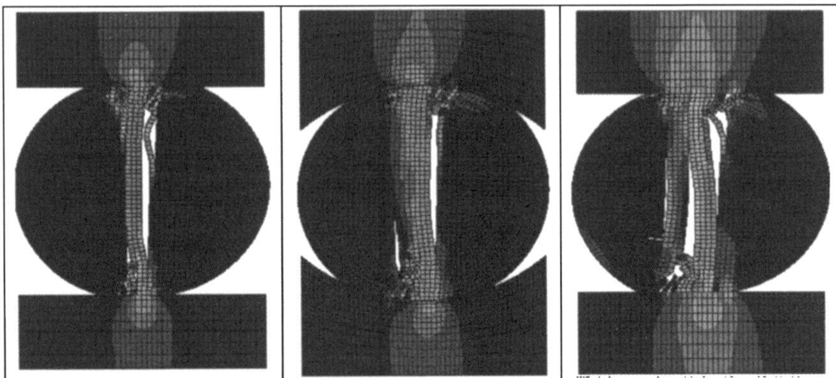

(a) Full disc with circular anvils	(b) Full disc with plane anvils	(c) Disc with flats and plane anvils

FIGURE 6: Final fracture appearances of the discs under diametral compression (3 times enlarged deformation)

Figure 7 gives the load-displacement curves for the three configurations. It shows that, similar to the results found by Awaji and Sato (1979), the full disc with plane anvils gives the lowest failure load, while the other two cases give a similar load but different displacements at final failure. The predicted failure stresses using the final loads for the three configurations are (a) 4.05 MPa, (b) 6.63 MPa and (c) 6.43 MPa, respectively. It can be seen that the disc either with flats or loaded by circular anvils gave predicted strengths which were similar to the input value of 6.3 MPa. The figure also shows a series of "pop-ins" taking place prior to the main failure. Lower predictions (3.68, 5.00 and 5.00 MPa) would be obtained if the loads at the point where the first load-drop occurred (circled in Figure 7) were used instead.

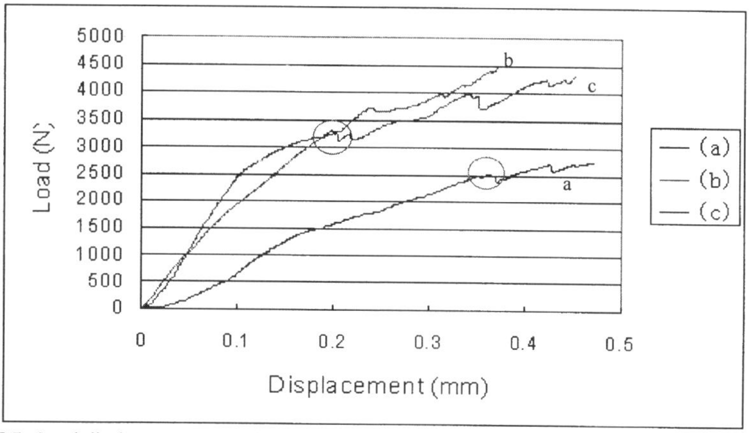

FIGURE 7: Load-displacement curves for (a) full disc with plane anvils; (b) full disc with circular anvils; and, (c) disc with flats and plane anvils.

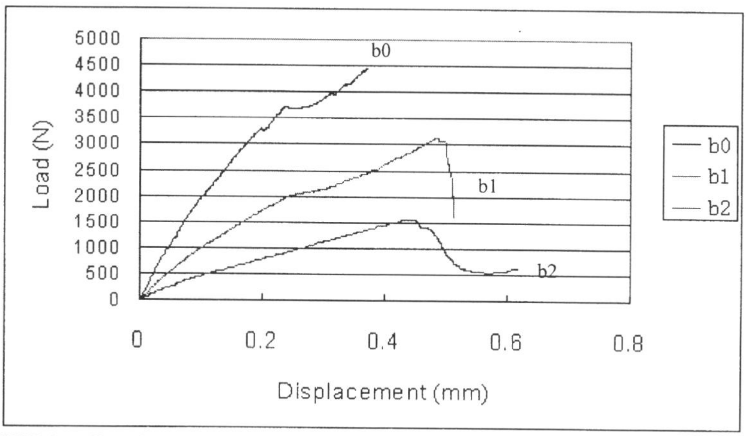

FIGURE 8: Effect of strength degradation on load-bearing capacity of disc loaded by circular anvils.

Figure 8 shows the predicted effects of material degradation on the load-bearing capacity of the disc loaded with circular anvils to simulate the effects of irradiation damage. Both the tensile strength and Young's modulus of graphite were reduced to simulate material degradation, the assumed values being (b1) E = 2.786 GPa and σ_t = 3 MPa, and (b2) E = 1.412 GPa and σ_t = 1.434 MPa. The case (b0) used the virgin values as listed in Table 1.

The predicted strengths for the cases with degraded materials are (b1) 4.4 MPa and (b2) 2.2 MPa, respectively, which are higher than their input values. It seems therefore that this particular configuration might cause an overestimate of the strength of degraded graphite. A possible reason for this strength overestimation could be the increased crushing at the loading points which resulted in a reduction of the maximum tensile stress at the centre of the disc for the same applied load.

CONCLUDING REMARKS

Finite element analysis has been carried out to investigate the influence of modifications to the disc in diametral compression test whose aim is to reduce the contact stresses at the loading points, thereby minimizing the amount of crushing damage. Both the introduction of flats to the disc and the use of circular anvils seem to be effective in reducing these contact stresses while keeping the stresses in the middle of the disc relatively unchanged. For the disc with flats loaded by plane anvils, high stress concentrations have been predicted at the edges of the flats. The significance of these stress concentrations is not clear and would require further investigation.

The CDM failure model has been used successfully to simulate the fracture process, producing fracture patterns and strength predictions which are similar to those expected. The results confirmed that the use of plane anvils and discs without flats could lead to underestimation of the strength of the material. Simulation using degraded material properties, on the other hand, pointed to the possibility of overestimating the strength of degraded graphite when circular anvils were used.

Despite the apparent success of the CDM model, a word of caution is necessary here. The fracture simulation results appeared to be mesh-sensitive to some extent. Also, possible local densification and yielding of the graphite material under high compressive stresses have not been considered. Further work is therefore necessary to refine the numerical analysis.

Acknowledgements

The financial support of Nexia Solutions is gratefully acknowledged. The views expressed in this paper are those of the authors and do not necessarily represent those of the sponsor.

References

Awaji, H. and Sato, S. (1979). Diametral compressive testing method. *Journal of Engineering Materials and Technology*, 101, 139-147.

HKS, ABAQUS/Standard User's Manual, Version 6.4, 2004.

Shi, L., Li, H., Fok, S. L. Marsden, B. J. and Yu, S. (2004). The effect of softening on the predicted strength of brittle materials using a continuum damage mechanics failure model, *The 2nd International Topic Meeting on High Temperature Reactor Technology*, Beijing, China, September 22-24, 2004.

Timoshenko, S. P. and Goodier, J. N. (1970). Theory of elasticity, 3rd edition, McGraw-Hill Inc.

Vardar, Ö. and Finnie, I. (1975). An analysis of the Brazilian disk fracture test using the Weibull probabilistic treatment of brittle strength. *International Journal of Fracture*, 11, 495-508.

Zou, Z., Fok, S. L., Oyadiji, S. O., and Marsden, B. J. (2004). Failure predictions for nuclear graphite using a continuum damage mechanics model. *Journal of Nuclear Materials*, 324, 116-124.

Irradiation Damage in Graphite from First Principles

Malcolm I. Heggie[1], Irene Suarez-Martinez[1], Gemma Haffenden[1], Gianluca Savini [1], Ahlam El-Barbary[2], Christopher Ewels[3], Robert Telling[4], Christopher Cousins[5]

[1]Department of Chemistry, University of Sussex, Falmer, Brighton, BN1 9QJ

[2]Ainshams University, Cairo, Egypt

[3]LPS, University of Paris-South, France

[4]DSTL, Poole, UK

[5]School of Physics, Exeter University, EX4 4PY

Email: m.i.heggie@sussex.ac.uk

Abstract

The authors use first principles total energy calculations, of a kind unavailable to anyone previously, to build up a picture of the structures, energies and mobilities of point and extended defects in graphite, including the interstitial, the vacancy and their clusters, boron impurities and basal and non-basal dislocations. It is clear from our calculations that the historic model of atomistic processes in radiation damage of graphite is incorrect and the authors offer a compelling and coherent alternative picture based on interplanar bonds, sub-threshold processes and buckling of graphite planes.

Keywords

Graphite, radiation damage, first principles

INTRODUCTION

Graphite is one of the best studied materials of the last six decades. Its properties have been extensively measured and reported in journals such as Carbon, Chemistry and Physics of Carbon and in books such as The Physics of Graphite (Kelly, 1980), amongst others. In nuclear applications a rigorous 'engineering' approach has been taken in which conservative predictions of graphite integrity are made from materials test reactor data spanning a wide range of graphite types.

The academic community has always sought to interpret these data in terms of atomistic models, in the hope of finding new 'physics' or 'chemistry'. Their interpretations have been logical but not always based on the correct premises. For example, it was always believed that the self-interstitial is highly mobile between layers even at low temperatures and this led to a description of the evolution of radiation damage similar to precipitation of new graphene planes from a continuously fed two dimensional gas of carbon.

In our time, computer modelling from first principles has been able to offer a unique and definitive picture of atomistic processes occurring with radiation damage. This picture does not undermine materials predictions that only concern macroscopic phenomena, but affords deeper insight into the fundamental origin of these phenomena and predictions. The picture is not yet complete, but so far affords a new insight into irradiation at lower temperatures, where the authors have shown, amongst other things, that the carbon self-interstitial, C_i, cannot be mobile. In this article, our approach will be: 1) to justify our choice of the local density functional method; 2) to summarise findings for the vacancy, the interstitial and their various complexes, including their relation to experiment; and, 3) to illustrate the consequences of these findings for physical property changes, particularly dimensional change.

The principal conclusion is that radiation induced point defects at temperatures below 250 °C are both immobile and pin planes together. Most importantly, these two facts cause pronounced buckling of layer planes below 250 °C. This buckling is expected to be small at normal reactor operating conditions (450-650 °C).

As far as normal operating conditions are concerned, the authors find that the essential processes occurring within radiation damage have activation energies of 1.2 eV or greater, offering an atomistic basis for the equivalent temperature concept, where materials test reactor data are corrected for a thermal annealing effect with a 1.2 eV barrier (Bell *et al.*, 1962).

APPLICABILITY OF LOCAL DENSITY FUNCTIONAL CALCULATIONS

It has become an article of faith that density functional theory is unable to properly describe graphite, because it does not contain the dynamic correlation that gives rise to van der Waals forces. Such forces are the only important interactions between, for example, inert gas atoms and were thought to control the interlayer separation in graphite. However, there has always been a thread of dissent, beginning with the objection that there should be no interlayer conductivity in this case (P. Thrower, private communication) and continuing with the pronounced success of the simplest density functional method (the local density approximation, LDA) in obtaining graphite structures and band structures (Charlier *et al.*, 1994).

This year, both LDA and its refinement (generalized gradient approximation or GGA) were applied to graphite elastic properties. Elastic constants are the second differential of energy with respect to strains. Their indices indicate which strains are involved, for example, C_{33} is the c-axis stiffness and C_{13} is the interaction between a-axis and c-axis strains. LDA was found to give good elastic constants, with the single exception of C_{13} (Mounet and Marzari, 2005), which turned out small and negative. GGA proved much worse largely because the interlayer separation was poor (c/a = 3.45 instead of 2.725). Below it is shown the results for LDA using the AIMPRO code (Jones and Briddon, 1998). The authors used optimised lattice parameters of 2.4519 and 6.6527 Å (c/a = 2.7133) which turn out to be within 0.3% of experimental values. Our numerical differentiation uses a polynomial fit to 5 energy points separated by strain increments of 5×10^{-5} and it is noted that special attention must be paid to the precision of optimisation of the c lattice parameter, and differentiation of the energy with respect to it.

The table omits C_{13} for which there is an experimental ambiguity. Strain gauge measurements give 15 GPa, while compressibility data (Zhao and Spain, 1989) give 7.8 GPa when properly corrected (Cousins and Heggie, 2003). At the time of writing the authors obtain 15 GPa, but this is subject to further verification.

TABLE 1: Elastic constants for graphite obtained from LDA compared with experiment.

Elastic constant	LDA (GPa)	Experiment (GPa)
C_{11}	1104	1060 ± 20
$C_{12}=(C_{11}+C_{12})-C_{11}$	178	180 ± 20
C_{33}	35	36.5 ± 1
C_{44}	4.8	5.0 ± 0.5

It is clear from Table 1 that the agreement between the LDA calculations and experiment is excellent and scarcely leaves any role for van der Waals interactions. As for the poor performance of GGA (Mounet and Marzari, 2005), it is possible that the simple form of LDA exchange-correlation functionals precludes numerical inaccuracies which the more sophisticated, but less theoretically rigorous, GGA functionals admit.

DEFECT PROPERTIES FROM LDA

The structural defects introduced by neutron irradiation are Frenkel pairs, which normally separate into isolated vacancies (V) and self-interstitials (C_i) that can subsequently recombine after thermally activated migration. The authors routinely use 64 atom supercells – these can lead to size effects and imprecision of several tenths of an eV in absolute formation energies. Where possible the authors have examined larger supercells to gauge the convergence with supercell size.

The Vacancy

The vacancy has a formation energy of 8.3 eV and migrates through a 1.7 eV higher split-vacancy saddle point with an almost planar four-coordinate carbon atom. In detail the vacancy is subject to a weak symmetry-breaking but dynamic Jahn-Teller distortion, which maintains its time-averaged three-fold symmetry at most temperatures.

It can dimerise strongly with vacancy in an immediately neighbouring site (as a 'first-neighbour' V_2). This is almost completely immobile but there is a pronounced barrier to its formation (3.5 eV), which leaves open the possibility of weaker, more distant dimerisations either between planes (Telling *et al.*, 2003) or within the same plane (as a third 'neighbour' V_2, which has the two vacancy sites separated by three C–C bonds). Both these weaker dimerisations lead to V-V binding energies in the range 1-2 eV, which, when added to the V migration energy, offer a plausible explanation for the reportedly high activation energy (3.1 eV) for vacancy motion (Thrower and Mayer, 1978). Importantly the interplane dimers have short and strong interplane bonds, leading to interplane pinning.

Further aggregation of vacancies to form V_n is into lines which can be shown to be lower in energy than the alternative aggregation into disks (prismatic vacancy loops) for $n \geq 6$ (El-Barbary, 2005). Such lines are self-healing and cause progressive basal plane contraction for temperatures at which vacancies are mobile.

The Self-Interstitial

The self-interstitial has a formation energy of 5.5 eV (and thus with the vacancy gives a Frenkel pair energy of 13.8 eV). It bonds strongly with two host atoms in each neighbouring plane, forming two almost-equilateral triangles of single bonds, where the planes of the triangles make an angle of 60° with each other. The five carbon atoms central to this defect have a bonding topology similar to the pentane isomer, spiro-pentane, so this is named the spiro-interstitial.

Alternative structures are at least 1 eV higher in energy: for example, a structure with two bonds to the upper plane, but only one to the lower, has a structure resembling a capped letter Y (the Y-lid structure) and a formation energy of 6.8 eV. Another has two bonds to only one plane (the grafted interstitial) and a formation energy of 7 eV.

Migration through these metastable states imply a migration energy of at least 1 eV, contradicting the experimentally inferred values of 0.02-0.4 eV (Thrower and Mayer, 1978). A detailed search for the saddle point for motion (proceeding through saddle points with three and with two bonds to host atoms) has so far yielded a 1.2 eV barrier to motion.

Aggregation of C_i proceeds via the di-interstitial and this was thought to be key to the dimensional change of irradiated graphite (*e.g.* Tanabe *et al.* 2001, Niwase *et al.*, 2002). Modelling of C_{ni} is in progress, but to date it can be reported that there is no pronounced *c*-axis expansion for the structures considered with $n = 1$ and 2. Interestingly, the authors have found a significant volume of formation, but this does not come from a large *c*-axis expansion around the defect. Many of these defects produce a basal shear between the layers they connect. In 64 atom unit cells the effective density of defects can be large enough to cause a homogeneous shear between layers equal to half a C-C bond length (0.7 Å). This allows the axis of the spiro-C_i to follow the ***c*** direction and optimal bonding with the host. Doubling both basal cell dimensions prevents the homogeneous shift happening, forcing the spiro-Ci axis to deviate from the ***c*** direction, but there is still a large local shear around C_i. This local shear causes a stacking error and it is this which pushes the planes apart. As an illustration of stacking disorder effects, AA graphite (planes eclipsing each other, rather than regular AB stacking) gives an interlayer distance of 3.6 Å. This formation volume cannot be responsible for dimensional change, since it does not have the correct relationship between the c axis expansion (δX_c) and the basal contraction (δX_a). Therefore the authors are inclined to rule out the idea that changes in interlayer spacing (d_{002}) measured by diffraction are caused by homogeneous distributions of C_{ni} where $n = 4 \pm 6$ as suggested by Martin and Henson (1964) who added the mysterious caveat that such nuclei could not nucleate new graphite planes.

Further aggregation of C_i can nucleate new graphite planes (interstitial prismatic loops) and independent authors (Brown *et al.*, 1969, Reynolds and Thrower, 1964) found a 1.2 eV activation energy for their evolution, where kinetic arguments were adduced to eliminate n>1 clusters as the mobile aggregating species (Brown *et al.* 1969). Whereas this presented early authors who believed in a very low C_i migration energy a problem, this is entirely consistent with our calculated C_i migration energy.

The Intimate Frenkel Pair

The interstitial and the vacancy might not separate completely during a collision (or might trap into a metastable state if they migrate together). This state is the intimate Frenkel pair (Ewels *et al.*, 2003) and as much as one third of Frenkel pairs are thought to be of this type according to Hennig and Hove (Kelly, 1980).

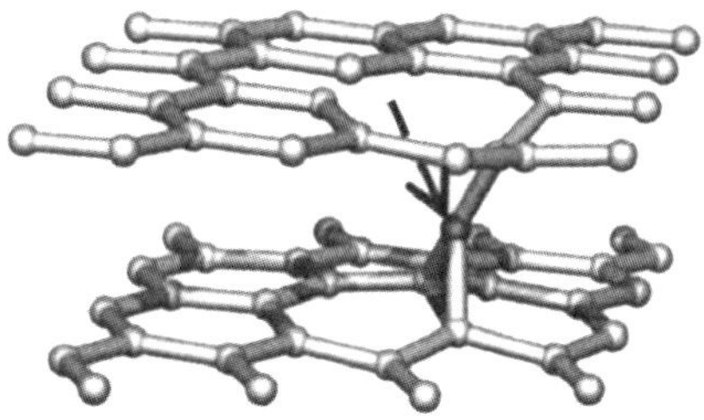

FIGURE 1: The intimate Frenkel pair from LDA. The arrow is the atomic displacement that forms the defect.

The LDA calculations of Ewels *et al* (2003) find that the iFP has approximately 10.8 eV formation energy, (giving a 2.9 eV binding between isolated V and C_i) and it experiences a 1.34 eV barrier to annihilation.

DISCUSSION

The foregoing results show that the defects in graphite are more complex than the early models reveal. The isolated defects do not produce pronounced *c*-axis expansions (in fact they appear locally quite dense compared with graphite, due to their interplanar bonding). They do however form strong shear-resistant links between graphite planes and need thermal activation to move. Such links present an elegant explanation of the role of a light neutron irradiation in stiffening the graphite interlayer shear constant, C_{44}, by eliminating reversible basal dislocation motion.

The interplanar links can also cause the graphite sheets to buckle and in the next section the authors show how this occurs. It is important to see what is happening at low temperatures, because it is from low temperature irradiations that the extremely low migration energy of Ci was deduced. After irradiation at 4K, annealing starts at around 77K (see, for example, Kelly 1980) and this was thought to correspond to movement of C_i. The authors have shown that this is impossible and therefore an alternative explanation for physical property changes and their annealing at low temperatures must be sought.

Buckled Graphite

Where graphite planes of different length are pinned together, buckling will occur – it is easier to compress the longer plane by buckling than extending the shorter plane by stretching. A model for this is an array of climb dipoles of basal edge dislocations. These are shown in Figure 2, where LDA has been used to optimise both atom coordinates and superlattice vectors. These are preliminary calculations with Monkhorst-Pack *k* point mesh of 1x1x2 and AIMPRO basis of pdpp ('pdpp' means each atom has three radial Gaussian basis functions expanded with spherical harmonics up to l =1 and one radial Gaussian expanded with spherical harmonics up to l =2). The result is a structure for which $\delta X_a/X_a$ is –10% and $\delta X_c/X_c$ is 63% – values which are typical of the dimensional changes observed for irradiations below 250 °C. This array of climb dipoles comprises perfect basal edge dislocations of line direction, $\boldsymbol{l}$ =<0,1,–1,0> and Burgers vector $\boldsymbol{b}$ = <2,–1,–1,0>, separated in the climb direction, <0,0,0,1>. Cell dimensions are chosen for computational convenience: 4 atoms were removed from one of the layers of an 80 atom, 4 layer cell. The resulting buckling adopted the fundamental mode with wavelength equal to cell width (22 Å), which is therefore a lower bound for this dipole array.

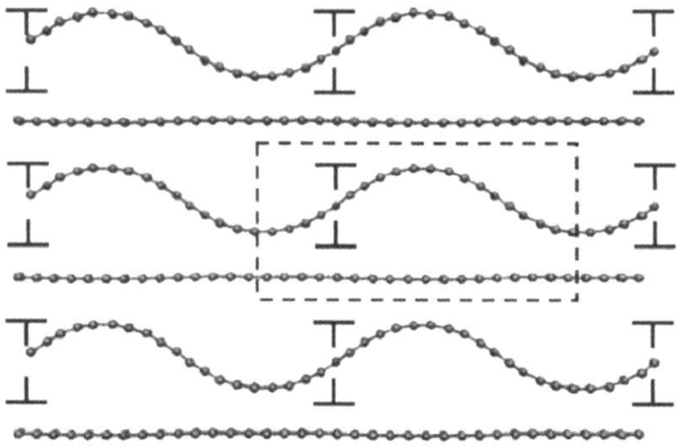

FIGURE 2: Climb dipoles of basal edge dislocations. Dislocation T symbols represent perfect dislocations of *l* = <0,1,–1,0> (into the plane of the page) and ***b*** = <2,–1,–1,0> (left-right on page). Climb direction is <0,0,0,1>. The 152 atom calculational orthorhombic supercell is denoted by the broken line.

It is indeed ironic that earlier workers (Lachter and Bragg, 1988) who found that ball milling of graphite produced comparable dimensional changes to neutron irradiation sought to explain the effects of ball milling in terms of the picture developed for neutron irradiation, *i.e.* in terms of creation of Frenkel pairs and a homogeneous distribution of C*ni* clusters. A more natural explanation is that ball milling generates basal dislocations which lock-in to produce buckled planes, and that the neutron irradiation effects follow this pattern too.

While the example of Figure 2 shows the principle of dimensional change through buckling, it is to a certain extent artificial: the structure of the supercell prevents the disappearance of the dislocations, since they lie on different planes. In practice the dislocations can glide in the basal plane and the glide is blocked by interplanar bonds (in V_2, C_i and iFP). By taking the original 76 atom unit cell, adding it to the same cell inverted and fixing the end of the straight planes together with spiro-Ci the authors created the constrained glide dipole of Figure 3.

As the interplanar shear stress from the buckling pushed against the spiro-Ci pinning point, the spiro-Ci axis moved off the c axis, but most of the buckling was retained. This calculation is presented as a proof of concept – in particular it should be noted that this is a two dimensional model of a three dimensional process.

It appears clear that different amounts of buckling arise from trapping of different numbers and distributions of basal dislocations. Higher basal dislocation concentrations lead to stronger buckling and higher internal interplanar shear stresses. The rate of escape of basal dislocations by thermally activated jumping of the pinning points is crucial but it is also stress-assisted. Dimensional change saturates when all the crystal is buckled and the interplanar shear stresses equal the ultimate shear strength of the pinning points. At 250 °C the thermally activated release from pinning points (with barriers *ca.* 1.2 eV) is rapid and internal shear stresses are small. At 4K, no thermal activation is possible and internal shear stresses reach a limit given by the density and shear strength of pinning points. Annealing occurs at intermediate temperatures by thermally activated release of pinning points which are already subject to interplanar shear stress.

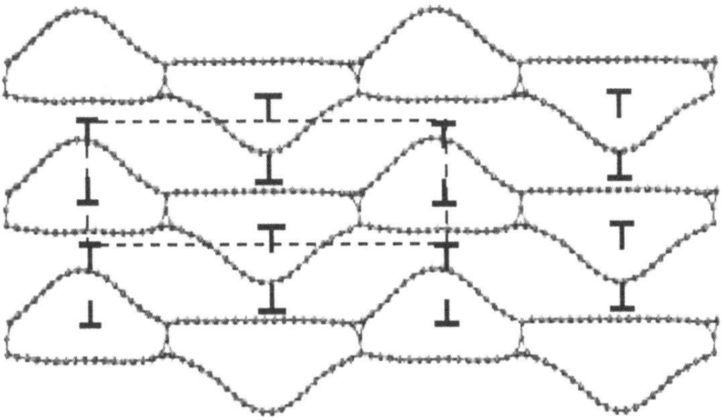

FIGURE 3: Climb dipoles of basal edge dislocations. Dislocation T symbols represent dislocation axes and dashed lines the calculational supercell. Pinning points are spiro-C_i.

It is useful to speculate how this buckling damage can arise, so the authors show a schematic in Figure 4. The cascade of damage leads to enormous and transient local heat rises (thermal spike) which cool quickly, but do so in the presence of defects which have interplanar bonds. These pinning defects lock in the buckling in a temperature dependent way – the higher the irradiation temperature, the lower the ultimate shear strength of the pinning defects and the smaller the internal stresses and buckling. Here the graphite planes are behaving like flexible inextensible sheets – which even during the thermal spike are experiencing ***c***-axis expansion and basal contraction.

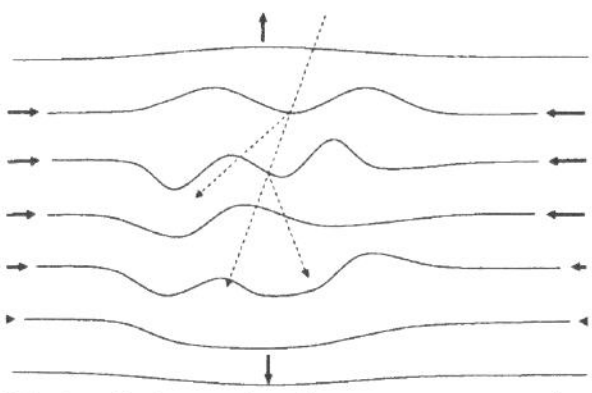

FIGURE 4a: Schematic diagram: energetic neutron and knock-on atoms excite planes which behave like flexible inextensible sheets, contracting basal plane, expanding c direction

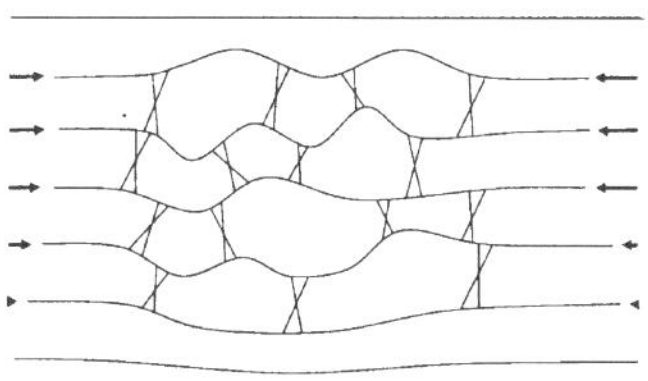

FIGURE 4b: During cooling, some of the deformation is pinned in place by interlayer bonds (here represented as spiro-C_i)

Iwata *et al.* (1974) showed that irradiation reduces ***c***-axis resistivity and that annealing in the range 85K-180K causes this reduction to continue (with a pattern of charge carrier and mobility changes characteristic of the irradiation process). Annealing would be expected to return the resistivity monotonically to near its unirradiated value. This hitherto unexplained and enigmatic result finds an elegant explanation in that buckled planes can have interplanar separations too large for the spiro-C_i, so adatoms result (Figure 5a). These convert to spiro-C_i upon gentle annealing, appearing to continue the radiation damage (Figure 5b). This occurs in the early stages of annealing because the conversion of adatoms to C_i, releases energy and has a driving force stronger than the simple unbuckling of perfect planes.

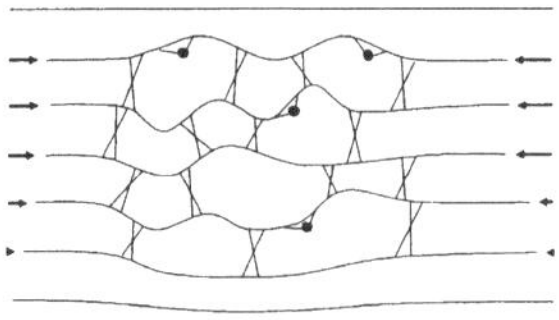

FIGURE 5a: Where interplanar distances are too large, adatoms form instead of spiro-C_i.

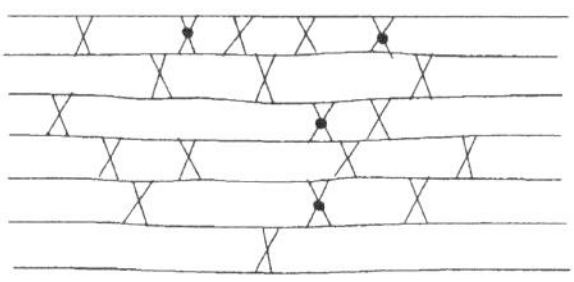

FIGURE 5b: After a gentle anneal, planes unbuckle as they slip past pinning points and new spiro-C_i from adatoms.

Thus interplanar pinning and buckling offers a better explanation of dimensional change and electrical resistivity changes than the current model, and this reinforces the finding that self-interstitial migration (which has a substantial role in the development of radiation damage at normal reactor operating temperatures) is approximately 1.2 eV and not 0.02 eV.

CONCLUSIONS

LDA calculations provide a sound basis for an atomistic description of radiation damage in graphite, including elastic properties. Defect calculations reveal that the historic picture of interstitials that are mobile at low temperatures (below 250 °C) is incorrect and radiation effects in this temperature regime are controlled by immobile defects which pin planes together, causing internal interlayer shear stresses and buckling of planes.

Acknowledgements

The authors would like to thank Dr. Mark Bradford for discussions and enthusiasms and British Energy Generation Ltd. for financial support. The views expressed in this paper are those of the authors and do not necessarily represent those of the British Energy Generation Ltd.

References

Bell, J. C., Bridge, H., Cottrell, A. H., Greenough, G. B.,Reynolds, W. N. and Simmons, J. H. W. (1962). *Phil. Trans. Roy Soc* **A254** 361.

Brown, L. M., Kelly, A., and Mayer, R. M., (1969). *Phil. Mag.*, **19** 721 (1969).

Charlier, J.-C., Gonze, X., and Michenaud, J.-P., (1994). *Europhysics Letters,* **28**, 403.

Cousins, C. S. G. and Heggie, M.I. (2003). *Phys. Rev. B*, **67,** 024109.

El-Barbary, A. A, D.Phil. Thesis, University of Sussex, 2005.

Ewels, C. P., Telling, R. H., El-Barbary, A., Heggie, M. I., and Briddon P. R., (2003). *Phys. Rev. Lett.,* **91,** 025505.

Iwata, T., Nihira, T., Matsuo, H. (1974). *J. Phys. Soc. Jap.,* **36** 123.

Jones, R., and Briddon, P. R. (1998). In:- Identification of Defects in Semiconductors, Vol **51A** Semiconductors and semimetals (ed. M. Stavola, Academic Press, Boston).

Kelly, B. T., (1980). Physics of Graphite (Applied Science, London).

Lachter, J., and Bragg, R. H. (1986). *Phys. Rev. Lett.*, **33,** 8903.

Martin, D. G., and Henson, R. W. (1964). *Phil. Mag.*, **9,** 659.

Mounet, N., and Marzari, N. (2005). *Phys. Rev. B*, **71,** 205214.

Niwase, K. (2002). *Phil. Mag. Lett.*, **82,** 401.

Reynolds, W. N., and Thrower, P. A., (1965). *Phil. Mag.*, **12,** 573.

Tanabe, T., Muto, S., Niwase, K. (2001). *Appl. Phys. Lett.*, **61,** 1639.

Thrower, P. A., and Mayer, R. M. (1978). *Phys. Stat. Sol (a)*, **47,** 11.

Telling, R. H., El-Barbary, A., Ewels, C. P. and Heggie, M. I. (2003). *Nature Materials,* **2,** 333.

Zhao, Y. X. and Spain, I. L. (1989). *Phys. Rev. B*, **40,** 993.

Development of a Code to Predict Fundamental Material and Fracture Properties of Nuclear Graphite

Chris Lynch and Gareth B Neighbour
Department of Engineering, University of Hull, Cottingham Road, Hull, UK. HU6 7RX
Email: c.lynch@eng.hull.ac.uk

Abstract
Polygranular graphite is used as a moderator, reflector and major structural component within Advanced Gas-cooled Reactors (AGRs) operated by British Energy Generation Ltd. In recent times cracking phenomena has occurred within the graphite bricks at an earlier stage than is to be expected. This paper aims to increase the understanding of graphite crack propagation by suggesting modelling routes to better understand quasi-brittle fracture processes. This paper also briefly evaluates previous methods for modelling fracture processes in quasi-brittle materials and presents a review of these methods with the aim of selecting the most important attributes to model the fracture behaviour of nuclear graphite using novel computational methods. A new code, written in C++, currently predicts the in broad terms the microstructure and mechanical properties of nuclear graphite with increasing radiolytic oxidation. The model also has the potential to model the effects of neutron irradiation simultaneously with radiolytic oxidation given a set of rules. This paper also focuses on presenting the initial set of results from the first stage of modelling and comparing these results with the classical Knudsen relationship for nuclear graphite.

Keywords
Fracture, modelling, degradation

INTRODUCTION

Polygranular graphite is used as a moderator, reflector and a major structural component within Advanced Gas-cooled Reactors (AGRs) operated by British Energy plc. In recent times, cracking phenomena has occurred within the graphite core bricks at an earlier stage than were expected. This paper reports work which is part of a project which aims to increase the understanding of graphite crack propagation by modelling quasi-brittle fracture processes. This paper briefly evaluates previous methods for modelling fracture processes in quasi-brittle materials and presents a review of these methods with the aim of selecting the most important attributes to model the fracture behaviour of nuclear graphite using novel computational methods (C++ code). This paper focuses on presenting the initial set of results from the first stage of modelling and comparing these results with the classical Knudsen relationship for nuclear graphite.

FAILURE AND THE COMPLICATIONS OF THE FRACTURE PROCESS

The quasi-brittle fracture response of polygranular graphite has been documented by Brocklehurst (1977), Sakai *et al.* (1983), Allard *et al.* (1991), Ouagne (2001) and Fazluddin (2002). The reasons behind this response, Figure 1, are complex, but there are five main categories that can be considered as complications. These are:

- Microstructure
- Crack / surface morphology

- Fracture process zone and sub-critical processes
- Scale
- Degradation

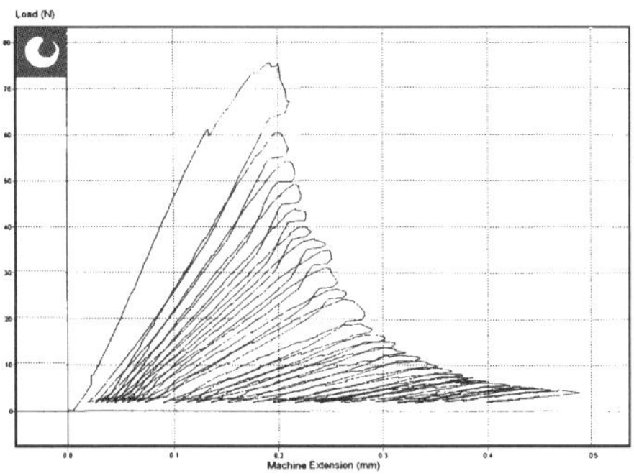

FIGURE 1: Cyclic load-displacement curve for AGR graphite (Chevron Notched Specimen) which shows stable crack growth and quasi-brittle characteristics.

Degradation is a major issue for nuclear graphite in-service within the reactor core. Radiolytic oxidation attacks both the open and ultimately closed pore network after breakthrough as the calcination cracks within the filler particles become interconnected with the volatile gas evolution & transport pores. For a detailed explanation on both neutron irradiation and radiolytic oxidation see Brocklehurst *et al.* (1970), Murdie *et al.* (1986) and Birch *et al.* (1990).

Fracture Models

A detailed review of current and previously used fracture models for quasi-brittle materials has recently been undertaken (Lynch and Neighbour, 2004). The outcome of the review was that no single one model was found to suitably describe the fracture of polygranular graphites. In the analysis of the various fracture models, for the purposes of polygranular graphite at least, it is thought appropriate to combine the key attributes of several models in the hope of generating a more realistic model. That is, the overall outcome is a combination of the ligament approach, the fictitious crack model and the crack band theory. These attributes of the various models which are thought to be beneficial to any future model are briefly described below.

Ligament Model

The central argument to the ligament model (Steer, 2003) is that "stiffness attracts load" which is a widely used and accepted paradigm in structural engineering. When microstructural elements of a solid are linked so that their displacements are equal, the elements resist loads in proportion to their stiffness. The model assumes that as each ligament reaches its maximum tensile strength it is removed from model producing the fracture response seen in Figure 2, closely simulating the softening response associated with polygranular graphite and that described by Sakai *et al.* (1983).

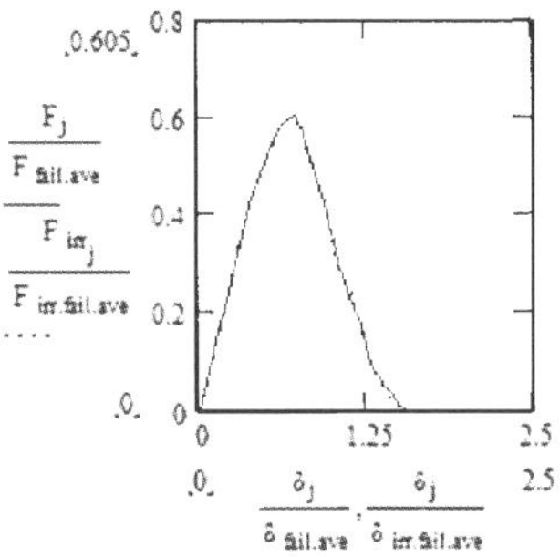

FIGURE 2: Normalised load-displacement results of the ligament model for as-received nuclear graphite (solid) and irradiated nuclear graphite (dotted) (reproduced with permission - Steer, 2003).

Fictitious Crack

The fictitious crack model proposed by Hillerborg *et al.* (1976) assumes a crack propagates when the stress at the crack tip reaches the maximum tensile strength (f_t). Once the crack has opened the stress transfer is not assumed to be constant, but gradually falls to zero as the crack width increases. The non-linear stress transfer, post crack initiation, gives the model the ability to simulate the softened response associated with quasi-brittle materials.

Crack Band Theory

Bažant and Oh (1983) originally proposed the crack band theory to overcome the problems of misrepresentation of crack paths, premature convergences and size effect mesh dependency associated with the smeared crack method. The model uses Hillerborg's fictitious crack model (previously described) and spreads it over a band of finite width. The width of the band, to which a micro-cracking area is smeared across, is related to the constituent parts, *e.g.* aggregate size in concrete. Bažant and Oh (1983) stated

> *"the line along which the microcracks are scattered is not straight (or smoothly curved) but is highly tortuous, deviating to each side of the straight line extension by a distance equal to about the aggregate size, as the crack is trying to pass around the harder aggregate pieces."*[5]

The crack band theory uses a three-parameter model to describe the material properties, and therefore takes into account the microstructure of the material.

Knudsen Relationship

The Knudsen equation is a widely used semi-empirical structure-property relationship, commonly fitted to a wide range of material properties that change exponentially with increasing porosity, such as strength and Young's modulus. The Knudsen equation has proved to be applicable to a wide range of materials (Rice, 1977) and is widely used for the prediction of the effects of oxidation on the graphite moderators (Burchell *et al.*, 1985). Traditionally, the Knudsen relationship is used to predict the decline in a mechanical property

---

[5] This statement may help explain the rough / tortuous fracture paths also seen in nuclear graphite, and bears some commonality with observations made by others such as Neighbour (1993) who discusses the crack travelling around filler particles.

with decreasing density in the nuclear industry. For example, it is common to describe the relationship between Young's modulus and oxidation of AGR graphite by the equation

$$E/E_0 = \exp(-bx), \quad [1]$$

where E is the Young's modulus of the graphite oxidised to a fractional weight loss x, E_0 is the Young's modulus of the unoxidised graphite and b is an exponent used to fit the data.

DEVELOPMENT OF THE ABSTRACT MODEL

There are a number of methods of modelling the relationship between microstructure and the mechanical properties of brittle materials. The Minimum Solid Area (MSA)[6] provides the most appropriate explanation of the empirically found Knudsen relationship. The Knudsen equation is a type of MSA model (based upon the contact between solid spheres), and therefore it is possible to fit the Knudsen equation to MSA models and extract values for the decay parameter *b*. In the development of the model, it is useful to start with a two-dimensional model and work up to a three dimensional model. To illustrate the point, first consider a simple model of a spherical pore within a unit cell of length unity. The MSA of a spherical pore with a radius *r*, in a cubic unit cell is thus defined as $1-\pi r^2$ and the volume of porosity is defined as $4/3\pi r^3$. That is to say MSA is a two dimensional parameter obtained from a three dimensional model. Equally, if a two dimensional model is used, *i.e.* a circle within a square, correspondingly, the one-dimensional parameter (equivalent to MSA, but called here the Minimum Line Length - MLL) for a circle with a radius *r*, in a square unit cell is defined as $1-2r^2$ and the area of porosity is defined as πr^2. As porosity increases, a similar trend is observed, but, initially at least, the 2D model is more pessimistic indeed it exhibits a more exponential behaviour than the MSA model. There are differences between the two models, with MLL showing greater exponential behaviour than MSA especially at low fractional porosities, though both models percolate at different fractional porosity; the worst case being the 3D model percolating at 0.55 which is a curiosity of the idealised system.

The Model

The application currently being developed is split into four components: graphical/user interface; microstructure; degradation; and, fracture. Each component has hierarchical dependencies, meaning although the components are independent in terms of code they are dependant upon the previously executed component for the input of subsequent calculations.

Graphical and User Interface

OpenGL is used for a means of displaying the model on a computer screen as this allows easy conversion to 3D in the future, and also reduces load on the main processor and RAM. For display reasons, at 100% magnification, one pixel equates to a physical measurement of 10 microns. The user interface is a dialogue based windows interface allowing the user to enter desired size of model and other variables such porosity distribution values.

Microstructure

The microstructure component is the core component for the application and is responsible for the base structure used throughout all other (physical and virtual) dimensions and

[6] Minimum solid area is defined as the minimum solid cross-sectional area of a plane normal to the applied stress.

subsequent calculations. The component uses dynamic multi-dimensional arrays to store information. If the array is truly 2 dimensional, the array can be simply visualised using the analogy of a stack of paper; if the array is based on 3 dimensions, then the array can be visualised as a layer cube. A central feature of the model is that it takes a pore-size distribution from an as-received nuclear graphite and allocated a percentage contribution to certain classes of Equivalent Pore Diameter (EPD); in the initial model there were 7 falling into the following bands: 0-50 microns (33%); 51-100 microns (18%); 101-150 microns (12%); 151-250 microns (22%); 251-350 microns (11%); 351-450 microns (0%); and, 451-550 microns (4%).

The basic pore shape chosen is one of a circle because circles are also the base shape for filler particles within AGR moderator graphite and circles are easily manipulated to produce complex shapes. This component of the model creates all three phases of the material - filler, binder and porosity. Currently the filler particles are not produced using a defined particle size distribution, but an assumption is made that a normal distribution exists around a mean (measured value), although it is intended to add a well-defined distribution in later versions. In essence, the model creates a dynamic matrix and populates with a marker for solid material. The filler particles are added with a certain percentage of inherent calcination cracks (initially closed porosity). Given that the total porosity is known, additional porosity is added using an algorithm using a probabilistic approach until all categories of porosity are full and the total porosity is approximately 18%, Figure 3.

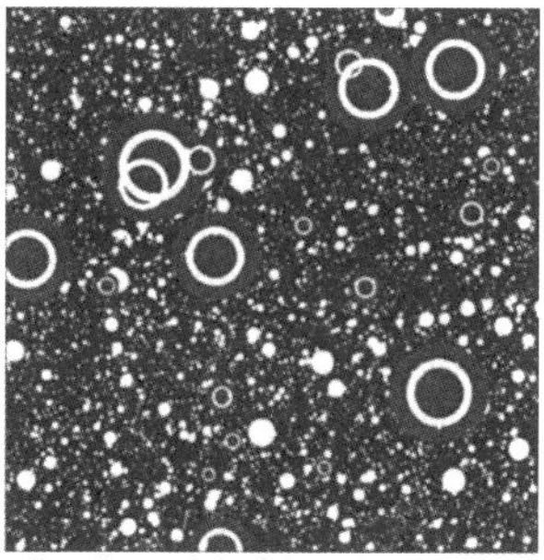

FIGURE 3: Example of an initial and early model of the microstructure showing a simulated 18% porosity.

Degradation

The degradation component of the model depletes the structure created by the microstructure component. The component will contain both neutron irradiation (yet to be incorporated) and radiolytic oxidation simulation. To simulate radiolytic oxidation, the degradation is applied to specific areas of the structure using a series of complex logic statements. That is, it mimics the real mechanism in attacking pore walls of open porosity within a certain distance of the creation of the oxidising species. In detail, the program statements determine whether or not the value picked by the number generator[7] is at the edge of a pore or not, if the selection is in the correct place then the cells within the vicinity are modified to simulate the pore growing in size. This is repeated until a user defined porosity / weight loss value is reached: an example microstructure can be seen in Figure 4. The effects do not have to be solely visual indeed any value within the multi-dimensional arrays can be dynamically modified to reflect

---

[7] A number generator is used to pick two random values, to create a co-ordinate for accessing the array.

simulated changes using a logic statement. Therefore this component has been created with expansion at its core.

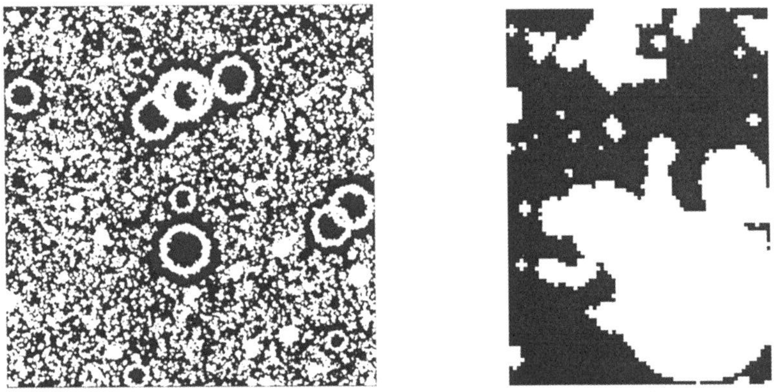

(a) (b)

FIGURE 4: A simulated microstructure of (a) 50% porosity, radiolytic oxidation graphite with (b) enlarged fine detail section of pores simulating the morphological effects of radiolytic oxidation.

Fracture

The fracture component is a component that can be run independently from the degradation component. The component will combine aspects of the ligament model, the fictitious crack model and the crack bad theory to create a path through the structure (as previously described). A displacement / load will be applied incrementally and after each increment various calculations are performed by the interrogation of the model. In this complex area of the model, some limitations of the main parameters can be made. Firstly, the model will concentrate on a band of finite width approximately the same size as the largest filler particle size. The second limitation is one applied per load cycle. The limitation will be the size of the area in which to scan. The component will seek out the most likely to break ligaments based on a set of rules, logic and calculations. The area used will correspond in size to the FPZ of AGR graphite. At this time the precise criterion have yet to be fully defined, however currently a simplified stress system combined with the ligament model is being used. To simulate the crack formation, those cells deemed to have broken will be removed and the remaining cells within the band will have their corresponding stress values modified accordingly.

The process will be as follows:

1. Loads / displacements are applied to the model.
2. Within the crack band there will be a scanning zone of the array which will be monitored using a simplified FEA process. As cells in this zone reach there maximum tensile stress they will be removed. The stress values in the zone will then be modified using the tension softening relationship predicted by the ligament model.
3. If once the tension softening calculation has been applied to a cell and the cell has reached its maximum tensile stress it will be removed and so on and so on until all the cells within the simulated FPZ have been scanned.

4. The centre of the simulated FPZ then moves to a cell deemed most likely to fail next and the load is then increased again and the process is repeated until the model has completely broken.

The fracture component can be used to directly modify the array being displayed or can be used to modify a corresponding array and called upon to be displayed separately or as an overlay (nth dimension); this is always for a virtual crack path to be viewed.

Initial Results

The results from the version development show a trend moving towards the Knudsen model, Figure 5, however it is still some way away from matching the approximation. This can be explained due to the symmetrical shape of the circular pores because they have an aspect ratio of 1, this means their size is distributed equally when considering MSA (in reality MLL). The degradation is not yet concentrated on the filler particles, this affects the aspect ratio of the degradation mechanism as it is spread throughout the material.

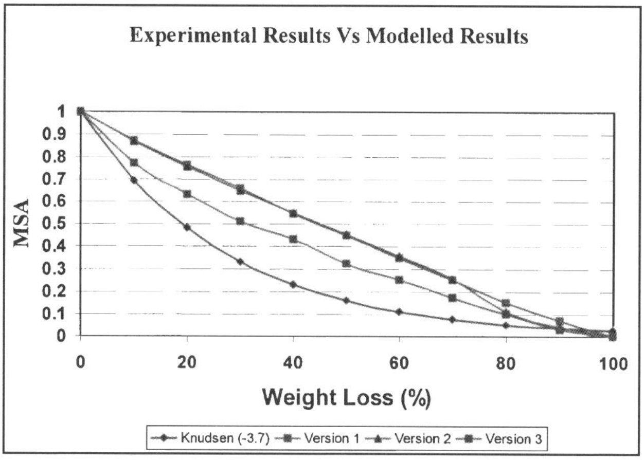

FIGURE 5: Comparison of the various developmental stage models against the ideal Knudsen relationship (see Miller and Brocklehurst, 1988).

To overcome the problem of a constant aspect ratio, the development of the circles within circles method was adopted, Figure 6. This method of pore shape formation is based upon Weisstein (1999), and packs circles using geometric relationships to enable the possible formation of highly complex shapes. Four diameter circles are used, Figure 7. D1 is the largest, then D2, D3 and finally D4. The diameters of the circles are found according to the ratio 1:0.3:0.22:0.14, respectively. The diameter of D1 is found using a random number generator picking numbers between the smallest and largest pore size when combined with the pore size distribution. The complex shapes are created using probabilities and a hierarchical Boolean method. The method can produce highly complex shapes with varying aspect ratios as shown in Figure 8.

FIGURE 6: Pore shape output from circles with circles approach.

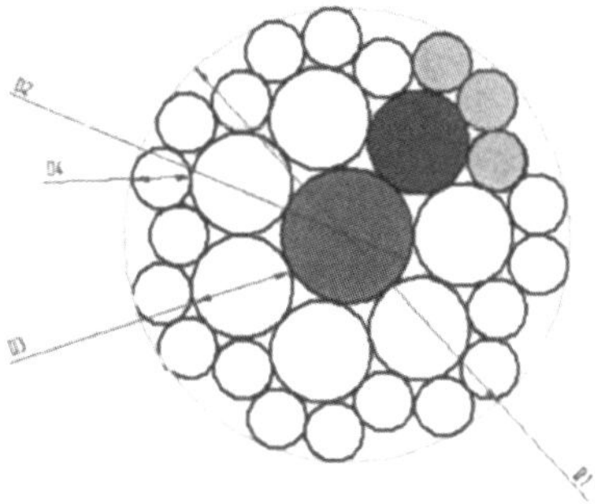

FIGURE 7: Schematic of the method used to pack circles for the circles within circles approach.

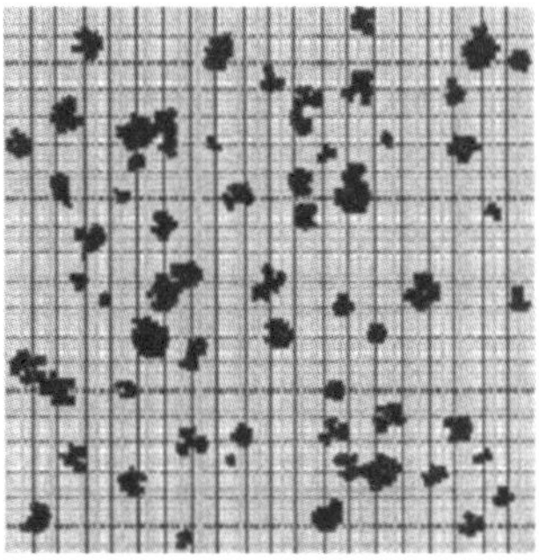

FIGURE 8: Results from manual simulation of the circle algorithm.

SUMMARY AND CONCLUSIONS

The application described in the paper aims to help provide a robust explanation for mechanical degradation and crack phenomena observed in AGR graphite moderator during operation. The design of the proposed computational code has been detailed and the initial stages of development shown. The paper has included:

- Findings of a detailed literature review of existing fracture models.
- A hybrid of three previously used fracture models. It has considered that the most suitable approach for the modelling of fracture for polygranular graphite is to use the ligament theory in conjunction with the fictitious crack and crack band models. The complex nature of the fracture process involved within quasi-brittle materials have been briefly outline and discussed. The application uses the ligament model to simulate the softening response of each phase of the material and applies this relationship to the cells within a dynamic array. A simplified FEA approach is proposed using a combination of both the fictitious crack model and crack band theory

and will spread the non-linear fracture processes over a band on finite width and thickness. The model therefore contains no predefined crack path.

- The application has been shown to be capable of building all the elements of AGR moderator graphite microstructure and also has been shown to be capable of including degradation mechanisms. Methods for the inclusion of further components and mechanisms, *e.g.* neutron irradiation, have been briefly discussed and the method for the creation of the fracture component has been shown.
- Initial development of the application has been shown and compared to existing empirical models.

Acknowledgements
The authors would like to thank Mr. A Steer and Dr J. Reed for their many examples of useful information and guidance. Thanks is also given to EPSRC and British Energy plc for financial support. The views expressed in this paper are those of the authors and do not necessarily represent those of the British Energy Generation Ltd.

References

Allard, B., Rouby, D., Fantozzy, G., Dumas, D. and Lacroix, P. (1991). Fracture behaviour of carbon materials. *Carbon*, [29], 457-468.

Bažant, Z. P. and Oh, B. (1983). Crack band theory for fracture of concrete, RILEM Material and Structures, 16, 155-177.

Birch, M., Schofield, P., Brocklehurst, J. E., Kelly, B. T., Harper, A. and Prior, H. (1990). The combined effects of fast neutron damage and radiolytic oxidation on the physical properties of Gilsocarbon graphite, *Carbon*, [90], 242-243.

Brocklehurst, J. E. (1977). Fracture in polycrystalline graphite. *Chemistry and Physics of Carbon*, [13],145-279.

Brocklehurst, J. E., Brown, R. G., Gilchrist, K. E. and Labaton, V. Y. (1970). The effect of radiolytic oxidation on the physical properties of graphite, *Journal of Nuclear Materials*, [35], 183-194.

Burchell, T. D., Pickup I. M., McEnaney B. and Cooke R. G. (1985). The relationship between microstructure and the reduction of elastic modulus in thermally and radiolytically corroded nuclear graphites. *Carbon*, [24], (5), 545-549.

Fazluddin, S. B. (2002). R-curve behaviour in nuclear graphite materials. PhD Thesis. University of Leeds.

Hillerborg, A., Modeer, M. and Petersson, P. E. (1976). Analysis of crack formation and crack growth in concrete by means of fracture mechanics and finite elements. *Cement and Concrete Research*, [6], 773-782.

Lynch, C. and Neighbour, G. B. (2004). Modelling crack growth in complex heterogeneous brittle materials. First Year Progress Report, HULL/ENG/NUC/R009.

Miller, W. N. and Brocklehurst, J. E. (1988). A statistical analysis of physical and mechanical property changes for graphite radiolytically oxidised in different coolants and recommended design rules for CAGR moderator and sleeve graphites. NRL-R-2009(S). AGR/GCWG/P(88)3.

Murdie, N., Edwards, I. A. S. and Marsh, H. (1986). Changes in porosity of graphite caused by radiolytic gasification by carbon dioxide. *Carbon*, [24], 3, 267-275.

Neighbour, G. B. (1993). Microstructural processes leading to fracture in nuclear graphites, PhD Thesis, University of Bath.

Ouagne, P. (2001). Fracture property changes with oxidation and irradiation in nuclear graphites. PhD Thesis, University of Bath.

Rice, R. W. (1977). Microstructure dependence of mechanical behaviour of ceramics. In:- MacCrone (Editor), Treatise on Materials Science Technology, Volume 11: Properties and Microstructure, 199-379, Academic Press, New York.

Sakai, M., Urashima, K., and Inagaki, M. (1983). Energy principle of elastic-plastic fracture and its application to the fracture mechanics of polycrystalline graphite. *Journal of the American Ceramic Society*, [66], 868-874.

Steer, A. G. (2003). A structural approach to the description and determination of its physical properties. Private publication.

Weisstein, E. W. (1999). "Circle Packing." from MathWorld-A Wolfram Web Resource. http://mathworld.wolfram.com/CirclePacking.html, CRC Press LLC, accessed January 15th 2006.

Modelling Graphite Ageing: Black Art or Forensic Science?

Mark A Davies and Mark R Bradford[#]
RPS Engineering and Safety, Birchwood Park.
[#]British Energy, Barnwood.
Email: mark.r.bradford@british-energy.com; mark.davies@pbmr.co.za

Abstract
The UK fleet of Advanced Gas-Cooled Reactors have been operating for a substantial period of time and rely on data obtained from MTR programmes dating back to the 60s and through to the end of the 80s to support current and future operation. Historically, an empirical approach to the irradiation behaviour of graphite has been used and, due to the nature of the available MTR data and the lack of a consistent set of data containing all the relevant measurements, this is still largely the case at present. Differences in interpretation of the data can have an impact on the assessed integrity of AGR core components. Consequently, a thorough review of the basis of the current models is being carried out, and new models are being developed as necessary. As an illustration, this paper presents some new interpretations of the available MTR data for irradiation creep.

Keywords
AGR, creep, integrity

INTRODUCTION

Modern finite element codes, used to analyse irradiated graphite components, incorporate a constitutive material model that is complex, including solution dependent feedback mechanisms and significant numbers of input parameters. Ideally, the derivation would be based on a substantial amount of experimental data but, in reality, data are sparse and limited in extent. Consequently, material modelling becomes almost a black art rather than a science, often relying on empirical relationships and accepted wisdom rather than fundamental understanding accompanied by objective validation. The precise interpretation of the experimental datasets has a bearing on the predictions of future core component integrity. As a result, a thorough, forensic, re-examination of the sparse and disparate Material Test Reactor (MTR) information is being undertaken in an attempt to develop improved models. This paper illustrates the process by considering recent findings on irradiation creep.

UK CREEP METHODOLOGY

UK Creep Model

Irradiation induced creep is defined as the difference in length between a stressed and an unstressed specimen of the same material properties subjected to the same irradiation environment. The currently employed formulation is based on (Kelly, 1989), who proposed:

$$\varepsilon_c = \left(\frac{\sigma}{E_0 SW}\right)\left(1 - \exp^{-4\gamma}\right) + \beta\left(\frac{\sigma}{E_0 SW}\right)\gamma \qquad [1]$$

The first term is primary creep and saturates at 1 initial elastic strain unit (esu), and is fully recoverable with continued irradiation following load release (Price, 1981). The secondary component is proportional to esu and dose with a secondary creep coefficient, β(Ti) of 0.23. It is considered to be irrecoverable. The creep modulus is affected by structural (S) and radiolytic oxidation (W) changes to the Young's modulus. Young's modulus is virtually unaffected by creep strain, however it has long been argued that creep strain modifies the CTE of graphite (Kelly, 1989), and the current methodology incorporates this effect as a linear dependence on the sum of the elastic and creep strain components.

Based on an earlier theory (Simmons, 1965), it was proposed that the change in CTE would, in turn, modify the underlying dimensional change rate of a stressed specimen (Kelly and Burchell, 1994). Thus, 'true' creep strain is the difference in length between an unstressed specimen and a stressed specimen with the same properties as the unstressed specimen and is the sum of the observed 'apparent' creep strain and the 'interaction' strain:

$$\varepsilon_c^i = \int_0^{\gamma} \left(\frac{\alpha_x^1 - \alpha_x}{\alpha_c - \alpha_a} \right) \left(\frac{dX_T}{d\gamma} \right) d\gamma \qquad [2]$$

where $\alpha^1{}_x$ - α_x is the difference between the CTE of the stressed and the unstressed specimens; α_c , α_a is the crystal thermal expansion coefficients parallel and perpendicular to the basal planes; and $dX_T/d\gamma$ is the crystal shape change parameter. Finally, according to the Kelly theory, the creep Poisson's ratio should be equivalent to the elastic Poisson's ratio for both primary and secondary creep and independent of the accumulated creep strain.

NEW ANALYSIS OF DATA

The Effect of Strain on CTE

It was noted above that the currently endorsed UK methodology incorporates the modification of the CTE by means of a linear dependence on the sum of the elastic and creep strain components, *i.e.* the change in CTE continually increases with the creep strain. However, there is evidence in both UK and US data that this treatment may not be correct. Figure 1 shows some of the UK data for creep strain as a function of dose (Kelly *et al,* 1977). Figure 2 shows the CTE of the same, isotropic graphite, specimens. It can be seen that, whilst the creep strain continues to increase with dose, the change in CTE appears to be saturating.

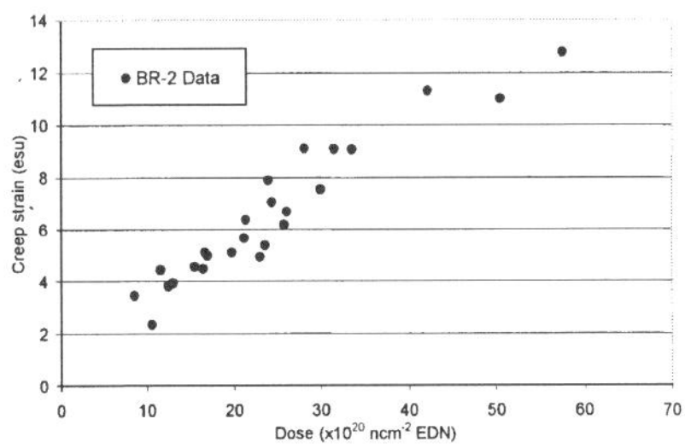

FIGURE 1: BR-2 Creep Data

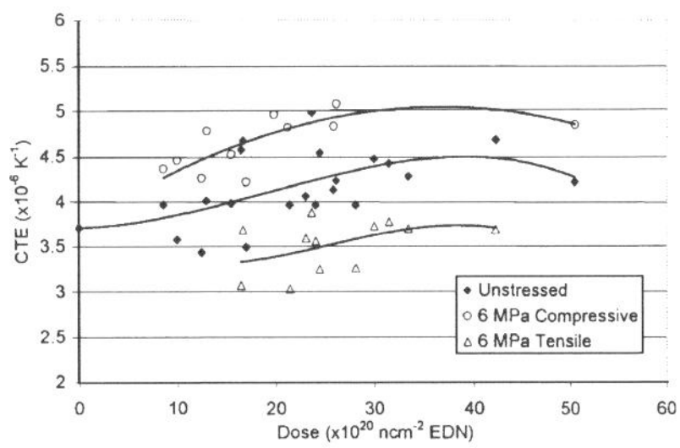

FIGURE 2: BR-2 CTE data

A similar behaviour is observed in US experiments on H451 graphite (Mobasharan, 1990), which employed two constant compressive loads of 13.8 MPa and 20.7 MPa. The data for samples irradiated at 600 °C is shown in Figure 3; similar behaviour is observed at 900 °C. Figure 4 shows the mean saturated change in CTE plotted against initial elastic strain derived from the unirradiated static modulus appropriate to the applied stress. This suggests a linear relationship between the two parameters, with some indication of temperature dependence. However, this dependence may, in part, reflect the different temperature ranges of the CTE measurements (the quoted CTE is the mean CTE over the range 20- (T_{irr}-100) °C).

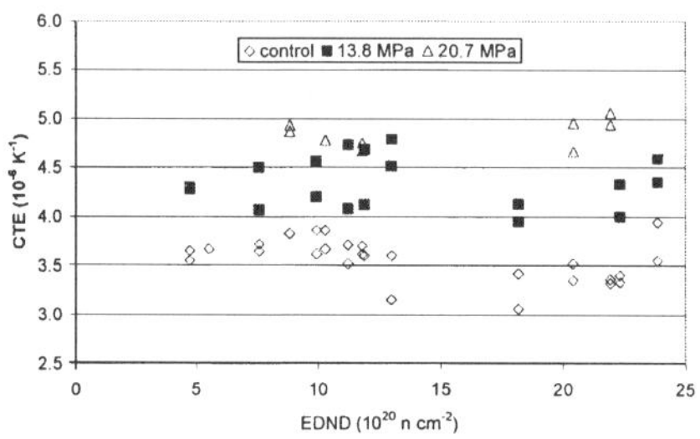

FIGURE 3: CTE change of H451 subject to compressive stress irradiated at 600 °C

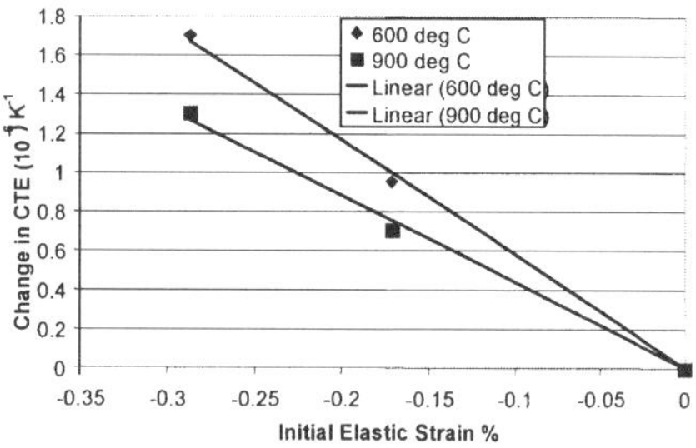

FIGURE 4: H451 CTE change as a function of initial elastic strain

The dose increment required to generate the change in CTE due to the applied stress appears to be similar to that required for the saturation of the "pinning" component of the irradiation-induced change to Young's Modulus. This may indicate that the mechanisms are related. Irrespective of whether this proves to be the case, there is very strong evidence that the change in CTE saturates at low dose, and may be related to initial elastic strain.

The Support for Interaction Strain

The validation of interaction strain was carried out on H451 graphite irradiated at 900 °C (Kelly and Burchell, 1994). A linear dependence of CTE on dose was assumed and, as a result, the interaction strain becomes a quadratic function of dose for a constant applied stress. The 'true' creep strain is a linear function of dose for a given applied stress, consequently the measured 'apparent' creep strain will increasingly diverge from the 'true' creep strain. The behaviour of the interaction strain, extrapolated out to 60×10^{20} n cm^{-2} EDND, is shown in Figure 5.

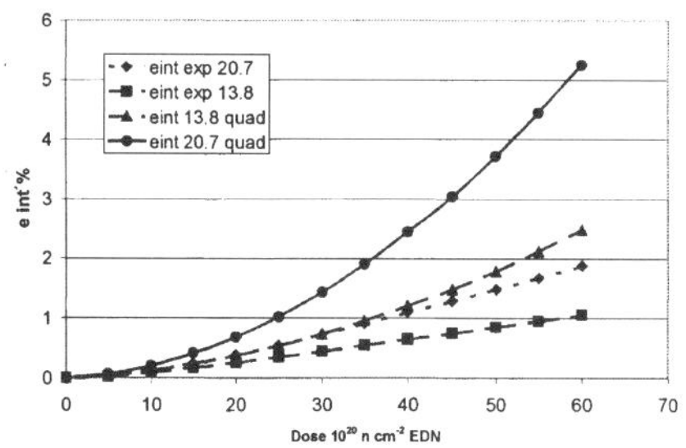

FIGURE 5: Interaction Strain for H451 graphite irradiated at 900 °C.

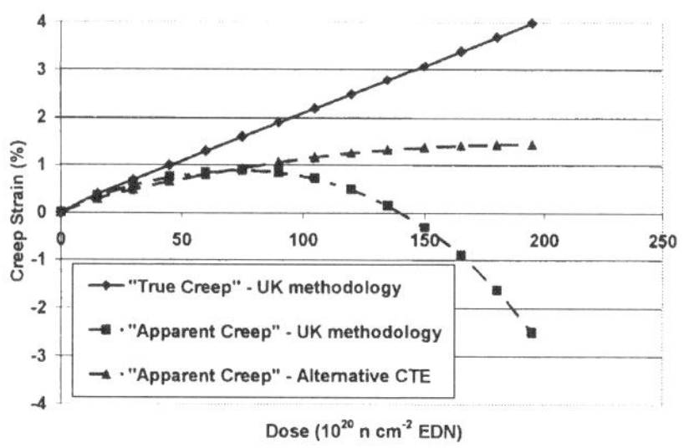

FIGURE 6: 'True' and 'Apparent' Creep derived from interaction strain.

Figure 5 also shows an evaluation of interaction strain assuming that the change in CTE saturates and is proportional to the elastic strain applied. This leads to a fixed interaction strain rate and a much smaller correction, which will affect the correlation with measurement presented in the original paper.

Figure 6 shows the 'apparent' creep derived from the endorsed interaction strain and from the modified interaction strain extrapolated to high doses. Both sets of results indicate that the interaction strain in its present form is significantly overestimated at high dose and leads to non-physical results for a constant applied load MTR tests. In an AGR block the situation is more akin to a restrained shrinkage test and equally non-physical behaviour could be predicted. The extrapolation of the interaction strain indicates the limitations of the current formulation.

By relating the interaction strain to the Simmons' theory, the current methodology implicitly assumes that the underlying driving force is related to structural changes. This would also imply that the change in CTE and dimensional change rate should also be accompanied by a significant change in Young's modulus. The available data show only a weak dependence of modulus on creep strain, which is further evidence against interaction strain as defined by Kelly and Burchell (1994).

The Effect of Thermal Annealing

Tensile and compressive specimens along with their unstressed "controls" were irradiated in BR-2 and subsequently thermally annealed (Brocklehurst, 1969). Both the CTE and the dimensional change of the samples were measured during the thermal anneal and it was found that the CTE in both the compressive and tensile samples recovered to the unstressed control CTE, albeit at slightly different annealing temperatures. A significant proportion of the supposedly irrecoverable secondary creep strain was also recovered. Figure 7 shows the residual strains following each stage of the thermal anneal[8].

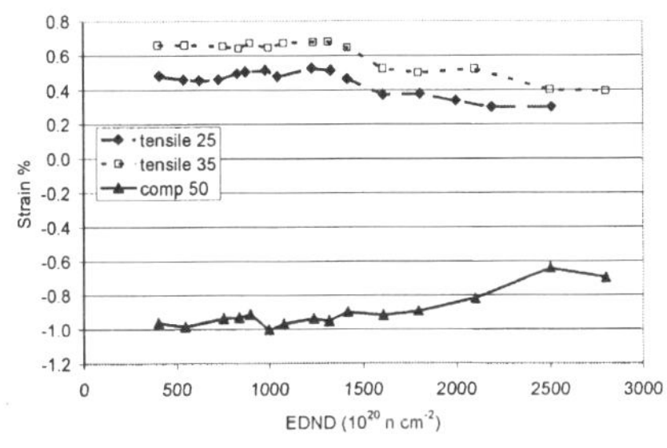

FIGURE 7: Residual strain following thermal anneal.

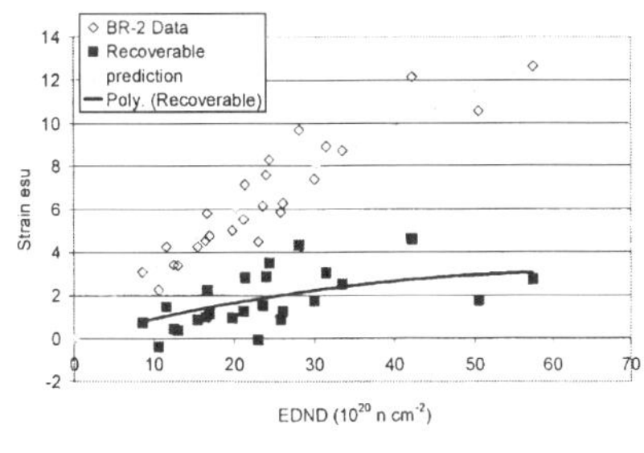

FIGURE 8: Recoverable strain derived from BR-2 data.

The thermal annealing experiments offer the possibility that a proportion of what has traditionally been assigned to secondary creep is, in fact, recoverable. This would have significant implications for the dose dependence of irradiation creep, particularly since the irrecoverable strain, defined as the residual strain post thermal anneal, appears to exhibit a linear response with dose, equivalent to a reduced secondary creep coefficient of ~0.15 (c.f.

---

[8] Unfortunately, the control sample for the compressive specimen was not thermally annealed and the residual strain has had to be obtained by scaling from the control dimensional change at 35×10^{20} n cm^{-2} EDN.

β=0.23). The implied associated recoverable strain has been derived from the BR-2 data (Figure 8) and suggests that it is saturating beyond 30×10^{20} n cm^{-2} EDN at approximately 3 initial esu. The trend shows a remarkable similarity to that for CTE (Figure 2) and suggests that the mechanism inducing the change in CTE is linked to the component of strain that can be recovered by thermal annealing. Further evidence can be found in the data for irradiation annealing.

Irradiation Annealing

A compressive specimen was irradiated to give significant creep strains, unloaded and re-irradiated to a high dose (Kelly, 1989). The dimensional change of the specimen and the control specimen is shown in Figure 9. The sample shows an immediate recovery followed by a slower recovery over a sustained dose. Having identified that the available MTR data may be incompatible with interaction strain, the behaviour points to irradiation annealing of a recoverable strain substantially in excess of primary creep. Unfortunately, the associated CTE data is not available on these samples and so it is not possible to confirm the CTE recovery that would be expected from the thermal annealing experiments.

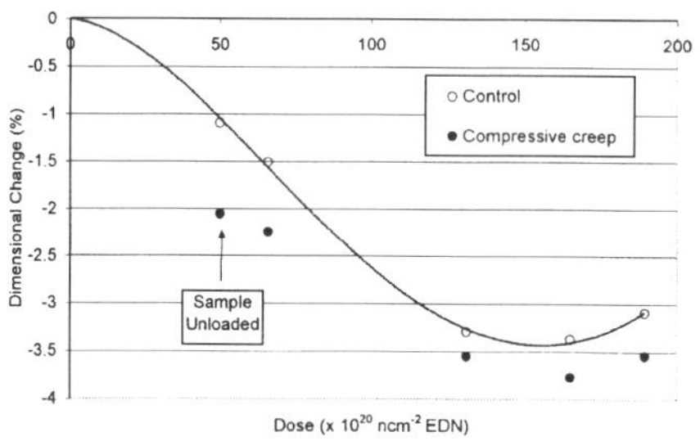

FIGURE 9: BR-2 irradiation anneal.

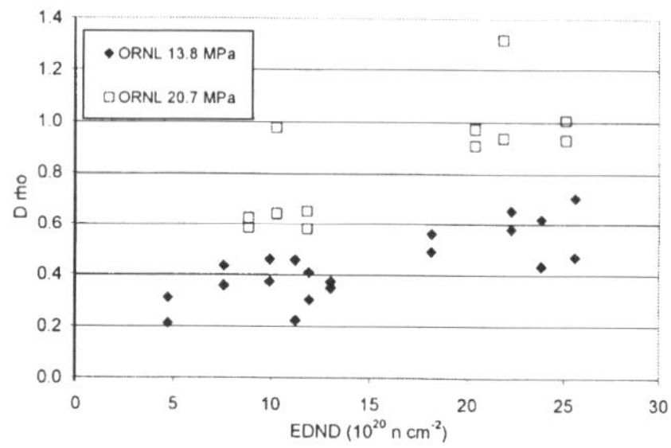

FIGURE 10: ORNL Volume Change (%) at 600 °C.

Volume Change

The endorsed creep model assumes that creep Poisson's ratio for both primary and secondary creep is equivalent to the elastic ratio and is independent of dose. Figure 10 shows density change data for 600 °C ORNL H451 data. Both sets of results indicate that the volume change occurs at low dose and appears to be saturating. This would suggest that secondary creep occurs at constant volume, which is supported by an analysis of US lateral creep strain information (Price, 1981).

REVISED CREEP MODEL

By drawing together the information from UK and US creep experiments, strong evidence has been found to suggest that:

- The strain induced change in CTE is not a function of secondary creep strain, but saturates after a dose of $\sim 30 \times 10^{20}$ n cm^{-2} EDN.
- The concept of interaction strain is not validated at doses beyond $\sim 30 \times 10^{20}$ n cm^{-2} EDN and appears to become non-physical when extrapolated to higher doses.
- There is evidence, from both thermal and irradiation annealing, for a recoverable strain several times that of primary creep, and a lower associated secondary creep

coefficient than has been hitherto assumed. The dose at which the recoverable strain saturates bears a striking similarity to that for the saturation of the CTE change.

An alternative creep model is therefore proposed comprising only primary creep, recoverable creep and irrecoverable creep, *i.e.*

$$\varepsilon_c = \alpha\left(\frac{\sigma}{E_0 SW}\right)\left(1-\exp^{-K_1\gamma}\right) + \omega\left(\frac{\sigma}{E_0 SW}\right)\left(1-\exp^{-(K_2\gamma)}\right) + \beta\left(\frac{\sigma}{E_0 SW}\right)\gamma \qquad [3]$$

There is no interaction strain. All components are proportional to esu and the effects of structural changes and radiolytic oxidation are also included. The rates of saturation of the primary and recoverable creep components are controlled by the dose constants K_1 and K_2. The irrecoverable creep is synonymous with secondary creep, but with a coefficient, β, reduced from that in the present model. As an illustration of the capability of this alternative formulation, Figures 11 and 12 show its predictions plotted against the BR-2 and ORNL (600 °C) data, illustrating good agreement without the need for any interaction strain.

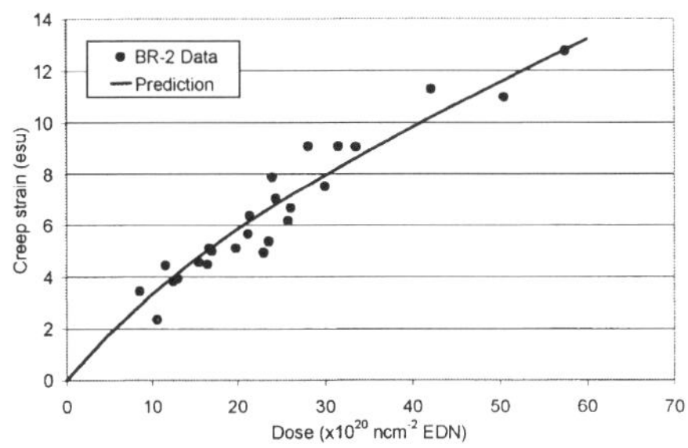

FIGURE 11: BR-2 – alternative creep formulation

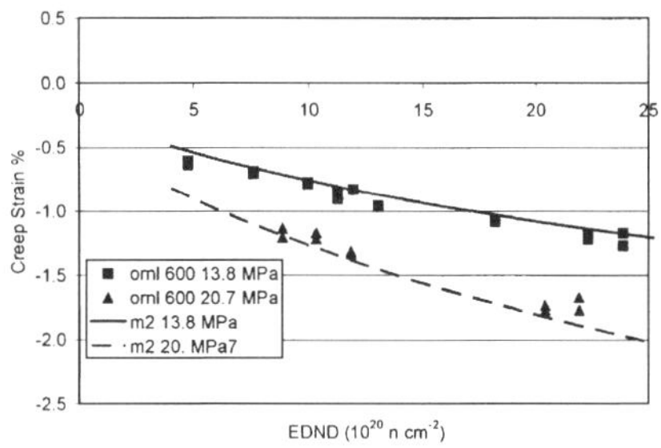

FIGURE 12: ORNL creep at 600 °C

As the model is still a work in progress, the constants in Equation [3] have been allowed to vary to achieve a reasonable fit to each dataset. However, when the same constants are applied to the creep recovery specimen (Figure 13) and ORNL volume change data (assuming secondary creep occurs at constant volume, Figure 14), it can be seen that the generalised form of the model predicts a wide range of observed behaviour very well.

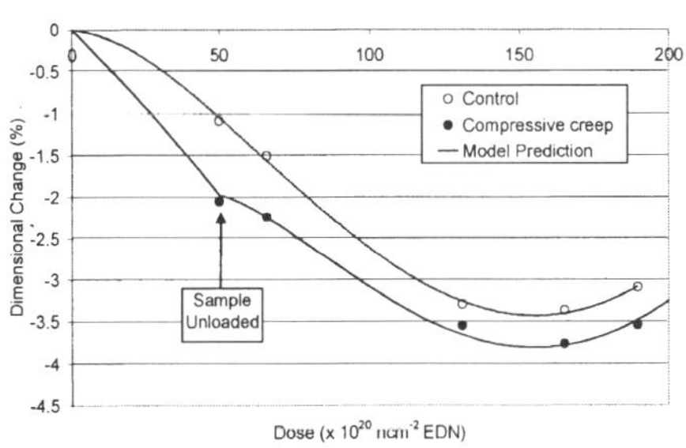

FIGURE 13: BR-2 irradiation anneal.

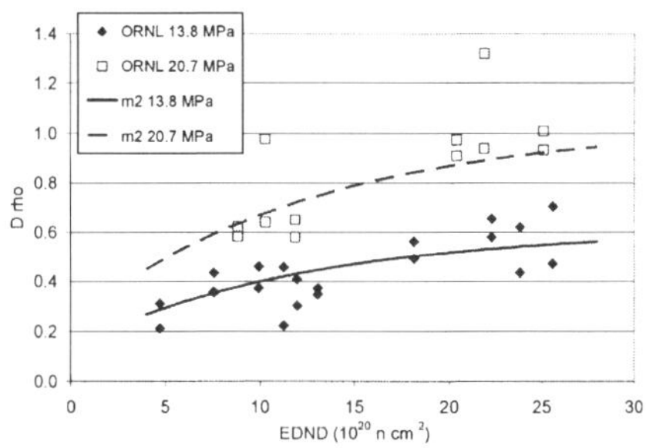

FIGURE 14: ORNL volume change (%) at 600 °C.

Overall, the level of agreement between the proposed alternative model and creep data from a broad variety of materials is very encouraging and suggests that it is a viable alternative to the methodology proposed by Kelly and Burchell (1994).

DISCUSSION

When considering the approach taken to the interpretation of graphite data in the past, the impression that can sometimes come across is that the modelling has, at times, been something of a "black art". Often there has been a piecemeal, sequential, adoption of new information within an established method. This is typified by the "bolt-on" inclusion of interaction strain within the UK creep methodology, as opposed to a reassessment of the totality of the available information. The present study has attempted to utilise a more holistic, forensic science approach by recognising that irradiation creep is universal and applies to all nuclear graphites and that, consequently, it is possible to incorporate all available data, however sparse and disparate in the search for commonality of trend. In this way, a more robust and widely applicable model can be derived. The proposed alternative creep formulation (Equation 3) is the result of such a forensic approach.

The alternative formulation is likely to have an impact on the predicted stress states and properties (particularly CTE) of AGR moderator fuel bricks and, when combined with companion studies (Bradford and Steer, 2005), this offers a valuable source of validation. The creep model affects the time evolution of stress, and so comparison of the stress variations in time with observations of the timing of brick bore cracking can potentially be used to support any claimed improvement in modelling based on the assessment of MTR data. Comparison of observed and predicted brick bore shapes can be used to validate shrinkage behaviour, and the CTE variations within trepanned samples may be used to validate the total methodology (due to the dependence of CTE on stress state). Therefore the next steps in the development process are:

- To determine the best estimate values of the constants in Equation [3], including any temperature dependence.
- To implement the model in the material property prediction and stress analysis codes.
- To verify that the coded model accurately reproduces the MTR specimen experiments from which it was developed, adapting the model and/or implementation as necessary.
- Finally, to run representative calculations for AGR brick geometries and to compare the predictions of the time evolution of brick shape, stress and properties, particularly CTE, with the results of AGR inspections.

In accordance with good scientific practise, the findings from these comparisons should then be fed back into the assessments to inform the next generation of models. In terms of the impact on the assessment of core components, definitive statements cannot be made until the process outlined, above is completed. However, a qualitative assessment has been carried out, based on previous studies that have investigated the sensitivity of the stress predictions of core components to variations in the coefficients of the existing models. On balance, it is judged that the revised model may perturb the predictions of stress evolution such that there may be an increased potential for brick bore cracking early in reactor life, and a reduced potential for keyway root cracking later in life. Neither outcome is expected to present a challenge to the evolving AGR core safety cases.

CONCLUSIONS

A re-evaluation of the information from UK and US creep experiments, has found that:

- The strain induced change in CTE is not a function of secondary creep strain, but saturates after a dose of ~$30x10^{20}$ n cm^{-2} EDN.
- The concept of interaction strain is not validated at doses beyond ~$30x10^{20}$ n cm^{-2} EDN and appears to become non-physical when extrapolated to higher doses.
- There is evidence, from both thermal and irradiation annealing, for a recoverable strain several times that of primary creep, and a lower associated secondary creep coefficient than has been hitherto assumed. The dose at which the recoverable strain saturates bears a striking similarity to that for the saturation of the CTE change.

An alternative creep model has therefore been proposed comprising only primary creep, recoverable creep and irrecoverable creep, which shows considerable promise. Although further work is needed before the full consequences of adopting the alternative model are assessed, present indications are that there is unlikely to be any challenge to the evolving AGR core safety cases.

Acknowledgements

The authors would like to thank British Energy Generation Ltd for permission to present this paper.

References

Bradford, M. R. and Steer, A. G. (2005). A holistic, structurally-based approach to modelling graphite properties, *Ageing Management of Graphite Reactor Cores,* Proceedings of a Conference held in Cardiff, United Kingdom, 28-30 November 2005 (this volume).

Brocklehurst J. E. and Brown R. G. (1969). Constant stress irradiation creep experiments on graphite in BR-2, *Carbon*, Vol. 7, 487-497.

Burchell T. D. (2005). Private Communication.

Kelly B. T. and Burchell T. D. (1994). The analysis of irradiation induced creep experiments on nuclear reactor graphite, *Carbon*, Vol. 32, No. 1, 119-125.

Kelly B. T. and Brocklehurst J. E. (1977). UKAEA Reactor group studies of irradiation-induced creep in graphite, *Journal of Nuclear Materials,* 65, 79-85.

Kelly B. T. (1989). A review of irradiation induced creep in graphite under CAGR conditions, UKAEA Report ND-R-1406.

Price R. J. (1981). Review of irradiation induced creep in graphite under HTGR conditions. General Atomic Report GA-A16402, August 1981.

Simmons J. H. W. (1965). Radiation damage in graphite, Pergammon Press, Oxford.

Mobasheran A. S. (1990). PhD Thesis, University of Tennessee, U.S.A.

A Holistic, Structurally-Based, Approach to Modelling Graphite Properties

Mark R Bradford and Alan G Steer
Engineering Division, British Energy, Barnett Way, Barnwood, Gloucester, UK.
Email: mark.r.bradford@british-energy.com

Abstract
In this paper, the authors describe the progress towards development of a holistic model of irradiated graphite properties, which is based on an understanding of the interconnection of the structure. The approach has enabled us to explain many of the correlations that exist between irradiated graphite material properties and to make predictions of the effect of radiolytic oxidation beyond the limits of the database that currently exists. The principles are illustrated with reference to the irradiation "structure term" of Young's modulus. It is shown that the currently used definition is actually a combination of three separate processes – pore closure driven densification, increased structural interconnectivity and (latterly) pore generation. It is the structural interconnectivity component that is most closely linked to the changes in other properties such as dimensional change rate and coefficient of thermal expansion.

Keywords
Irradiation, graphite properties, structure

INTRODUCTION

Traditionally, the approach that has been taken when modelling the irradiation behaviour of graphite is to treat each property independently and to define empirical and semi-empirical models for each. However, this approach has led to some notable inconsistencies when attempting to include the effects of influencing factors for which data are sparse. A notable example is the long-standing assumption that radiolytic oxidation affects the irradiation behaviour of dimensional change but not that of the coefficient of thermal expansion (CTE). However, both properties are concerned with the transmission of crystallite driven strain throughout the material microstructure – shrinkage strain in the case of dimensional change and thermal strain in the case of CTE. Therefore, it is reasonable to expect that radiolytic oxidation would either affect both properties or neither, but not one in isolation of the other. It was inconsistencies such as those outlined here, that has led the authors to reinvestigate the basis of the existing models. The aim of the study is to define models with self-consistent assumptions, based upon a better understanding of the material behaviour as a whole. This will yield improved model reliability when extrapolating beyond the limits of the databases used to derive them. This paper gives an overview of the approach by discussing the role of structural connectivity in the irradiation behaviour, with particular emphasis on the so-called "structure term" of Young's modulus.

A STRUCTURAL APPROACH TO GRAPHITE PROPERTIES

Evidence for the Influence of Structure in the Irradiation Behaviour of Graphite

When subjected to fast neutron irradiation, the bulk properties of graphite are modified. In near-isotropic graphites, such as that in the AGRs, bulk shrinkage is accompanied by an increase in Young's modulus, whilst the CTE initially rises then steadily falls (Figure 1).

However, under similar irradiation conditions, the properties of the individual crystals[9] do not undergo any significant change, with the exception of dimensional change. In this latter case, growth and shrinkage in the crystallographic c-axis and a-axis orientations, respectively, is observed (Figure 2).

Attempts to interrelate the properties by treating the material as an assemblage of crystals (*e.g.* Simmons, 1965) found some early support in apparently explaining a link between initial shrinkage rate and the CTE. However, this approach breaks down once the material approaches dimensional change "turnaround" as the change in shrinkage behaviour runs contrary to that predicted from the CTE (Kelly *et al.*, 1966). More recently Kelly *et al.* (1994) attempted to model the various property behaviours *via* the crystal strain rates and a "structure term". Unfortunately, for isotropic material the method relies on the Simmons methodology and was therefore impractical.

In principle, given a computer model that is sufficiently large and representative, and suitable input properties, it should be possible to numerically model the behaviour of the bulk material. One example of such an approach that has met with moderate success is the work of Hall (2005). Unfortunately these approaches are computationally expensive (Hall's studies were restricted to two dimensions) and may ultimately struggle for a lack of representative crystal properties - even the most highly oriented Compression Annealed Pyrolytic Graphites are still made up of imperfectly aligned multiple crystals.

The authors have therefore chosen to adopt a different approach to the problem – that of using analogies and simplified conceptual models to understand how the various components of the structure interact with

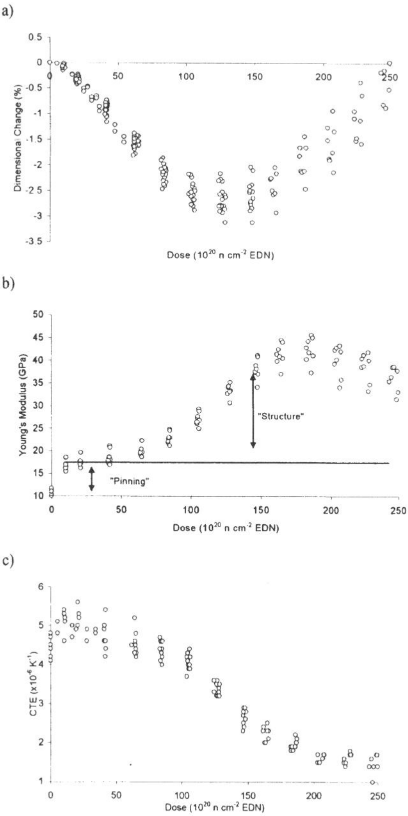

FIGURE 1: The behaviour of AGR Gilsocarbon graphites with fast neutron irradiation: a) dimensional change; b) Young's modulus; c) CTE. The irradiation temperature is 430 °C.

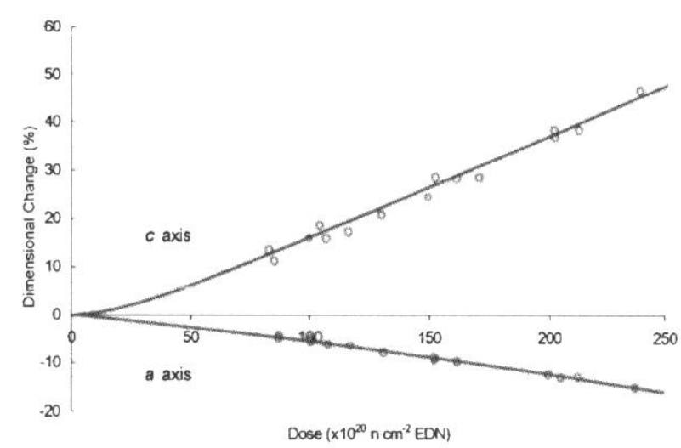

FIGURE 2: The dimensional change behaviour of HOPG with fast neutron irradiation (400-700 °C). Data taken from Kelly *et al.* (1971).

[9] Deduced from the behaviour of highly oriented pyrolytic graphite, HOPG – see Kelly and Brocklehurst, 1971 and Seldin and Nezbeda, 1970.

one another, and hence the ways in which properties are interrelated. The approach is similar to that adopted by Simmons, but extended into three dimensions. In this way the authors aim to define the key relationships that will enable improved semi-empirical models to be developed.

The Structure of Unirradiated Graphite

The graphite used in the Advanced Gas-cooled Reactors (AGRs) is comprised of a Gilsocarbon filler and flour, and a pitch-based binder. The target density was achieved by liquid pitch impregnation. Post-graphitisation, the crystallites appear to be relatively similar in size across the material, whether associated with the filler, binder or impregnant. There are differences in local orientation and ordering of the crystallites within the various regions (Figure 3) but, on a macroscopic scale, the crystallites are randomly oriented, giving a near-isotropic material. Interspersed throughout the graphitic phases, there are three principal classes of porosity:

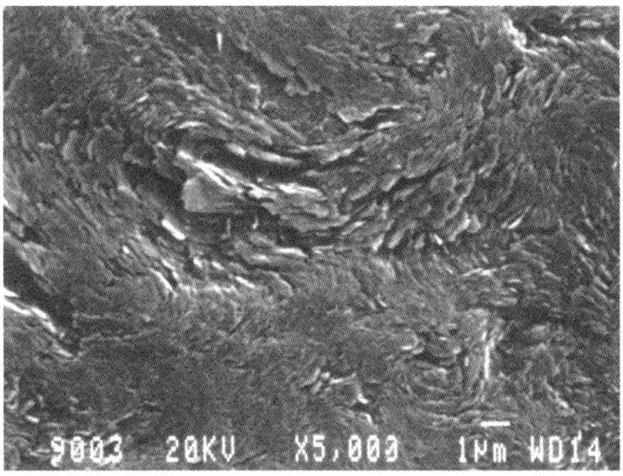

FIGURE 3: Scanning electron micrograph of binder phase in Gilsocarbon graphite showing typical crystallite size and the existence of short range order and medium-range disorder.

- *Calcination cracks.* Large, lenticular pores formed during the initial coking process. The "onion-skin" cracks in Gilsocarbon particles are classic examples of calcination cracks.
- *Gas evolution and transport pores.* During the baking of the "green stock", hydrocarbons and other gases are driven off and form these typically cylindrical aspect-ratio pores.
- *Thermal shrinkage cracks.* At graphitisation temperatures, the atomic motion is facile and the structure is relatively "fluid", thermal energy allowing bonds to break and reform readily. During cool-down, the structure "freezes", whereupon the anisotropic crystal CTE, particularly the high c-axis CTE, causes local stresses to form. If the stress is high enough, microcracks are formed. These are likely to be of high aspect ratio and, due to the nature of their formation, will be generally aligned with the crystal c-axis. Microcrack initiation at different temperatures during the cooling will result in a distribution of crack sizes and there will still be some residual stress at room temperature in regions of the material where cracking has not occurred.

The Conceptual Model

Within the proposed conceptual model, the authors represent the microstructure of the material by a network of ligaments, connected by joints, interspersed with voids/porosity. The ligaments are graphite crystallites or, at a coarser level, domains; the joints are the chemical bonds between the crystallites or domains. The measured bulk properties will be determined by the properties of the component crystals, the nature of the connections between them and, in some cases, by the extent and type of porosity. Where the properties of individual ligaments differ, as in the case of anisotropic graphite crystals of varying alignment, the assessment of the bulk behaviour from the crystal properties must also take into account the action of one ligament upon another. This is a principle well understood in mechanical engineering.

Application of the Conceptual Model to the Irradiation Behaviour of Graphite
The application of the concepts of the structural model to the irradiation behaviour of nuclear graphite may be illustrated by considering the "2½ dimensional" structure in Figure 4, which represents the graphite crystals in their most connected state. Although highly simplified, this is adequate for the present purposes of demonstrating the construction of bulk behaviour from that of individual components, and the response of the structure to irradiation and oxidation. In this representation, the bulk graphite is composed of two parallel plates connected by three ligaments – the inner ligament having a differing material property from the outer two because of the differing alignments of graphite planes.

Illustration of the Basic Principles

Young's Modulus: The bulk modulus of the composite structure is defined by the strain response for a given load. If a load is applied to the plates in Figure 4, then all three ligaments will experience equal strain and the greatest stress will be experienced by the ligament(s) of highest modulus. This is a clear illustration of the engineering principle that "stiffness attracts load". Further, the 2:1 split of ligament types (crystal alignment) in the illustration enables us to show that the load distribution, and hence the bulk modulus of structure of Figure 4, will be different according to which of the ligament types is the higher modulus. It therefore follows that the number and relative degree of connectivity of the stiffest ligaments will dominate the bulk modulus of the composite structure. In graphite, the highest modulus is in the plane of the crystals – *i.e.* the crystallographic a-axis.

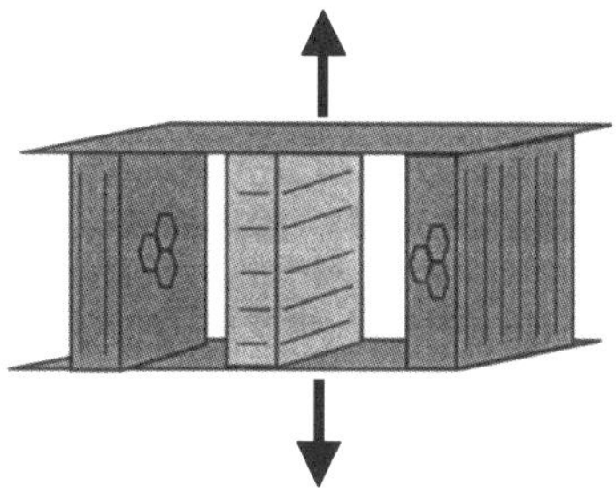

FIGURE 4: Schematic representation of a unit within the structural model. The different shades represent ligaments of differing material property.

The illustration also enables us to show the effect of increasing the porosity within the structure. As the internal porosity is increased, the ligament thickness is reduced. This reduces the internal load bearing areas, causing the stress in any given ligament to increase. If the underlying ligament moduli are unchanged, then the increased ligament stress will result in an increased strain – yielding a reduction in the bulk modulus. The bulk modulus of a graphite specimen will therefore be determined by the underlying crystal moduli, the degree of interconnectivity of the structure and the degree of porosity (*i.e.* density).

Coefficient of Thermal Expansion: The CTE of the composite structure reflects the equilibrium distribution of internally generated thermal stress driven by the differing expansion or contraction of the individual ligaments. In graphite, high thermal expansion is observed in the crystallographic c-axis, whilst there is negligible expansion (or even contraction) in the a-axis. Consequently, in the Figure 4 illustration, the outer ligaments contract whilst the central ligament expands, with the result that the outer ligaments will be placed in tension whilst the inner ligament will be compressed. The net thermal strain of the structure will then depend on the response of each ligament to the stress – *i.e.* it will be a function of the ligament moduli. Since the c-axis has significantly lower modulus, it follows that, although the crystal is trying to expand in the c-axis direction, the compressive load imposed upon it by the parallel a-axis ligaments constrains the expansion. As a consequence, the bulk CTE of graphite is strongly influenced by the a-axis connectivity.

In standard continuum models, CTE is independent of degree of porosity, provided that at least one continuous strand of the network remains intact (Hacker *et al.*, 1999). Figure 4 is consistent with this, in that the bulk CTE will be unaffected if the interconnectivity is left intact. The bulk CTE of a graphite specimen will therefore show a strong dependence on the underlying crystal CTE and moduli and the degree of interconnectivity of the structure, but is not likely to be a strong function of the degree or type of porosity.

Illustration of the Application to the Irradiation Behaviour

It has been noted above that irradiation does not fundamentally change the crystal CTE or moduli[10]. It does, however, change the size and shape of the crystals (ligaments) and, consequently, the porosity. It also affects the connections, allowing existing bonds to be broken and reformed and new connections to be made when porosity is closed. It therefore restores to the structure some of the fluidity that exists at graphitisation temperatures. It is this dynamic behaviour that allows the material to undergo irradiation creep when subjected to a stress, whether externally applied or internally generated.

Dimensional Change: The highly simplified representation in Figure 4 shows a structure with complete connectivity between the individual ligaments and the plates. In other regions of the microstructure, the thermal shrinkage stresses associated with cooling from graphitisation would have led to microcracking, generating c-axis aligned porosity. It is this alignment that enables thermal shrinkage cracks to "accommodate" the crystal c-axis expansion shown in Figure 2, with the result that the a-axis contraction initially dominates, giving the bulk behaviour of Figure 1a.

In the early stages of irradiation, crystal dimensional change rates are slow and there will be little, if any, closure of porosity and increase in connectivity. However, residual internal stress will be relaxed by creep processes. As irradiation proceeds, the rate of c-axis growth will increase and the thermal stress cracks will be closed – slowly at first but then at an increasing rate, which some HOPG irradiation data (Kelly *et al.*, 1971) suggest may saturate after an irradiation dose of $\sim 50 \times 10^{20}$ n cm^{-2} (Figure 2). Once the pores are closed, the structural interconnectivity increases and continued growth begins to strain the surrounding material. However, the irradiation also allows the connecting bonds to be broken and reformed, allowing the structure a degree of freedom to rearrange and rebalance the strain. As more aligned cracks are closed, the c-axis expansion begins to dominate. The shrinkage rate then slows and reverses – see Figure 1a - which is referred to as "turnaround".

An accompanying effect of the continued expansion of the c-axis post-pore closure is the generation of new porosity. However, due to the distribution of the pore and crystallite sizes, this will not occur in all regions of the material simultaneously. Consequently some regions of the bulk material will still be undergoing pore closure whilst others undergo pore growth. Interestingly, the scale at which these two processes occur is likely to be different. Thermal shrinkage microcrack closure will occur throughout the structure, whilst pore growth, being

---

[10] Seldin and Nezbeda (1970) show an increase in c44, but the increase is an order of magnitude less than that required to cause a significant change in the bulk modulus of a structure such as that shown in Figure 4.

driven by relief of stresses caused by crystal strain, may tend to favour extension of existing cracks (*e.g.* lenticular calcination cracks).

Young's Modulus: Upon irradiation, the crystal moduli do not change sufficiently for this to be reflected in a change in the bulk modulus. The observed changes with irradiation must therefore be the result of structural change. At low dose, there is no significant change in porosity. However, irradiation creep relief of residual stress and rebalancing/stiffening of initial connections between crystals and domains should affect the bulk modulus, and experimental evidence suggests that it does – the rapid initial increase known as "Pinning". With increasing dose, the crystal shape change drives a closure of porosity and an increase in the proportion of the internal load bearing area relative to the bulk, hence the modulus increases slowly. With further irradiation, the accommodation porosity is consumed, and the interconnectivity increases in 3-dimensions. Although not immediately apparent from the highly simplified representation of Figure 4, this further increases the a-axis contribution to the bulk modulus and the rate of rise is more rapid. However, continued c-axis expansion then begins to generate new porosity and the bulk modulus will first peak, then fall, as the effect of pore generation outweighs that of increased interconnectivity, which will saturate as accommodation porosity is consumed. This consideration therefore indicates that the traditional "structure term" (Figure 5) is, in reality, comprised of three components: – pore closure (densification); increased connectivity, which will asymptote as accommodation porosity is consumed; and crack/pore formation at higher dose. The authors find that a very good representation of the behaviour is obtained if the densification is modelled as the reverse of radiolytic oxidation and the pore generation is modelled in a similar manner to thermal oxidation.

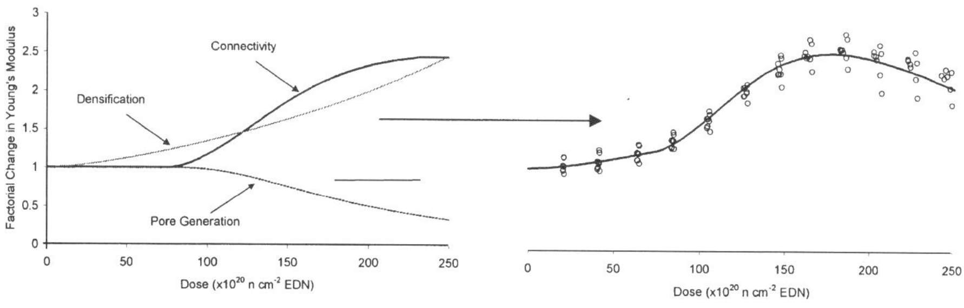

FIGURE 5: The construction of the Young's modulus "structure term" from its separate components.

Coefficient of Thermal Expansion: It is known that externally applied elastic strain modifies the bulk CTE (S D Preston, personal communication), and so it follows that the observed increase in CTE at low dose (Figure 1c) may be the result of the relief of residual cooling stress. With increasing dose, the crystal shape change drives a closure of porosity without increasing the structural interconnectivity. The CTE will therefore, to first order, be unaffected[11] during this phase. As the accommodation porosity is consumed, and the interconnectivity increases in 3-dimensions, the influence of the a-axis will increase and the bulk CTE is expected to fall. By contrast, pore generation due to continued c-axis expansion

[11] Interestingly, a consideration of the relative changes of the shapes of the crystals, and hence contact areas, suggests that the CTE should fall slightly during the early densification phase as relatively more of the contact area is made up of a- axis aligned material, increasing the contribution of that component.

is likely to have little effect on bulk CTE. The high dose increase in porosity is likely to be dominated by growth of existing cracks and pores and, since the connectivity in material around the pores will remain largely unaltered, the continuum theories of CTE behaviour will apply.

Implications for the Modelling of Graphite Properties

Correlations Between Properties

From the consideration of the structural behaviour of graphite outlined above, it follows that:

- The turnaround in dimensional change and the rapid rise in modulus in the mid-dose region are both due to the crystal dimensional change driven closure of accommodation porosity. Pore generation follows as a consequence. Therefore, it should be expected to find a strong correlation between the dimensional change pore generation rate and the underlying modulus structural connectivity term. This agrees with experimental data (Figure 6).
- The behaviour of bulk CTE is only significantly affected by the 3-dimensional structural connectivity. Therefore, the authors expect to see a strong correlation between the CTE and the underlying modulus structural connectivity term. Again, this is in full agreement with the experimental data (Figure 7). The interconnectivity model also predicts that the fall in CTE should not be continual, but should asymptote, as can be seen in Figure 1c.
- The authors also expect to see a correlation between the CTE and the modulus after the initial relaxation of cooling residual stress, and before the pore closure-induced increase in modulus occurs. This, also, is validated by the available data (Figure 8).

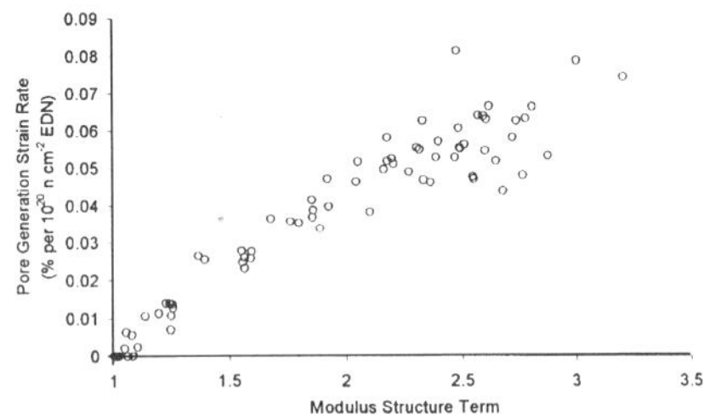

FIGURE 6: The correlation between the dimensional change pore generation rate and the Young's modulus structural connectivity term

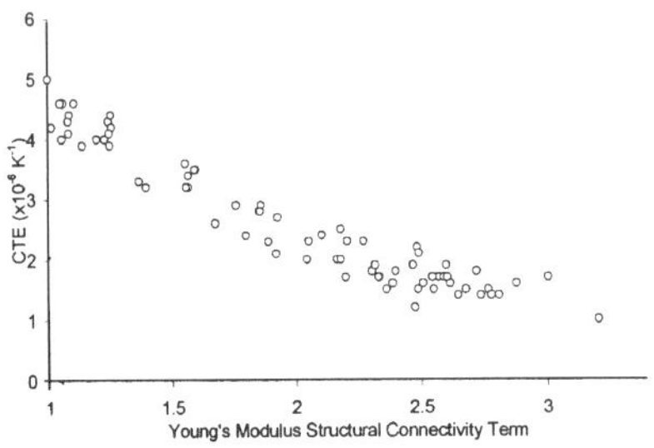

FIGURE 7: The correlation between the CTE and the Young's modulus structural connectivity term.

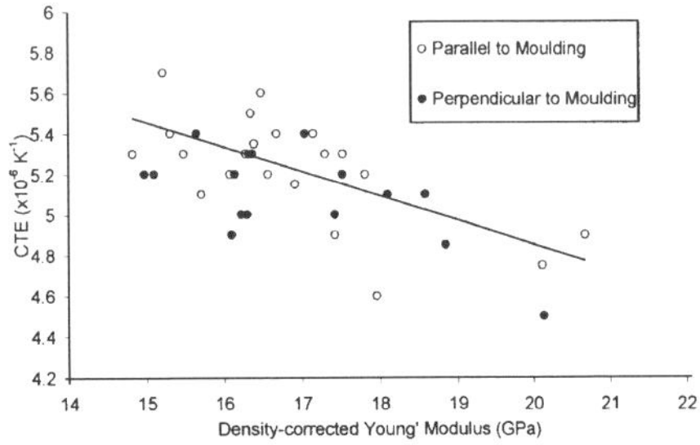

FIGURE 8: The correlation between CTE and Young's modulus at low dose.

The Influence of Radiolytic Oxidation

In contrast to its thermal counterpart, which is predominantly limited to the gas transport porosity, radiolytic oxidation acts throughout the whole of the open porosity within the structure. The processes that control radiolytic oxidation act in such a way that it is most efficient in small, slab shaped pores, and favours the corrosion of the outermost planes of a crystal (Best *et al.*, 1985). Both of these aspects come together to suggest that open thermal shrinkage cracks will be oxidised efficiently in such a way as to increase the available accommodation porosity. From the discussion presented here, it can be seen that this would lead to a delay in shrinkage turnaround, for which there is clear evidence in support from Material Test Reactor experiments (Kelly *et al.*, 1974). However, since it is the closure of the accommodation porosity that leads to the rapid increase in modulus and the fall in CTE, radiolytic oxidation is now expected to delay both of these phenomena as well.

Application in the Assessment of AGR Component Behaviour

The structural connectivity term identified within this paper can be used as a surrogate for the closure of accommodation porosity. The correlations that have been identified can then be used, in a pragmatic manner, to define simplified semi-empirical models for bulk dimensional change, modulus and CTE for use in component stress analysis. Since these models are now founded in a physical understanding of the material behaviour at a microstructural level, and since there is now complete internal consistency in their treatment, the authors believe they increased confidence in their behaviour with oxidation and to high dose.

CONCLUSIONS

A conceptual, holistic model of the behaviour of irradiated graphite has been developed, based on an understanding of the interconnection of the crystallites within the structure. The approach has enabled the identification of the most significant correlations that exist between certain of the properties. It has been shown that the currently used definition of the "structure term" of Young's modulus can be considered to be a combination of three separate processes – pore closure driven densification, increased structural interconnectivity and (latterly) pore generation. The structural interconnectivity component of modulus is closely linked to the changes in other properties such as dimensional change rate and coefficient of thermal expansion and may be used to define self-consistent models for the simultaneous irradiation and oxidation behaviour of these three properties.

Acknowledgements

The authors would like to thank British Energy Generation Ltd for permission to present this paper.

References

Best, J. V., Stephen, W. J. and Wickham, A. J. (1985). Radiolytic graphite oxidation. *Progress in Nuclear Energy*, 16, [2], 127-178.

Kelly, B. T. and Brocklehurst, J. E. (1971). High dose fast neutron irradiation of highly oriented pyrolytic graphite. *Carbon*, 9, 783-789.

Kelly, B. T., Brocklehurst, J. E. and Ashton, B. W. (1974). The interaction of radiolytic oxidation and fast neutron irradiation in the moderators of CO_2 cooled reactors, *UKAEA report TRG 2589(S).*

Kelly, B.T. and Burchell, T. (1994). Structure-related property changes in polycrystalline graphite under neutron irradiation. *Carbon*, 32, [3], 499-505.

Hacker, P. J., Neighbour, G. B. and McEnaney, B. (1999). The coefficient of thermal expansion of graphite. *Bath Nuclear Materials Group report BNMG/REP/024, HSE Reference FC/GNSR/44.*

Hall, G. (2004). Microstructural modelling of nuclear grade graphite. *EngD thesis, University of Manchester.*

Kelly, B. T., Martin, W. H. and Nettley, P. T. (1966). Dimensional changes in pyrolytic graphite under fast-neutron irradiation. *Phil Trans A*, 260, 37-49.

Seldin, E. J. and Nezbeda, C. W. (1970). Elastic constants and electron-microscope observations of neutron-irradiated compression-annealed pyrolytic and single-crystal graphite. *J App. Phys.* 41 [8], 3389-3400.

Simmons, J. H. W. (1965). Radiation damage in graphite, Pergamon Press, Oxford.

X-ray Tomographic Observations Applied to Porosity Models for the Thermal Properties of Oxidised Nuclear Graphite

L Babout, T J Marrow, P Mummery and B J Marsden[§]
School of Materials, Materials Performance Centre, University of Manchester
[§]School of Mechanical, Aerospace and Civil Engineering, Nuclear Graphite Research Group, University of Manchester
Email: james.marrow@manchester.ac.uk

Abstract
The porosity network in thermally oxidised samples has been classified in terms of shape of isolated pores and fractions of isolated/continuous porosities using computed X-ray microtomography. A model had been developed and tested against measurement of thermal conductivity. The model, which is based on the Euler approach, takes into account the different families of pores and shows a good agreement with the experimental data.

Keywords
Tomography, thermal conductivity, porosity

INTRODUCTION

Polygranular graphite is used in UK gas-cooled reactors mainly because it is an effective moderator for fast neutrons. During reactor operation, fast neutron irradiation and radiolytic oxidation lead to significant graphite dimensional and material property changes. In particular, radiolytic oxidation can lead to the generation of significant porosity. Despite the fact that radiolytic oxidation and thermal oxidation are known to have different effects on microstructure evolution (*e.g.* Burchell, 1986), to avoid difficulties associated with the examination of irradiated graphite samples, thermal oxidation is often used to simulate changes caused by irradiation. These studies on thermally oxidised graphite are useful to demonstrate the level of information that can be obtained from trepanned irradiated samples. The monitoring of graphite density variations is usually performed using balance or immersion measurements. These measurements can give a satisfactory measure of the bulk or average weight loss and bulk density, but do not allow local density variations to be investigated. X-ray tomography is a good means of studying changes in local density, allowing three-dimensional density maps to be obtained (Baruchel 2000, Maire 2003, Burch 2002, Philips 1979). In particular X-ray tomography measurements allow density and structural variations to be evaluated over a sample at a range of important length scales:

- The length scale of the sample itself, for direct comparison with other bulk density measurements.
- The macro-scale, corresponding to density changes across the sample.
- The microstructural feature scale, for studying the variation of density from one phase to another and establishing their relative sensitivity to oxidation.
- The element or voxel scale, essentially the smallest element in the volume resolvable defined by the tomographic instrumental spatial resolution.

The techniques for density measurement in thermally oxidised Pile Grade A (PGA) graphite specimens at the sample, macro and element scales have been presented in a previous paper (Babout, 2005). The aim of this paper is to demonstrate how X-ray tomography data can be applied to the semi-empirical modelling of material properties, in particular thermal properties.

MEASUREMENT OF THERMAL PROPERTIES

The thermal diffusivity, α, of the specimens was measured using the laser flash technique (Parker, 1961). The technique has been used successfully to measure the thermal properties of carbon and graphite. Here, the temperature rise is measured, as a function of time, on the back face of a specimen heated by a laser pulse on the front face (Figure 1a). A typical result is shown schematically in Figure 1b. The pulse source is a solid state Nd-glass laser (λ = 1.067 μm) with a beam diameter of 16 mm. The energy output is in the range 5 – 95 J, and the pulse dissipation time is nominally 0.6 ms. The specimen is mounted in a graphite holder placed inside a graphite susceptor. A lid, having a hole with the same diameter as the sample, serves to cut off the excess laser beam. The susceptor is located within an induction coil positioned in a vacuum chamber. Radiation from the back face of the sample is collected via a lens and mirror system, and transmitted to the detector assembly external to the vacuum chamber by an optical system. The radiation detector is an InSb detector, sensitive to 5.5 μm, having a response time of 1.5 μs. The diffusivity, α, is calculated from the half-rise time, $t_{1/2}$, and thickness, L, by

$$\alpha = 0.136 \frac{L^2}{t_{\frac{1}{2}}} \qquad [1]$$

The thermal conductivity, k, can be calculated from the diffusivity given knowledge of density, ρ, and specific heat capacity, C_p, of the material, using the relationship $k = \alpha\rho C_p$.

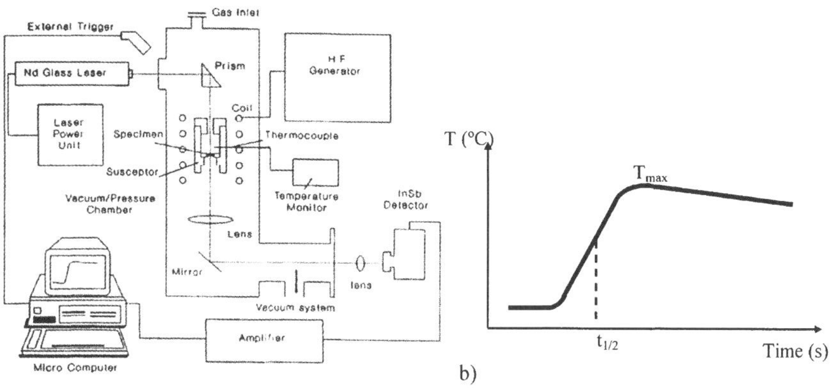

a) b)

FIGURE 1: a) Schematic of laser flash equipment, and b) definition of half-rise time, $t_{1/2}$

MODELLING OF THERMAL PROPERTIES

In order to compare reliable observations of porosity at different weight losses, it is necessary to use the same standards for all samples. This is done by applying the same set-up for X-ray tomography scans and data reconstruction, and the same parameters for data analysis, such as thresholding and segmentation. Further details of application of X-ray tomography to measure the density variations in nuclear graphite are given elsewhere (Babout 2005 & 2006).

Calculation of the relative fractions of continuous and isolated porosity is required in order to use the different contributions of these two populations to predict the thermal properties of the sample. The technique developed in this work is novel since it uses, to our knowledge for the first time, the data extracted from X-ray tomography to calculate these fractions. The method, which consists of calculating the number of intersection between a growing three-dimensional box and the porosity, is very sensitive to the irregularity and tortuousity of pores. This is important in thermally oxidised graphite where coalescence of pores is observed, particularly in the matrix (Figure 2). Consequently, the three-dimensional connectivity of porosity may be evaluated from three-dimensional tomography data.

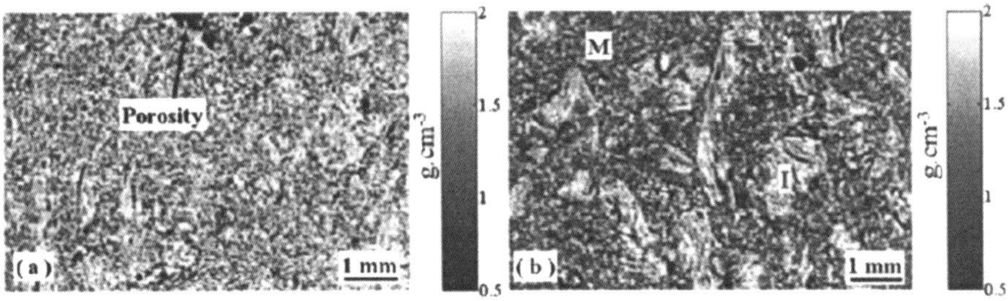

FIGURE 2: Reconstructed tomography data showing longitudinal slices of (a) PGA sample at its virgin state; and (b) after thermal oxidation corresponding to 22.5% of bulk weight loss. The extrusion direction is vertical in the images. (I): Filler Particles, (M): Matrix. The density variation has been calibrated (Babout, 2006).

The pores are identified and labelled by thresholding and segmentation of the tomography data. The characterisation into families of isolated and percolated or continuous pores has been performed using a criterion based on the complexity of the pore shape. The method consists of calculating the average number of intersections of the pore with a set of cubes, increasing in size incrementally from the pore centroid to the pore's bounding cube. This is illustrated, for two dimensions, in Figure 3, for which the number of intersections with an isolated pore is shown to be lower than that for a continuous pore. The criterion assumes that any pore with an average number of intersections larger than 6, which corresponds in three dimensions to the maximum for a sphere, is classified as continuous porosity. The volume fraction of each family of pores may then be calculated. It is recognised that the spatial resolution and the selected threshold for the pore segmentation may influence the actual content of both categories, and the calculated pore data is therefore an approximation. Standardising the classification method, however, enables a description on the effect of thermal oxidation on the relative proportions of isolated and continuous pores.

A further step of the classification is to separate the isolated pore family into sub-groups of different morphology. The aim is to implement the morphological parameter resulting from this new classification into a porosity model for material properties. The sub-groups are three ellipsoid shapes, *i.e.* spheroid, prolate and oblate (F).

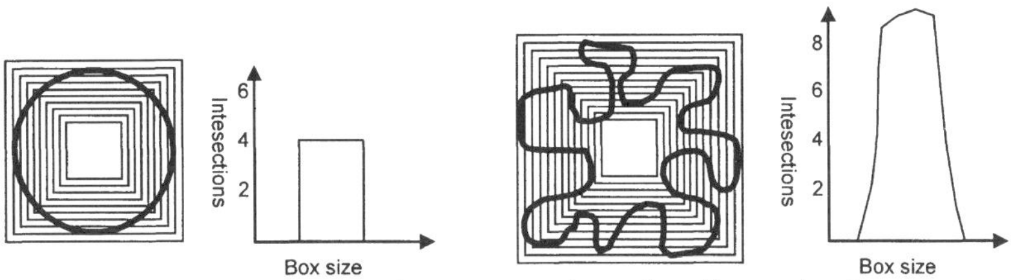

FIGURE 3: Description of the character of a pore in terms the number of intersections between a growing box and the pore surface (2D example). The average number of intersections for all boxes increases significantly for percolated pores.

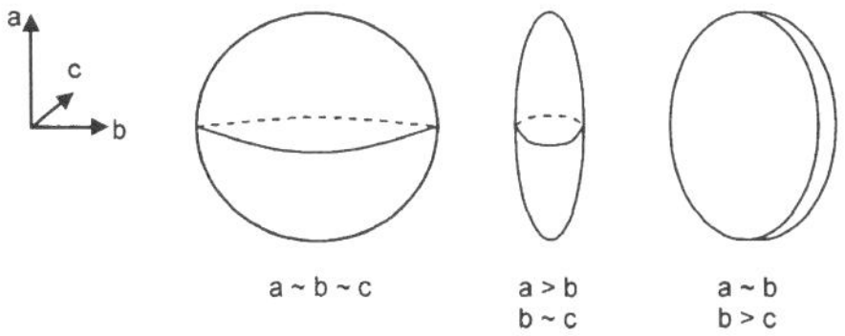

FIGURE 4: Spheroid, prolate and oblate ellipsoids with the corresponding dimension characteristics.

The classification of the observed isolated pores into the three sub-classes, with respect to their aspect ratios b/c and a/b, is shown in Figure 5 for virgin PGA. The classification involves two boundaries to divide the porosities in three different sub-groups with the following criteria:

- Prolate: b/c "small" (b/c < 2.5) and a/b "large" (a/b > 2.5).
- Oblate: b/c "large" (b/c > 2.5) and a/b "small" (a/b < 2.5).
- Spheroid: a/b and b/c "small".

The boundaries were chosen in accordance with the distribution of aspect ratios shown in Figure 5 and are both arbitrarily set to 2.5. The relative fractions of continuous and isolated porosity, and the different fractions of isolated pore types are summarised in Table 1 as a function of thermal oxidation weight loss. The spheroid sub-group is the largest component of the isolated pores. The relative fractions of the subgroups do not vary systematically with weight loss, indicating that thermal oxidation is not selective with pore shape for the measured porosity.

ANALYTICAL MODEL FOR THERMAL CONDUCTIVITY

The pore characterisation by X-ray tomography was applied to an analytical model for thermal conductivity. Three different approaches have been considered:

- The Euler model which takes into account the fraction of continuous and isolated pores (Euler, 1957, Brocklehurst, 1970).
- The Meredith and Tobias' model which takes into account the morphology of pores (Ceruschi, 2004).
- The extended Meredith and Tobias approach, used by Cernuschi, which combines iteratively the different families of pores (Ceruschi, 2004).

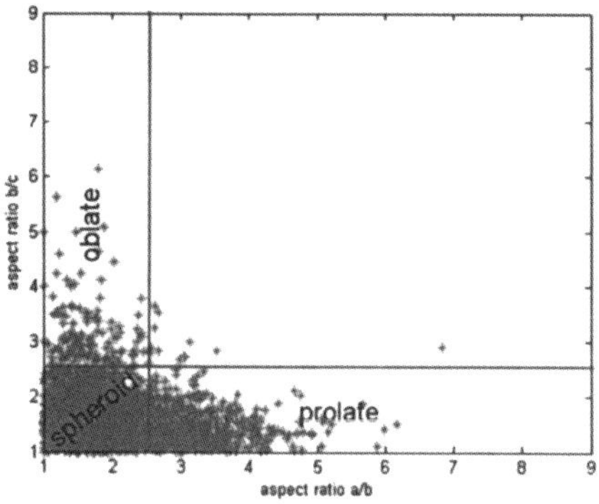

FIGURE 5: Classification of isolated pores with respect to their aspect ratios for virgin PGA.

Table 1: The effect of weight loss (WL) on thermal conductivity (k), measured by the laser flash method in thermally oxidised PGA graphite. The relative fractions of isolated (f_{iso}) and continuous (f_{cont}) porosity determined by X-ray tomography are given. The mean aspect ratio (Aspect) and fractions of the different sub-groups of isolated porosity (f_1, f_2, f_3) are also given, relative to the total fraction of isolated porosity. The mean aspect ratio for oblate sub-group is expressed inversely so all ratios are defined in accordance with the revolution axis (revolution axis for spheroid and prolate: along "a" axis – revolution axis for oblate: along "c" axis).

Weight Loss	*Thermal Cond.*	*Porosity Ratio*		*Spheroid*		*Prolate*		*Oblate*	
WL frac.	***k*** (W m^{-1} K^{-1})	f_{iso}	f_{con}	**Aspect** (a/c)	**Fraction** (f_1)	**Aspect** (a/c)	**Fraction** (f_2)	**Aspect** (c/a)	**Fraction** (f_3)
0.000	178	0.837	0.163	2.341	0.755	4.346	0.135	0.149	0.109
0.045	143	0.481	0.519	2.355	0.711	4.368	0.193	0.163	0.096
0.126	103	0.402	0.598	2.313	0.789	4.267	0.139	0.145	0.072
0.141	86	0.325	0.675	2.325	0.815	4.304	0.126	0.153	0.062
0.197	46	0.042	0.958	2.381	0.786	4.204	0.143	0.155	0.071
0.233	21	0.006	0.994	2.394	0.667	4.832	0.167	0.167	0.167

Euler's approach, used also by Brocklehurst (1970) to model thermal conductivity of irradiated graphite samples, considers two bounds which correspond to two extreme cases. A periodical array of isolated solid spheres represents continuous porosity, whereas an array of spherical pores represented the isolated pore condition. The corresponding bounding limits for the change in thermal conductivity k/k_o, denoted by Y_1 and Y_2 are:

$$Y_1 \qquad k/k_0 = \pi\left(r^2 - 0.25\right) \qquad [2]$$

where r is the solution of

$$1 - f = \pi\left(-2.67r^3 + 3r^2 - 0.25\right) \qquad [3]$$

Y_1 is valid for the volume fraction of porosity: $\pi/6 \le f \le 1$

$$Y_2 \qquad k/k_0 = (1 - f)/(1 + 0.5f) \qquad [4]$$

Y_2 is valid for the volume fraction of porosity: $0 \le f \le \pi/6$

Meredith and Tobias's approach considers the shape of the porosity and volume fraction f directly in an analytical expression.

$$\frac{k}{k_0}=\left[\frac{2-f}{2+(W-1)f}\right]\left[\frac{2(1-f)}{2(1-f)+Wf}\right] \qquad [5]$$

$$\text{where } W=\frac{1}{3}\left(\frac{1}{2F}+\frac{2}{1-F}\right) \qquad [6]$$

Equation [6] is function of the shape factor F which is directly correlated to the aspect ratio of the pores. Cernuschi's approach develops the Meredith and Tobias model and allows the different categories of porosity shapes to be taken into account in a two-phase model. The model has been defined with a unique iterative approach in order to obtain thermal conductivity expressions independent from the order of mixing of the different pore types within the graphite, as expressed in Equation [7]:

$$\begin{aligned}\frac{k}{k_0}=\frac{1}{6}\Bigg\{&\Phi\left[\frac{f_1}{1-(f_2+f_3)}\right]\Psi\left[\frac{f_2}{1-f_3}\right]\Theta[f_3]+\Phi\left[\frac{f_1}{1-f_3}\right]\Psi\left[\frac{f_2}{1-(f_1+f_3)}\right]\Theta[f_3]\\&+\Phi\left[\frac{f_1}{1-f_2}\right]\Psi[f_2]\Theta\left[\frac{f_3}{1-(f_1+f_2)}\right]+\Phi\left[\frac{f_1}{1-(f_2+f_3)}\right]\Psi[f_2]\Theta\left[\frac{f_3}{1-f_2}\right]\\&+\Phi[f_1]\Psi\left[\frac{f_2}{1-f_1}\right]\Theta\left[\frac{f_3}{1-(f_1+f_2)}\right]+\Phi[f_1]\Psi\left[\frac{f_2}{1-(f_1+f_3)}\right]\Theta\left[\frac{f_3}{1-f_1}\right]\Bigg\}\end{aligned} \qquad [7]$$

Where Φ, Ψ and Θ correspond with the expressions of Equation [5] for the different fractions of pore type (f_1, f_2, f_3, where $f_1+f_2+f_3=f_{iso}$). The isolated pore fraction, f_{iso}, and the isolated pore aspect ratio values are shown in Table 1. The approach that has been implemented here attempts to incorporate the maximum amount of information that can be extracted from X-ray tomography. The method combines the Euler approach for continuous pores and the Cernuschi approach to the different families of isolated pores. The bounds are combined with a weighting law, which depends on the fractions of continuous (f_{cont}) and isolated (f_{iso}) porosity, to determine the thermal conductivity,

$$\frac{k}{k_0}=f_{cont}Y_1+f_{iso}\left(\frac{k}{k_0}\right)_{Cernuschi(MT)} \qquad [8]$$

Where $(k/k_0)_{Cernuschi(MT)}$ is given by Equation [5]. The predicted thermal conductivity is also compared with a more simple expression, taking into account the Euler Bounds only, neglecting the pore shape,

$$\frac{k}{k_0}=f_{cont}Y_1+f_{iso}Y_2 \qquad [9]$$

The parameters Y_1 and Y_2 are calculated using the total porosity volume, which was determined using the bulk weight loss, assuming an initial porosity of 20%. The comparison of the model predictions with experimental data is shown in Figure 6. The initial conductivity, for 0% porosity, k_0, was set to 260 W m^{-1} K^{-1} by fitting to the data. The in-plane thermal conductivity of highly oriented pyrolytic polycrystalline graphite (HOPG), which is close to the theoretical density of graphite, varies between 160 and 390 W m^{-1} K^{-1} (Nightingale 1962 and Bokros 1969). This demonstrates a variation with structure, but suggests that the value obtained for fully dense graphite is reasonable.

Strictly, the solution for connected porosity (Y_1) is not valid over the range of total porosity volume observed. However, there is good agreement between experimental data and both models, indicating that a transition may occur between the upper bound (isolated porosity) and lower bound (connected porosity) with increasing weight loss. The data does not show a very strong sensitivity to the isolated pore morphology, although this may be due to the lack of variation in the samples observed, and also the fact that most of the isolated porosities have a spheroidal shape (Table 1). The figure shows that estimating the fraction of continuous porosity by X-ray tomography may be a suitable method for modelling the thermal conductivity variation with oxidation. Work is in progress to validate the proposed models.

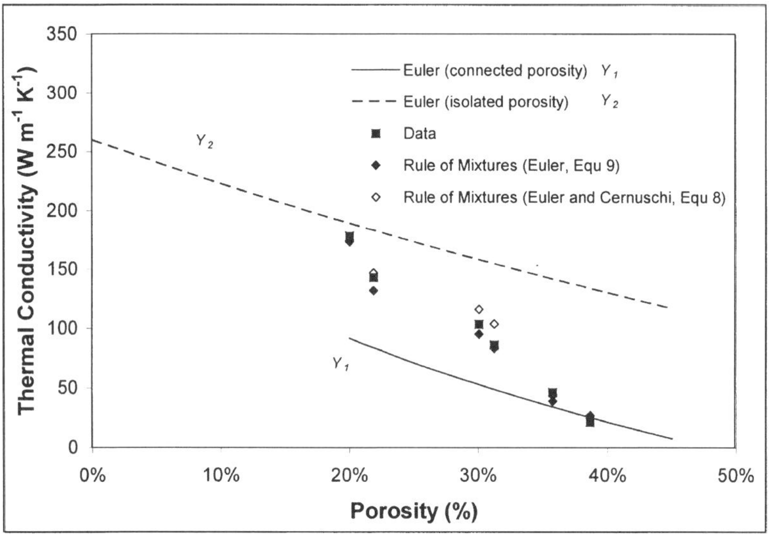

FIGURE 6: Comparison between models and experimental data at different porosity volume fractions for thermally oxidised PGA samples.

CONCLUSIONS

- The porosity data obtained from thermally oxidised samples can be classified in two different families of continuous and isolated porosity using X-ray tomography. The isolated porosity can be also be classified according to shape. These data have been implemented into a porosity-model for thermal conductivity.

- The comparison between the experimental data of thermal conductivity measurements and this model is good, indicating that the thermal conductivity is strongly sensitive to the relative fractions of isolated and continuous porosity in the graphite matrix.
- A transition in thermal conductivity may occur between an upper bound (isolated porosity) and lower bound (connected porosity) with increasing weight loss.

Acknowledgements
The authors would like to thank the Health and Safety Executive (NII) for financial support, and British Nuclear Group for provision of materials. The opinions expressed in the paper are those of the authors, and do not necessarily represent those of the research sponsors. The assistance of Dr G. Hall of the Nuclear Graphite Research Group, The University of Manchester, in the preparation of this paper is gratefully acknowledged.

References

Baruchel, J., Buffiere, J. Y., Maire, E., Merle, P. and Peix, G. (2000). X-ray tomography in materials science. 2000, Paris: Hermes Science. 204.

Babout, L., Mummery, P. M., Marrow, T. J., Tzelepi, A. and Withers, P. J. (2005). The effect of thermal oxidation on polycrystalline graphite studied by X-ray tomography. *Carbon*, [43], 765-774.

Babout, L., Marrow, T. J., Mummery, P. M. and Withers, P. J. (2006). Mapping the evolution of density in 3D of thermally oxidised graphite for nuclear applications. *Scripta Materialia.* Accepted for publication.

Brocklehurst, J. E., Brown, R. G., Gilchrist, K. E. and Labaton, V. Y. (1970). The effect of radiolytic oxidation on the physical properties of graphite. *Journal of Nuclear Materials*, 35(2): p. 183-194.

Burch, S (2002). Measurement of density variations in compacted parts using X-ray computerised tomography. *Metal Powder Report*, 57(2): p. 24-28.

Burchell, T.D., Pickup, I. M., McEnaney, B. and Cooke, R. G. (1986). The relationship between microstructure and the reduction of elastic modulus in thermally and radiolytically corroded nuclear graphites. *Carbon*, 24(5): p. 545-549.

Cernuschi, F., Ahmaniemi, S., Vuoristo, P. and Mantyla, T. (2004). Modelling of thermal conductivity of porous materials: application to thick thermal barrier coatings. *Journal of the European Ceramic Society*, 24(9): p. 2657-2667.

Euler, F (1957). Simple geometric model for the effect of porosity on materials constants. *Journal of Applied physics*, 28(11): p. 1342-1345.

Maire, E., Fazekas, A., Salvo, L., Dendievel, R., Youssef, S., Cloetens, P. and Letang, J. M. (2003). X-ray tomography applied to the characterization of cellular materials. Related finite element modelling problems. *Composites Science and Technology*, 63(16): p. 2431-2443.

Parker W. J., Jenkins R. J., Abbott G. L., Butler C. P. (1961). Flash method of determining thermal diffusivity, heat capacity, and thermal conductivity. *J. App. Phys.*, 32: p. 1679-1684.

Phillips, D. H. and Lannutti, J. J. (1997). Measuring physical density with X-ray computed tomography. *NDT & E International*, 30(6): p. 339-350.

Nightingale R.E. (1962), Nuclear Graphite, Academic Press, p.436, 1962.

Bokros J.C. (1969). Deposition, structure, and properties of pyrolytic carbon (in *Chemistry and Physics of Carbon*, Volume 5, edited by P. L. Walker), Marcel Dekker, p.99, 1969.

PART C

Advances in the Understanding of Graphite-Component Behaviour and its Assessment

Crack Propagation Resistance and Damage Mechanisms in Nuclear Graphite

A Hodgkins, T J Marrow, P Mummery, B J Marsden[§] and A S L Fok[§]
School of Materials, Materials Performance Centre, University of Manchester
[§]School of Mechanical, Aerospace and Civil Engineering, Nuclear Graphite Research Group, University of Manchester
Email: a.hodgkins@postgrad.manchester.ac.uk

Abstract
Crack propagation in isotropic nuclear graphite has been studied using high resolution X-ray tomography to determine the mechanisms responsible for R-curve behaviour (rising fracture resistance with increasing crack propagation). The observations show a physical bridging zone, characterised by small ligaments in the binder/fine filler matrix, in the vicinity of the crack tip. The tomography observations have also shown a microstructure damage process zone ahead of the crack tip and in its wake, together with frictional bridging from fracture surface roughness up to 40 mm behind the crack tip. The measured R-curve is a consequence of both microstructural damage and crack bridging.

Keywords
Isotropic graphite, fracture, R-curve

INTRODUCTION

Rising resistance to crack propagation is observed in a range of polygranular graphites (Fazluddin, 2002). A common measure of this effect is the R-curve, which is the energy dissipated by crack propagation as the length of the crack increases. The R-curve behaviour of Gilsocarbon graphite, which is the grade used to moderate UK Advanced Gas reactors (AGR), has been reported previously (Ouagne, 2002). It was deduced that crack initiation was preceded by the formation and subsequent growth of a micro crack process zone in the material ahead of the crack tip. However, although stress induced micro cracks have been observed ahead of a stationary crack tip in fine grained isotropic graphite (Sakai, 1988) no direct evidence of micro cracks has been reported in the coarse grained Gilsocarbon graphite. Re-notching experiments showed that the crack wake was important, and observations of crack bridges (Ouagne, 2001) led to the deduction that bridging by particles in the wake was the dominant R-curve mechanism. In fine grained grades these bridges are caused by filler particles (Sakai, 1988). However, the bridges observed in the coarser grained IM1-24 (Ouagne, 2001) were significantly smaller than the mean filler particle dimensions, and appeared to be restricted to a region close to the crack tip (Fazluddin, 2002). This bridging was associated with a sharp rise in the R-curve, referred to as stage I. A subsequent plateau of constant resistance with increasing crack extension is characteristic of stage II of the R-curve (Ouagne, 2002); and this has been proposed to form once the micro crack process zone and the bridging zone have developed to a stable size. The extent of stage II is limited by the dimensions of the specimen and gives way to stage III. This was deduced to occur when there is an interaction between the rear face of the specimen, and the leading edge of the micro crack process zone. With this interaction, the resistance to crack propagation is reported to fall.

Linear elastic fracture mechanics methods are commonly employed to determine the fracture properties of graphite, although elastic-plastic methods have been adopted by some workers (Sakai, 1983). The R-curve for Gilsocarbon grade IM1-24 has been reported by several authors, using different methods, and significant variations have been observed (Fazluddin, 2002, Ouagne, 2001). Possible explanations for these variations include:

- Sensitivity to load configuration; the design of the specimen has a direct bearing on the way loads are transferred through it. Compact tension and single edge notched beam specimens can develop of high *T* stresses (*i.e.* shearing), which does not occur in tensile tests. The R-curve may therefore differ with specimen geometry (Fett, 1998).
- The implicit assumption that the fracture process zone is very small relative to the dimensions of the test specimen. This assumption may not be valid in graphite which may exhibit a size effect similar to that found in geological materials and concrete (Bazant, 1990).

The aim of the work reported in this paper was to positively identify the mechanisms that produce R-curve behaviour in coarse grain size polygranular isotropic nuclear graphite and to define the dimensional scale over which they take place. X-ray micro-tomography has been used to look beneath the surface of the specimens, for the first time enabling in-situ observations of the microstructure damage and crack morphology.

EXPERIMENTAL METHOD

Compact tension (CT) specimens, Figure 1, were loaded in tension under displacement control, at 0.05 mm/min, until crack propagation was detected; after which the specimen was unloaded. Repeated load cycles were used to obtain the R-curve (Sakai, 1983). The standard specimen design was modified to improve the resolution of X-ray micro-tomography and to control the crack path and crack shape. Earlier work had shown that side grooving on one side was necessary to prevent crack bowing and out-of-plane crack propagation, Figure 2. Crack extension was detected using electronic speckle pattern interferometry (ESPI), by observation of changes in the correlation fringe patterns, Figure 3. The crack length $a_{(i)}$, was determined by calculation from the unloading compliance (Ouagne, 2002), direct current potential drop (DCPD) and direct surface measurement by optical imaging and surface strain mapping by ESPI (Jones, 1989) to determine the location of the crack tip strain concentration.

X-ray micro-tomography (XRμT) observations were made using an X-Tech conical beam scanner with a tungsten target, filtered using 0.1 mm thickness of pure copper and the images were collected using a 1024 x 1024 pixel CCD camera. The sample to detector distance ranged from 480 mm to approximately 520 mm, giving a voxel (*i.e.* 3-dimensonal pixel) size of 0.08 - 0.12 mm. The data was analysed using AMIRA[12] visualisation software to obtain two dimensional sections and three-dimensional surfaces by threshold-based segmentation.

The specimens were then sectioned, perpendicular to the crack plane and polished using 0.25 μm diamond paste for optical microscopy examination. Features observed in XRμT were subsequently correlated to the optical images of the crack pathway at the same location.

[12] Amira Ver 3.1.1 – Mercury Computer systems GmbH

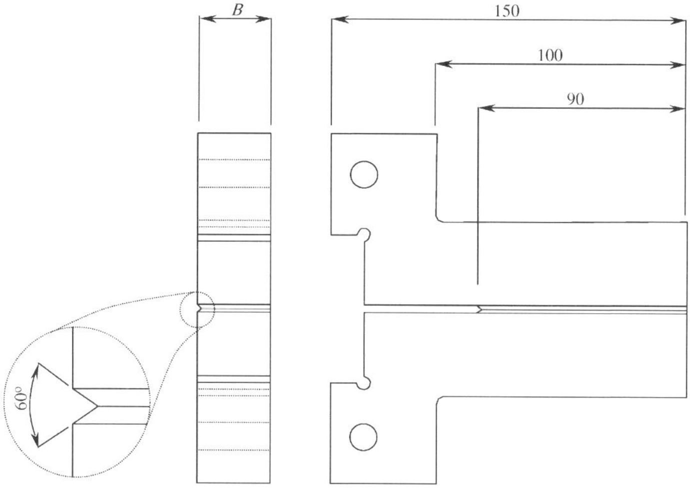

FIGURE 1: The compact tension specimen configuration used.

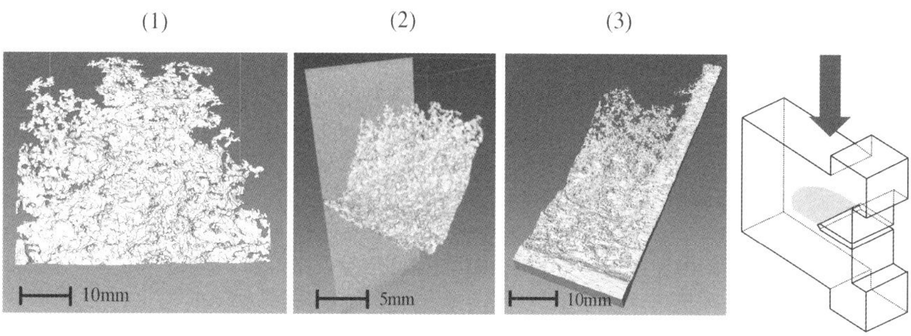

FIGURE 2: Images of the test specimens produced using XRµT showed that the shape of the crack is dependent on the thickness (1 & 2). The direction in which the crack propagates was controlled using side grooving (2 & 3).

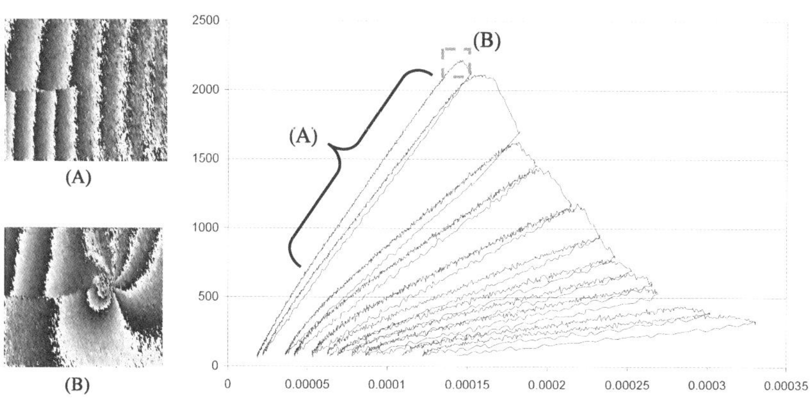

FIGURE 3: The onset of a significant change in the correlation fringes obtained by ESPI (from A to B) was used to identify the load at which crack propagation occurred. The correlation fringes change due to in-plane displacements of the specimen surface.

RESULTS

A summary of a typical R curve is given in Figure 4. The crack growth resistance was calculated from the load-displacement data using Sakai's method. There is an initial sharp rise in resistance from below 100 Jm^{-2} to approximately 300 Jm^{-2} as the crack extends up to 10 mm. Following this the resistance increases more gradually to 400 Jm^{-2}. The XRμT revealed discontinuities in the resolved crack surface, Figure 5. These can be observed in high concentrations close to the crack tip. Isolated discontinuities are also observed in the crack wake, up to 45 mm from the crack tip.

Optical examination of the discontinuities shows that those close to the tip are predominantly formed within the matrix of binder and fine filler particles. Some the isolated discontinuities can be shown to coincide with frictional contact, Figure 5. In most cases, however, the isolated discontinuities appear to be the result of locally small crack opening displacements, which are below the resolution limit (*i.e.* voxel size) of XRμT, Figure 6. XRμT observations of a sample subjected to successive load cycles showed that although the compliance change and ESPI correlation fringes indicated crack extension, there was no detectable change in the position of the crack tip. A marked reduction in the density of the isolated long range discontinuities was observed, however, Figure5 (1 & 2).

A zone of low X-ray attenuation was observed ahead of the crack tip and in the crack wake, Figure 7. Optical investigation of the same region shows micro cracking, Figure 8, which is deduced to be the cause of the reduced X-ray attenuation. The size of this zone is similar to the residual strain field observed, using ESPI, in samples, which had been loaded, without crack extension, and then unloaded. ESPI also shows that the region is present prior to primary crack initiation in specimens loaded to approximately 80% of the fracture load, Figure 7. The strain field in these images was determined by differentiation of displacement differences in the specimen surface between the unloaded states, before and after loading. The most significant cause for this residual strain is therefore deduced to be permanent deformation resulting from micro-cracking, although a change in crack opening due to contact of fracture surface asperities may also contribute.

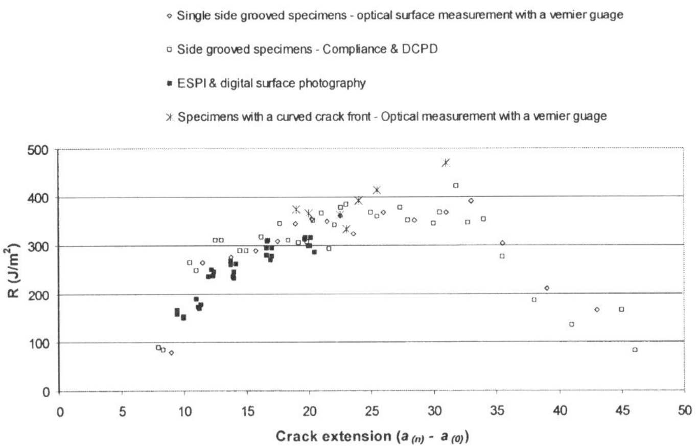

FIGURE 4: Crack propagation resistance, with increasing crack length, obtained for compact tension of various thickness and different crack monitoring methods. The data is for specimens without side-grooves, unless noted.

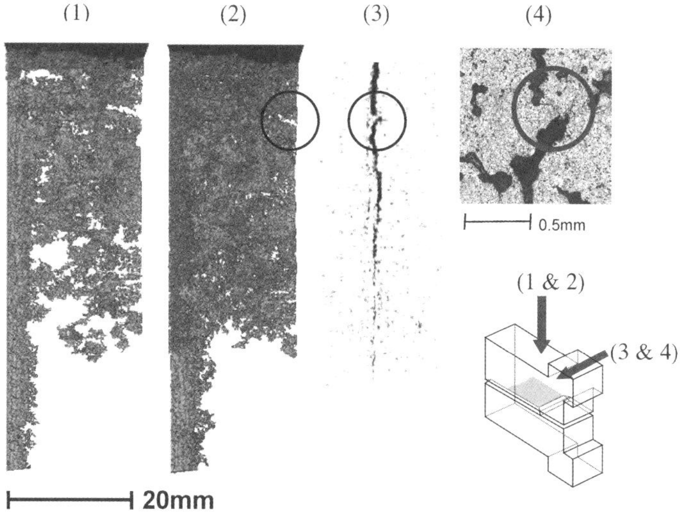

FIGURE 5: Discontinuities in the tomography image are observed over the crack path (1 & 2). A 2D section of the tomography data is shown in 3, with an optical microscopy image within the same region in 4. The same location is circled in each image; section 3 is orthogonal to 1 & 2.

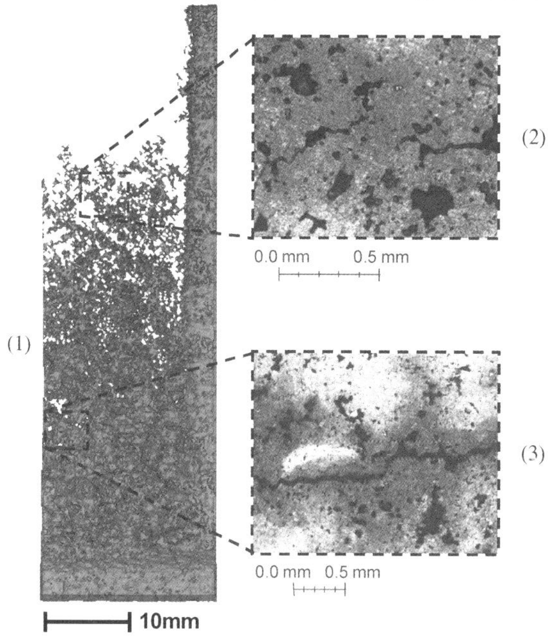

FIGURE 6: The region close to the crack tip is commonly very discontinuous in the computer tomography images (1). Optical microscopy images of this region show that bridging in the binder phase produces this effect (2). Low COD, below the resolution of the XRμT system is the most common cause of the isolated discontinuities at longer crack lengths (3).

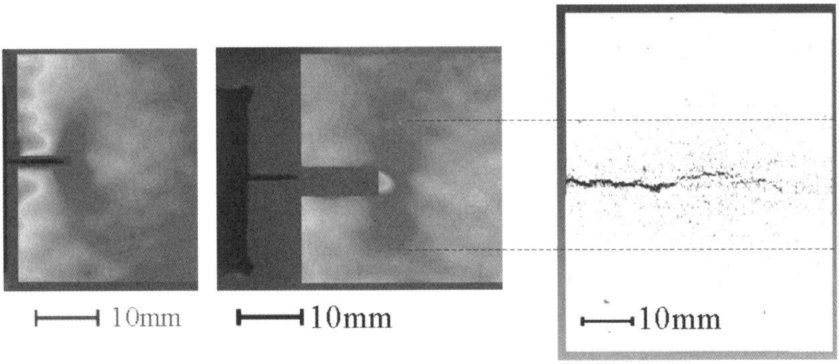

FIGURE 7: Comparison of the low attenuation region with the residual strain contours observed using ESPI before and after primary crack initiation.

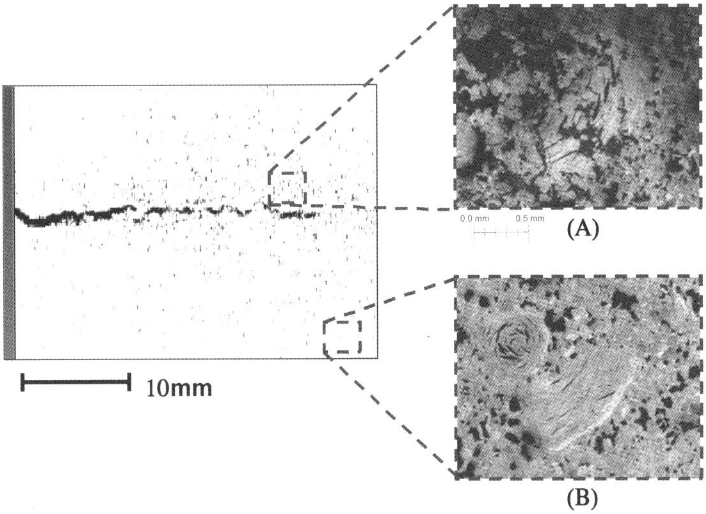

FIGURE 8: Optical microscopy shows that the low attenuation around the crack may be caused by extensive micro cracking (A). Outside the region there was no evidence of micro cracking (B). The arrow indicates the direction of observation.

DISCUSSION

The R curve for isotropic Gilsocarbon graphite is characterised by initially sharply rising resistance, Figure 4, followed by a further, shallower rise. Optical microscopy shows that some of the discontinuities that were identified by tomography, Figure 5, corresponded to frictional contacts between the crack faces. R-curve behaviour due to frictional contact has been observed in other materials, such as polycrystalline alumina (Marrow, 1993). Crack bridging by frictional contact is consistent with previous observations, using re-notching, that crack wake processes were responsible for a significant fraction of the R-curve behaviour in nuclear graphite (Sakai, 1988). This mechanism is expected to be sensitive to specimen geometry effects on the crack opening. The measured R-curve is therefore not likely to be independent of specimen geometry.

X-ray tomography also observed a region of reduced attenuation ahead of the crack tip and in its wake, Figure 7. This is caused by micro-cracking, which is due to heterogeneous deformation in the microstructure (Joyce, 2005). Micro-cracking is a toughening mechanism in some materials (Marshall, 1985). The development of the micro-crack process zone may therefore be responsible for some part of the measured R-curve behaviour in nuclear graphite.

Further work is in progress, using optical strain mapping (ESPI) to quantify the crack tip strain distribution and X-ray tomography to correlate this with the crack shape. The aim of this work is to confirm the relative contributions of wake and tip processes.

CONCLUSIONS

- Tomography has been used to observe the mechanisms responsible for R-curve behaviour in polygranular nuclear graphite.
- Crack bridging from frictional contact between the fracture surfaces is observed.
- A zone of reduced X-ray attenuation, due to micro-cracking, is also observed around the crack tip and in its wake.

Acknowledgements
The authors gratefully acknowledge the support of the Health and Safety Executive (NII) and British Energy Generation Ltd. The opinions expressed in this paper are those of the authors, and do not necessarily reflect those of the research sponsors.

References

Bazant, Z. P. and Kazemi, M. T. (1990). Determination of fracture energy, process zone length and brittleness number, from size effect, with application to rock and concrete. *International Journal of Fracture*, **44**: p. 111 - 131.

Fazluddin, S. (2002). Crack growth resistance in nuclear graphite. PhD Thesis, University of Leeds: Leeds. p. 252.

Fett, T., (1998). Friction induced bridging effects caused by the T-stress. *Engineering Fracture Mechanics*, **59**(5): p. 599 - 606.

Jones, R. and Wykes, C. (1989). Holographic and speckle interferometry. 2nd ed. Cambridge studies in modern optics (6), Cambridge: Cambridge University Press.

Marrow, J., Luprano, V. and Roberts, S. (1993). Scanning acoustic microscopy observations of high temperature crack bridging mechanisms in Alumina. *Journal of the American Ceramic Society*, **76**(11): p. 2915 - 2918.

Marshall, D. B. and A.G. Evans, A.G. (1986). *Fracture mechanics of ceramics*, R. Bradt, et al., Editors. 1986, Plenum Press: New York. p. 1 - 15.

Marshall, D. B., Cox, B. N. and Evans, A. G. (1985). The mechanics of matrix cracking in brittle fibre composites. *Acta Metallica*, **33**(11): p. 2013 - 2021.

Ouagne, P. (2001). Fracture property changes with oxidation and irradiation in nuclear graphites, PhD Thesis, Bath University: Bath.

Ouagne, P., Neighbour, G. B. and McEnaney, B. (2002). Crack growth resistance in nuclear graphites. *Journal of Physics, D: Applied Physics*, **35**: p. 927-934.

Sakai, M., *et al.* (1988). R-Curve behaviour of a polycrystalline graphite: microcracking and bridging in the wake region. *Journal of the American Ceramic Society*, **71**(8): p. 609 - 615.

Sakai, M., Urashima, K. and Inagaki, M. (1983). Energy principle of elastic-plastic fracture and its application to the fracture mechanics of polycrystalline graphite. *Journal of the American Ceramic Society*, **66**: p. 868-874.

Graphite Irradiation Testing at NRG Petten

Jaap G van der Laan and Arjan (J A) Vreeling
Materials, Monitoring & Inspection, Nuclear Research & Consultancy Group,
P.O. Box 25, 1755 ZG PETTEN, The Netherlands.
Email: vanderlaan@nrg-nl.com

Abstract
The High Flux Reactor at Petten is a major tool in a number of research and qualification programmes of materials for nuclear installations. The HFR has been the main tool in the irradiation programmes like DRAGON and the German HTR Modul. Furthermore graphite materials have been irradiated for the European fusion technology programme (1984-today), gradually focusing on ceramic composites. Currently some European framed projects on technology development for high and very high temperature reactors are implemented, including irradiation of present day graphites. NRG's Hot Cell Laboratories allow measurement of key properties like dimensional change, thermal expansion, thermal diffusivity, elastic modulus and strength. These facilities together with the irradiation facilities at the High Flux Reactor form the backbone of the current European graphite irradiation programme.

Keywords
Irradiation testing, post-irradiation examinations, qualification

INTRODUCTION

The Nuclear Research & Consultancy Group

The Nuclear Research and Consultancy Group (NRG) was established as a Partnership Firm in 1998 through the merger of ECN's (70% share) and KEMA's (30% share) business activities in the nuclear fields. ECN is the Energy Research Centre in the Netherlands, which was renamed from the Reactor Centrum Nederland founded in 1955, whereas KEMA was originally founded by the Dutch electricity utilities. At present NRG has about 350 staff. The successive mergers have created a centre of excellence in nuclear technology, integrating all relevant disciplines. NRG develops knowledge, products and processes for safe applications of nuclear technology in energy, environment and health. NRG offices and nuclear facilities are located in the Netherlands, in Petten (West) and Arnhem (East).

The High Flux Reactor

The High Flux Reactor at Petten is a major tool in a number of research and qualification programmes of materials for nuclear installations. The High Flux Reactor (HFR) in Petten is owned by the Joint Research Centre (JRC) of the European Commission in Petten. The reactor is operated and maintained by NRG-staff. In Spring 2005, the operating license was transferred to NRG. The HFR is crucial for NRG's activities in research, industrial applications like isotope production for health, and innovations in nuclear technology. The HFR activities are presently organised in so-called supplementary programmes, and a follow-up is being prepared.

The HFR is a tank-in-pool multi-purpose test reactor with a thermal power of 45 MW. Its core arrangement of 9 x 9 positions contains 33 fuel elements, 6 control rods and 22 beryllium reflector elements, and provides effectively 20 in-core positions. The pool-side facility (west), outside the core box, contains 12 experimental positions. Presently 4 of the 12 horizontal beam tubes are in use for medical and scientific projects. Peak flux levels are $3.0*10^{18}$ $m^{-2}s^{-1}$ for thermal and $2.1*10^{18}$ $m^{-2}s^{-1}$ for fast neutrons, respectively. Nuclear heating for in-core positions ranges from 1 to 12 W/g. The HFR is currently operated in cycles of 25 days, with an accumulated 280 to 290 full power days per year.

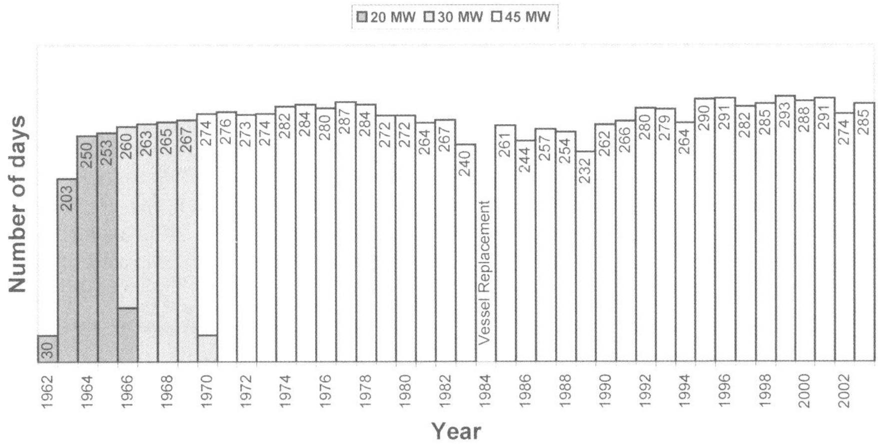

FIGURE 1: Availability record in Full Power Days per year for the High Flux Reactor since 1962.

Following the recent international initiative on research reactor fuel cycles, the High Flux Reactor started conversion to Low Enriched Uranium fuel effectively in October 2005 (Wijtsma 2005), and will be completed by mid 2006. Due to the different performance of the LEU fuel as compared to the HEU fuels used up to now, the nuclear characteristics of the core will change, like a little lower thermal flux in some areas, and a reduction of nuclear heating rate. After conversion the reactor cycle will typically be 31.5 days of which 28 full power days, for optimised fuel consumption. This implies a nominal 10.3 cycles per year.

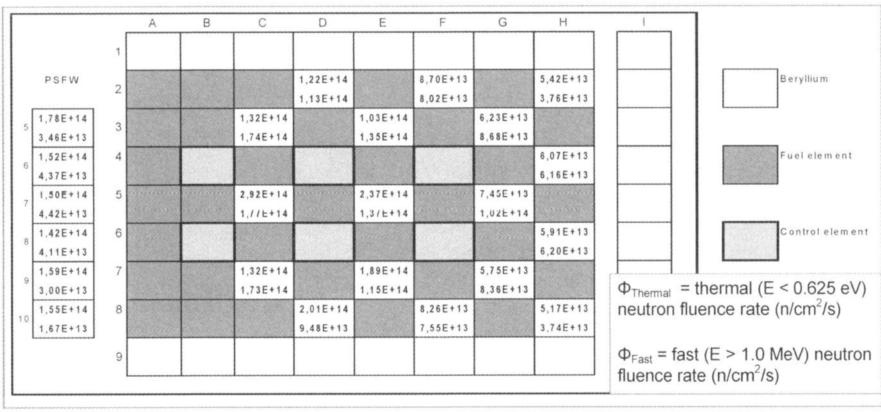

FIGURE 2: Core lay out of High Flux Reactor with fluence rate characteristics.

The HFR has been the main tool in the irradiation programmes for gas cooled reactors like DRAGON and the German HTR-Modul. Furthermore graphite materials have been irradiated for the European fusion technology programme (1984-today), eventually focusing on ceramic composites. Currently some European framed projects on technology development for high and very high temperature reactors are implemented, including irradiation of present day graphites.

The Hot Cell Laboratories

Dismantling of irradiation rigs and post irradiation examinations are typically performed in NRG's Hot Cell Laboratories (HCL). The HCL comprise 5 concrete, and 12 lead shielded cells, with few m^3 to few 10 m^3 working volume. Three lines of lead shielded cells are dedicated to materials testing. Test types available include tensile, flexural, fatigue, creep, impact, fracture toughness, (re-) welding, thermal diffusivity, dilatometry, and density. In 2005 this has been extended with a glove box line, installed for graphite post-irradiation examinations in support of HTR/VHTR development programmes. NRG facilities allow measurement of key graphite irradiation properties like dimensional change, thermal expansion, thermal diffusivity, elastic modulus and strength. These facilities together with the irradiation facilities at the High Flux Reactor form the backbone of the current European graphite irradiation programme.

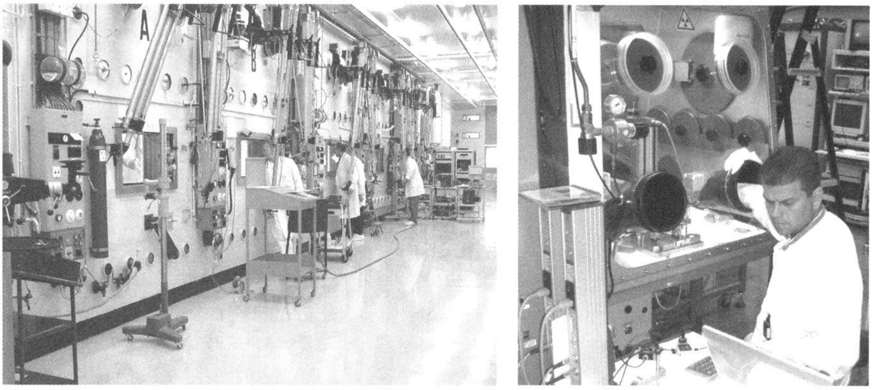

FIGURE 3: Front view - Hot Cell Lab interior (left) and glove box extension for graphite P.I.E. (right).

IRRADIATION PROGRAMMES

Materials applied in nuclear installations generally suffer from degradation of properties in long term operation. Depending on the neutron spectrum, displacement damage effects and nuclear reactions change the composition and crystalline ordering of the material. High fluence neutron irradiation causes hardening and embrittlement of metallic structural materials by extensive formation of lattice defects. Crystalline ceramics typically show changes in electrical and physical properties. Environmental effects can occur as well, and some synergistic effects are well recognised, like IASCC in LWR and the radiolytic corrosion of graphite in gas-

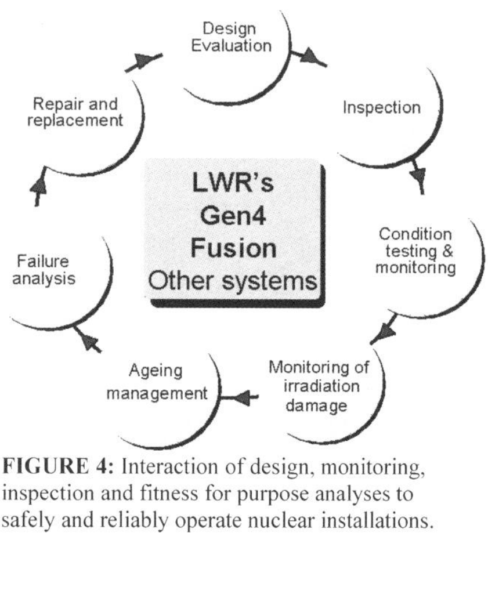

FIGURE 4: Interaction of design, monitoring, inspection and fitness for purpose analyses to safely and reliably operate nuclear installations.

cooled power plants like Magnox and AGR.

The design of core structures for nuclear installations, the plant licensing and operation and maintenance strategy typically rely on the good predictability of property evolution during the planned operational life. Material performance is mostly determined in dedicated experimental programmes in material test reactors. Irradiation programmes are tailored for material selection processes, acquisition of engineering design curves or support life-extension projects. NRG is contributing, for example, to the qualification of materials and sub-components for ITER, the international project to demonstrate the physics and technology of fusion power. These projects provide data for engineering design. In parallel, material development for reactor application requires to screen their high dose behaviour, and have feedback from conceptual design activities.

The Euratom Framed HTR Projects

Recently, there has been a worldwide renewal of interest in High Temperature Reactors. Two demonstration units, one in Japan (HTTR) and one in China (HTR-10), are operating. The development of the Pebble Bed Modular Reactor (PBMR) in South Africa and the Gas Turbine Modular Helium Reactor (GT-MHR) are at an advanced stage. The European Commission is supporting research projects for the development of High Temperature Reactor (HTR) technology. For materials issues the multi-partner projects HTR-M and HTR-M1 were implemented, with the aim to create the technological requirements for designing a European HTR by 2010. Despite much research that has been done on the behaviour of graphite grades in a nuclear reactor environment and vast amounts of data that exist, these graphite grades are no longer commercially available and the raw materials do not exist anymore. In addition, most data from the past are from low temperature experiments (<550 °C), whereas for HTRs the graphite temperatures will generally be higher than 550 °C, up to about 1000 °C. It was decided to start building a database that contains the data of the materials behaviour of presently available graphite grades under neutron irradiation at HTR relevant temperatures, in order to build a design base for a European HTR. As the most appropriate starter the irradiation project was focused on a dose of 7-8 dpa at 750 °C, with the view to add a double dose in the follow-up. In Spring 2005 the so-called RAPHAEL Integrated Project was launched, that is focused on the technology for a Very High Temperature Reactor, with helium outlet temperature well above 900 °C to potentially allow system integration with a chemical hydrogen production unit. In the frame of this project NRG will perform another 3 irradiation campaigns to get by about 2011 full curves for graphites at 750 and 950 °C, up to targets of 24 and 16 dpa, respectively.

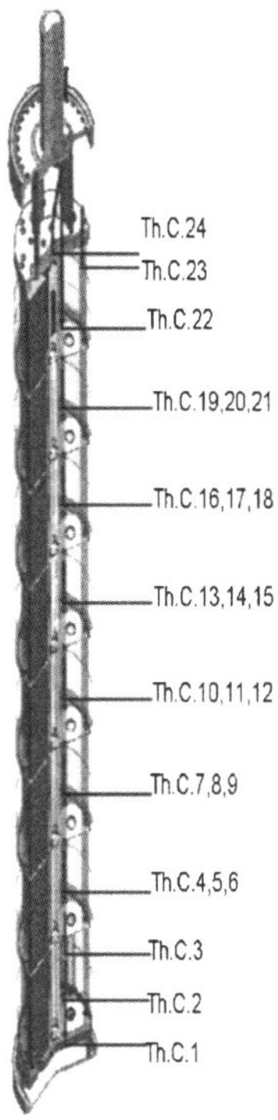

FIGURE 5: Irradiation geometry for the INNOGRAPH series.

The INNOGRAPH Irradiation Programme

The irradiation series denoted 'INNOGRAPH' is designed with the experience gained from the earlier graphite irradiations, and the fusion materials programme. The new design consists of well-instrumented modules, consisting of metallic cylinders with either 3 or 4 channels for specimen stacks, allowing to achieve temperatures in the range from 600 to 950 °C. Its total active stack height is optimised to profit from a larger flux variation (~65-100% of peak flux). A new specimen was designed to perform for all non-destructive tests and enabled maximization of the test matrix size, with an effective stack height of 6 or 12 mm. The basic design should allow the loading of active specimens as well, which necessitated taking every reasonable measure to prevent any additional activation induced during irradiation and handling. It was chosen to have high purity helium purge, and graphite foils wrapped around specimens to prevent metallic contaminations. The initial results of INNOGRAPH-01 confirm the contact dose rate after 4 months cool down are 2-250 mSv/h, and at 0.1 m: 1-23 mSv/h. Purified versus unpurified material showed a 10 times lower activity.

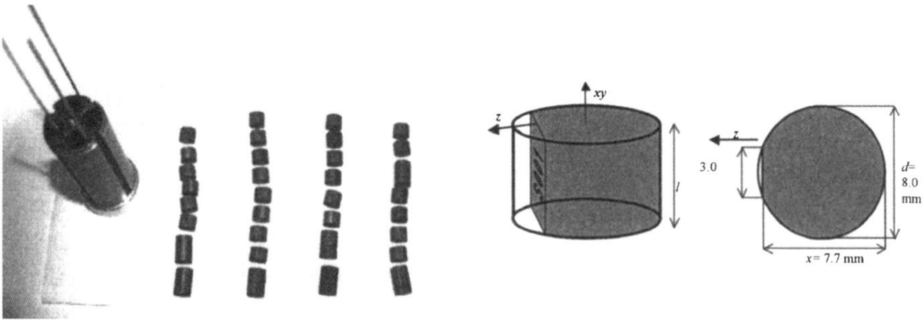

FIGURE 6: Left: INNOGRAPH Basic unit for irradiation of graphite specimens with 4 columns of stacked specimens, manufactured by Electro Discharge Machining from Mo-alloy or steel. Right: specimen geometry.

The following post-irradiation examinations are foreseen:

- Dimensions, Density, RT.
- Dynamic Young's modulus, RT.
- Coefficient of Thermal Expansion, RT-Tirr.
- Coefficient of Thermal Diffusivity, RT-Tirr.
- Crystallite sizes Lc, La (X-Ray Diffraction).
- Micro-structural evolution.
- Diametral and/or Axial Compression.

The first five measurements have been applied in the pre-irradiation characterization. Typically the effect of property scatter is reduced by performing pre- and post-irradiation tests on the same specimens, with the same facilities. Sometimes 'twinning' is applied as well: perform pre- and post-irradiation tests on neighbouring samples as cut from block. As the first step, the irradiation experiment INNOGRAPH has been performed in the High Flux Reactor (HFR) in Petten from February 2004 to April 2005. The graphite grades in INNOGRAPH can be divided in two levels. Firstly, the 5 main grades for which full design correlation curves will be obtained. Secondly, 3 minor grades for which an indication of the irradiation behaviour will be obtained. The test matrix addresses variation of the following parameters:

- Pitch versus pet coke filler material.
- Extrusion versus iso-moulding, and vibration moulding.

- Two EU based and 1 Japanese manufacturer, with industrial products.
- Edge versus block centre (block homogeneity).
- Orientation: Across-Grain vs. With-Grain.
- Purification.

TABLE 1: Main graphite grades in current EU programme (NBG grades by SGL Carbon, Pxyz grades by Graftech and IG grades from Toyo Tanso).

Manufacturing process	Pitch Coke	Pet Coke
Extrusion	NBG-10, PPEA	NBG-20, PCEA
Iso-moulding	IG-430	NBG-25, PCIB-SFG, IG-110
Vibration-moulding	NBG-18, NBG 17	

TABLE 2: Key properties and ranges covered by graphite grade selection in Table 1.

Property	Unit	Minimum	Maximum
Density	g/cm^3	1.713	1.862
DYM	GPa	8.2	12.4
CTE30-120	10^{-6} 1/K	2.8	4.7
CTE30-750	10^{-6} 1/K	4.0	5.8
λ (27 °C)	W/mK	114	169
λ (700 °C)	W/mK	59	87

In addition, some piggy backs are included in the experiments like Gilsocarbon graphite, Enhanced Conductivity (Highly Oriented) Pyrolytic Graphite (ECPG).

Initial Results

The project structure is to separate the P.I.E. stages in screening/trending and full curves: after completion of the first irradiation step only a fraction of the specimens will be measured prior to the re-irradiation in the second step. The test equipment installed allows for low activation specimen measurements in glove boxes, but they all are compatible with hot cell use as well. Figure 7 and 8 show preliminary results for dimensional change measurements of a number of grades, kept anonymous here, as dosimetry data were not validated at the moment of writing this paper.

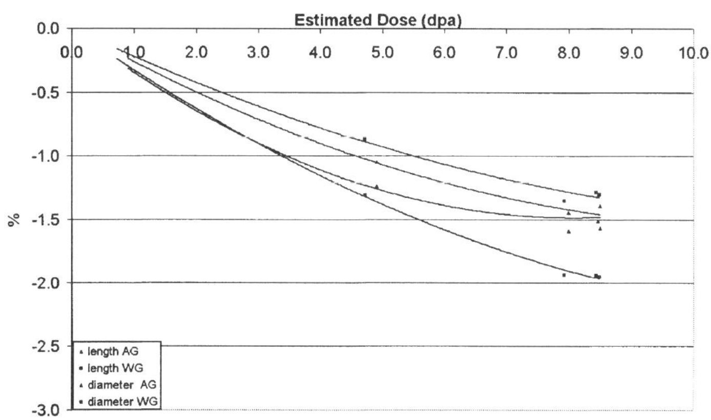

FIGURE 7: Preliminary data plot for length and diameter change of one of the extruded graphites after irradiation in INNOGRAPH-01, at 750°C nominal temperature. Curves are only drawn to guide the eye.

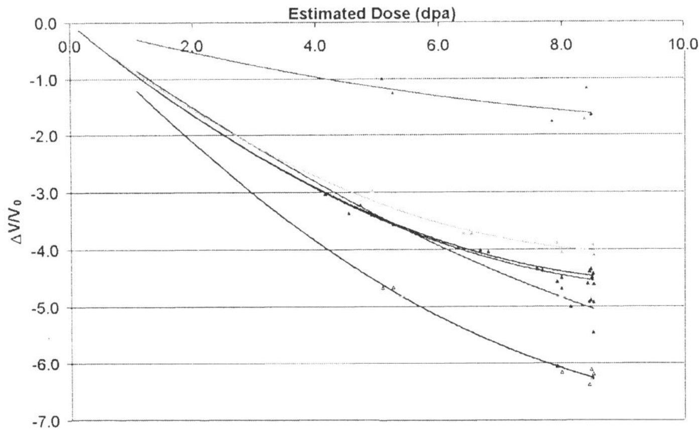

FIGURE 8: Preliminary data plot for volumetric change of six grades of graphite after irradiation in INNOGRAPH-01, at 750°C nominal temperature. Curves are only drawn to guide the eye.

The volumetric shrinkage behaviour at the maximum dose of about 8 dpa, is found between 1.5% and 6.3%, however for 4 out of 6 grades it is between 4% and 5%. Further data evaluation will have impact on the finalisation of the test matrix for the continued irradiation to 24 dpa.

Ongoing Developments

Microstructural Modelling

NRG collaborates with NGRG & the University of Manchester on modelling of irradiated graphite behaviour, like dimensional change, modulus, CTE and conductivity. It concerns, for example, the application of 3-D tomographic imaging of graphite, a three-phase material, for (a) use in the generation of a finite element mesh directly from the real microstructure; (b) prediction of the stress distribution from compressive loading; and (c) understanding a number of piggy backs with single crystal specimens are arranged in all HTR irradiations, with hot micro-tomography specifications to be detailed later.

Gradient Plasticity

NRG explores the applicability of the gradient plasticity model to graphite. This model incorporates local material behaviour in global material behaviour, which has been shown to work well for materials where the behaviour at micro-level influences the behaviour at macro-level (for example: concrete, and geo-materials such as rock and (sand)stone). Because strain localisation is included in the analysis by incorporating the strain gradients as separate degrees of freedom, results of finite element calculations are mesh-independent. NRG is exploring to extend the modelling work by setting up a customised 2D element for graphite. The next steps are then to model graphite, and match with mechanical tests at first for unirradiated material and subsequently for irradiated material from the INNOGRAPH irradiations, *i.e.* extend to 3D modelling and include anisotropy.

Irradiation creep

As effects of irradiation creep and stress relaxation play a significant role in the evolution of graphite thermal-mechanics, the acquisition of appropriate data is elaborate and costly.

Further developments concern in specimen miniaturisation their possible integration with near future irradiations, in particular for high temperature materials in HTR/VHTR environments.

Strength & Fracture

For a selection of present graphite grades various specimen geometries and loading configuration in three and four point bending are compared, along with diametral compression. Further experience is being gained with Single Edge Notched Beams and grooved or slotted cylinders under diametral compression.

CONCLUSIONS

The European R&D programme on graphites is focused on the irradiation of presently available grades, with emphasis on screening and reasonable design curves, allowing EU HTR design by about 2011. State of the art irradiation technology and post-irradiation test methods have been implemented at NRG Petten. Most of the current facilities and expertise built is of relevance to an MTR programme on graphite ageing, to support the UK AGR life extension projects. This availability also covers characterisation of trepanned material from the leading stations, and (re-)fabrication of specimens for (re-)irradiation in stressed or unstressed condition. The HFR is a key element of the Petten infrastructure, its conversion from HEU to LEU Fuel will be completed by mid 2006. Further innovations are ongoing, like building a new B-laboratory, and taking the PALLAS initiative for a new reactor for the period beyond 2015.

Acknowledgements

The authors would like to acknowledge the contributions of M.A. Davies, G. Haag and the partners in the HTR-M1 and RAPHAEL IP projects for their contributions to the work described here. In particular thanks to SGL Carbon (D/F), Graftech (F) and JAERI, Japan for supplying target material for the current programme. The work of R. den Boef, S. de Groot, D.S. d'Hulst, D.J.Ketema, M.A.C. van Kranenburg, W. Molijn, P. ten Pierick, and other contributing staff at NRG is very well appreciated. The work described is being performed with financial support from the European Union, and the Netherlands Ministry of Economic Affairs. The views expressed in this paper are those of the authors and do not necessarily represent those of the European Union, nor those of the Dutch Government.

References

Ahlf, J., Zurita, A. (Ed.), "High Flux Reactor (HFR) Petten – Characteristics of the installation and the irradiation facilities", Report EUR 15151 EN, 1994.

Buckthorpe, D., Couturier, R., van der Laan, J., Hegeman, H., Vreeling, A., Riou, B., Rantala, J., Ennis, P. J. Haag, G., Kuehn, K., Buenaventura, A., Friedrich, B-C. (2003). "Progress of HTR-M Projects for the HTR". In: Proceedings of the 17th Intern. Symposium on Structural Mechanics in Reactor Technology (SMiRT 17), IASMiRT, 17-22 August 2003, Prague (CZ).

May, R. (Ed.) (2006). A Joint Undertaking for HFR, EUR Report to appear January 2006.

NRG Annual Report 2004, http://www.nrg-nl.com/docs/library/2005/nrgannualreport2004.pdf

van der Laan, J. G. (Ed) (2005). "Fusion Technology Tasks at NRG, Progress Report 2004", NRG Report 21099/05.65013, Petten, 2005.

Vreeling, J. A., Haag, G., Davies, M. W., d' Hulst, D. S., Ketema, D. J., van Kranenburg, M. A. C., van der Laan, J. G., ten Pierick, P., Molijn, P. W. (2005). An update on the status of irradiation experiment INNOGRAPH-01, 6th International Graphite Specialists Meeting, Chamonix (F), September 2005.

Wijtsma, F. J. (2005). Status report on HFR conversion and relicensing project, 9th RRFM, Budapest, April 2005, http://www.euronuclear.org/meetings/rrfm2005/presentations/Wijtsma.pdf.

The Development of Measurement Techniques for Mechanical Properties Applicable to Small Reactor Graphite Samples

Nassia Tzelepi, Matthew Brown, Peter C Matthews, John F B Payne and Martin P Metcalfe

Nexia Solutions Ltd, Berkeley Centre, Berkeley, Gloucestershire, UK.

Email: martin.p.metcalfe@nexiasolutions.com

Abstract

Young's modulus and strength are key mechanical properties in the evaluation of component stresses and the assessment of component integrity. The development of measurement techniques for dynamic Young's modulus and strength and their validity in the context of international standards are described. The influence of sample size on measured properties is discussed. A major limitation on the applicability of such techniques is radiolytic weight loss. The findings of an ongoing technique development programme to address this key issue facing monitoring schemes for graphite cores late in reactor life are presented.

Keywords

Dynamic Young's modulus, ultimate tensile strength, trepanned

INTRODUCTION

The structural integrity of the gas-cooled reactor cores currently comprises a significant element in the graphite core safety cases, so that there is a need to have high confidence in mechanical properties trends with dose and weight loss. At the reactor design stage, it was envisaged that the monitoring of graphite mechanical properties in operating reactor cores would be addressed by measurements on pre-characterised installed samples designed for specific tests, mainly tensile and compressive strength. As the number of installed samples is generally limited and unrepresentative of behaviour at the fuel channel wall, trepanning specimens from the operating core components was also investigated and subsequently incorporated in the outage programme in the late eighties. Due to the invasive nature of the method of sampling, as far as component integrity is concerned, the trepanned samples are limited in size. Nevertheless, at the time of the trepanning rig development, the specimens were intended for measurements of physical and chemical properties only, for which the extracted specimen size was clearly sufficient. In the mid nineties, the possibility of measuring Young's modulus and strength of Magnox and AGR (Advanced Gas-cooled Reactors) trepanned samples was investigated. For both Young's modulus and strength, the associated ASTM (American Society for Testing and Materials) and British Standards were considered. On the former, the difficulty of measuring the strains and displacements of small samples excluded the standard static method. Of the two available dynamic methods, the sonic resonance (ASTM C747-93, 1993) would provide a more accurate answer but was not applicable on the available specimen geometry, *i.e.* discs of ø12 × 6 mm (Magnox) and ø19 × 6 mm (AGR). The adopted DYM (Dynamic Young's Modulus) technique was therefore based on the ultrasonic Time-of-Flight (ToF) method of ASTM C769-89 (1989). Although the preferred option was to use a standard testing method for tensile strength, the Magnox sample geometry was not compliant with any of the standards. As a consequence, the disc compression technique was developed and fully validated on virgin PGA (Pile Grade 'A') graphite. Due to the more favourable geometry of the AGR specimens, a 3-point bend test (ASTM C1161-90, 1990) on beams (6 × 6 × 18 mm) cut out of AGR discs was developed and fully validated on virgin Gilsocarbon. The first DYM and tensile strength measurements on Magnox samples were carried out in 1996 -1997.

DEVELOPMENT PROGRAMME

The adopted DYM and tensile strength techniques were used on a limited number of Magnox and AGR samples, as part of the graphite monitoring campaigns. The aim was to validate the MTR (Materials Testing Reactor) data used in the safety cases against reactor measurements. The high oxidation of the Oldbury core and other emerging issues in the last few years, however, required the techniques to be more accurate and repeatable. This led to an extensive strength and DYM technique development programme, with the objective to identify one or more measurement techniques producing accurate and reproducible results on trepanned specimens. Since there are more challenges associated with the PGA samples and this programme was commissioned by Magnox (and co-funded by BEGL), it focused on sample geometries associated with Magnox trepanned samples. There are five sample and/or testing characteristics that make the mechanical properties measurements on PGA trepanned samples very difficult to implement and to produce high confidence test data. These are discussed below.

Sample Size

The PGA trepanned discs measure ø12 × 6 mm, whereas the AGR discs measure ø19×6 mm. Although in theory, the trepanned cores (ø12 × 20 mm for Magnox and ø19×30 mm for AGR) from which the discs are machined, could be used instead, the density/property gradient along their length and the reduced number of specimens for these and other tests, do not make it a favourable option. The larger samples from the installed sets also have a density gradient along their length. Furthermore, there are only a limited number of these samples and the tests do not provide data on the actual reactor components. In addition, there are further difficulties with extracting, handling and dismantling installed set carriers.

Sample Microstructure

PGA graphite is a non-isotropic, non homogeneous material, with the larger filler particles measuring around 1 mm. The initial porosity of virgin graphite is ~20% and increasing with radiolytic oxidation. It is clear that apart from the specific technique requirements on sample size, the coarse grain structure alone requires bigger samples, so that the measurement is representative of the material. The microstructure in combination with the specific sample size also affect the reproducibility of the measurements and make the precision of a testing technique more difficult to demonstrate.

Empirical Models

Although there are predictive models based on trends from previous experimental data up to 35% weight loss, the predicted materials properties have not been validated for higher weight losses. It is therefore difficult to demonstrate that the measured values are within the required accuracy range.

Required Accuracy and Reproducibility

Because of previous achievements on technique development, the required accuracy and reproducibility of the measurement techniques on irradiated graphite are now higher, although the material is more degraded and more difficult to measure. In addition, the measurement accuracy cannot be directly demonstrated on irradiated material.

Remote Testing

The tests are carried out in a glove box or hot cell. This makes other experimental parameters, such as specimen handling and precise positioning harder to control. Considering the above limitations, a comprehensive development programme on strength and DYM measurements has been undertaken, described in the next sections. In collaboration with the Manchester University Materials Science Centre, CERAM (former British Ceramics Research Association), Serco and Frazer-Nash Consultancy (FNC), Nexia Solutions has been able to assess the applicability of each technique and carry out a validation exercise for the most promising candidates.

Dynamic Young's Modulus

There are two ASTM standards for measuring the Young's modulus of carbon and graphite materials. Both standards require a homogeneous and isotropic material and have specific specimen requirements. Specifically, the Ultrasonic ToF method (ASTM C769-89, 1989) states that it is applicable on specimens with a grain size of less than 0.8 mm and a "large diameter and length compared to the wavelength of sound in the material under test". On the other hand, the sonic resonance method (ASTM C747-93, 1993) requires a slender rod with parallel faces, *i.e.* a length-to-diameter ratio between 5 and 20. Since the Magnox and AGR trepanned samples would not satisfy the ASTM conditions on specimen condition and geometry, the DYM technique development programme covered the investigation of all available techniques.

Impulse Excitation

This test method is covered by a number of national and international standards including ASTM C1259, ASTM C623, ASTM C848, ENV 843-2, and ISO 17561. Impulse excitation exploits the fact that any object when struck resonates at a number of characteristic frequencies that are dependent on the properties of the material and geometry of object. Impulse excitation was considered for this study, as it can potentially measure both Young's and shear moduli and can be used on rods and discs in a hostile environment. This method can be applied on two specimen geometries (slender rod and thin disc), both of which were investigated using virgin PGA cylinders of different aspect ratios. The tests on the slender rod geometry showed that, although reliable measurements can be made on specimens with aspect ratios smaller than those recommended in the standards, the minimum sample length required for this measurement is $\geq$ 30 mm. The tests on the thin disc geometry showed that, although the method gave satisfactory results on the 1.5 mm disc, this was not the case for the 3 mm discs. This technique was therefore not pursued, since there are legitimate concerns over the machining difficulties of highly degraded material to these dimensions and the validity of a measurement made on such a thin disc that may only contain one filler particle across its width.

Sonic Resonance

Knowing the limitations of this technique from the ASTM standard, a fundamental theoretical understanding and a study of the effects of different experimental parameters was pursued.

This was achieved by FEA (Finite Element Analysis) and by experiments on virgin and thermally oxidised PGA specimens of different geometries. The tests on virgin graphite cylinders (minimum sample size: ø12 × 20 mm) are reproducible to better than 2.5% and accurate, with a maximum discrepancy (between the CERAM and Berkeley laboratories) of ~11% in DYM. The tests on thermally oxidised (in air) samples (ø12 × 20 mm) have shown very low scatter and, although the accuracy of these measurements cannot be assessed, the DYMs calculated from these frequencies are in very good agreement with the predicted values for this range of thermal weight losses. Although this technique could not be used on the PGA trepanned slices, its application on the bigger samples from the installed sets could be used as a validation of the Ultrasonic ToF test on irradiated samples.

Laser Time-of-Flight

This technique was investigated due to its perceived advantages over the conventional ultrasonic method on small specimens, which is strongly affected by parameters such as coupling and contact quality, frequency dependent attenuation and diffraction. Laser ToF uses the same principles as the ultrasonic ToF, but in this case a short pulse of laser light is focused onto the surface of the object of interest to generate the ultrasound. It has been successful in measuring high weight loss (~50% thermal) discs and is not strongly affected by geometry. The measurements of thermally oxidised samples, however, have not been reproducible and are found to be dependent on sample orientation, which was not observed with the virgin samples. Since further interpretation of the data and correlation of the results with microstructure characteristics would be required before using this test on irradiated PGA samples, the Laser ToF method was abandoned.

Ultrasonic Time-of-Flight

Previous work had provided a good understanding on the limitations of this technique. Hence, the aim was to explore new technologies and more sophisticated data analysis.

Ultimate Tensile Strength (UTS)

There are two ASTM standards on tensile and flexural strength testing of graphite and carbon materials. The flexural strength test requires that "the minimum dimension of the specimen" > 5 × "the largest particle dimension" and that "the length to thickness (minimum dimension) ratio" ≥ 8. The tensile strength test method has similar requirements on specimen dimensions, but the specimen has to be machined to a dog-bone shape. Since the trepanned samples would not satisfy the ASTM conditions on specimen condition and geometry, the UTS technique development programme covered the investigation of all available and innovative techniques.

Disc Bending

This test uses the same principles as a conventional beam bending test, but is applied on a typical PGA trepanned specimen geometry (ø12 mm × 6 mm). This technique was considered a strong candidate due to the minimal specimen preparation. The objective was to study theoretically and experimentally the effect of this geometry on the stress pattern and measured strengths. The tests and FEA show that disc bending provides a measure of tensile strength and that the behaviour of a disc (ø12 × 6 mm) in 3-point bend is similar to that of a rectangular beam of similar dimensions. However, the measured tensile strengths are significantly higher than expected and there appears to be a geometric dependence. This method was not investigated further in favour of the Ring Burst and Extended Tensile Strength methods.

Ring Compression

Although, this is a well known tensile test method, it is usually applied on thinner rings. Therefore, the applicability of this test on Magnox trepanned specimens, was investigated through theoretical and experimental work on specimens with different internal diameters. The tests showed failure at lower loads than would be expected from uniaxial tensile tests. In addition, the ring failure has a complex sequence with stable crack nucleation early in the test, which modifies the stress distribution and suggests that the calculated tensile strengths contain a high level of uncertainty.

Extended Tensile Test

Using the same principles as a conventional tensile test and based on a UKAEA method used on small graphite samples in the past, this technique involves the gluing of metal fittings to graphite samples and pulling under tension in a conventional testing frame. Tests on virgin cylindrical (ø5 × 10 mm) and rectangular (5 × 5 × 10 mm) specimens showed fracture in the graphite and not at the glue-graphite boundary with results in good agreement with reported values.

Ring Burst

Another way of measuring the strength of the PGA trepanned samples is by pressurising a ring internally and measuring the pressure at which the ring fails. The advantage of this technique is that the strain to failure can be derived by measuring the displacements of the ring periphery.

Three-Point Bend (AGR)

The 3-point bend test was considered the best candidate for application on beams machined from AGR discs. Since these specimens are a lot smaller than would normally be used, the objective of this work was to investigate theoretically and experimentally the effect of sample size and geometry on the measured strength. Analytical and FEA methods have been used to identify the most accurate expression for calculating the UTS. Effects related to the small specimen size and the small specimen fraction that is highly stressed strongly affects the measured UTS. Specifically, the ratio of UTS measured by this method to UTS measured by the uniaxial test was 1.51 ± 0.03 for virgin graphite. This method is being successfully applied on the AGR monitoring campaigns.

RESULTS

This development programme has successfully identified one technique for measuring the DYM and two very strong candidates for measuring the tensile strength of small trepanned samples. This section focuses on the results from the validated/strong candidate techniques.

Dynamic Young's Modulus

Ultrasonic Time-of-Flight

This development programme has led to the Mk3 DYM equipment, which uses piezo-composite transducers (ø16 mm) with novel rubber delay lines and a different data analysis that includes corrections for the rubber delay line distortion. The review by NDT Solutions of the existing (Mk2) measurement technique found that the current data analysis is accurate and fit for purpose, although a more sophisticated method could be developed if measurements on high weight loss samples have very low signal-to-noise ratio. A programme of proving trials was carried out, in order to demonstrate the accuracy and reproducibility of the Mk3 DYM rig on virgin PGA graphite and its reproducibility on irradiated graphite samples. Figure 1 shows the Mk2 and Mk3 measurements on 3 NPL (National Physical Laboratory) certified samples. This graph demonstrates that the Mk3 DYM rig gives accurate measurements on virgin PGA samples, with a maximum error in wave speed of 3.7% and, is in good agreement with the Mk2 DYM rig. Furthermore, all the measurements were repeatable within 0.6%. Finally, the Mk3 equipment was used to measure oxidised samples with weight losses up to 61% irradiated in the Idaho National Laboratory Advanced Test Reactor (Figure 3).

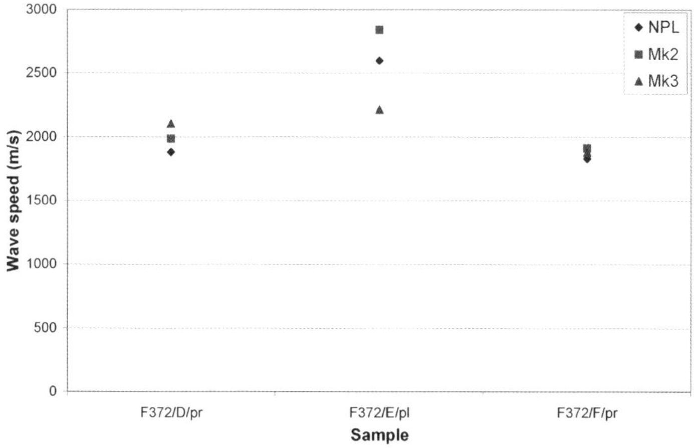

FIGURE 1: Assessment of the accuracy of the Mk2 and Mk3 rigs, by comparing the measured wave speeds with those certified by NPL.

Figure 2 demonstrates the good agreement between the Mk3 and Mk2 measurements and the DYM predictions on the Oldbury 2003 samples with radiolytic weight losses up to 33%.

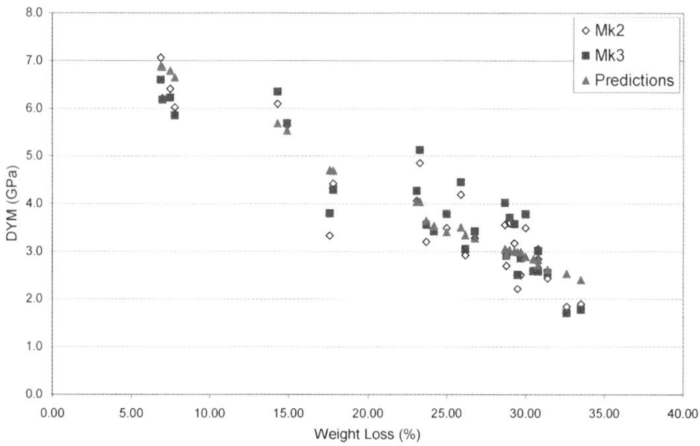

FIGURE 2: Comparison of the Mk3 with the Mk2 equipment and with the DYM predictions on the irradiated Oldbury 2003 samples

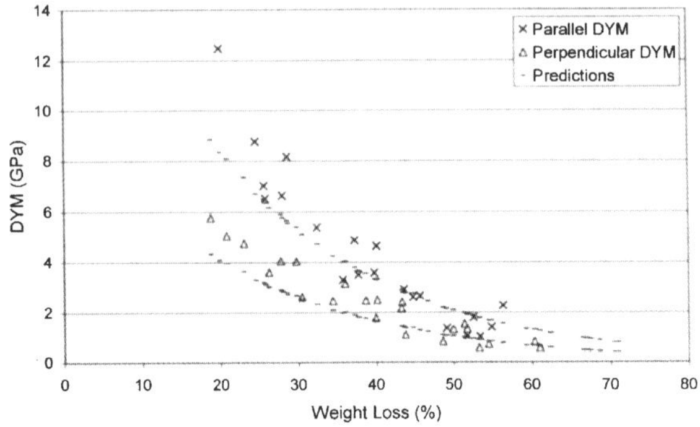

FIGURE 3: Mk3 DYM measurements and predictions on the high weight loss material produced in the Advanced Test Reactor.

Ultimate Tensile Strength (UTS)

Ring Burst

Results from tests on virgin Gilsocarbon and PGA rings showed good agreement with accepted values, with the perpendicular values being bounded by the perpendicular and parallel values obtained by uniaxial testing. In addition, thermally and radiolytically oxidised samples with weight losses up to 30% were successfully machined and tested. It is worth noting that the ring burst technique provides a UTS value which is a measure of the weakest part of the ring ligament and only lower bound UTS values can be obtained. Error analysis has shown a total potential underestimate in UTS of up to 20% and machining and increased sample handling will only further decrease the potential strength of the graphite samples. Figure 4 shows the results from tests on Oldbury 2003 samples and the assumed trend with weight loss, the scatter arising principally from imperfections in the narrow ring ligament.

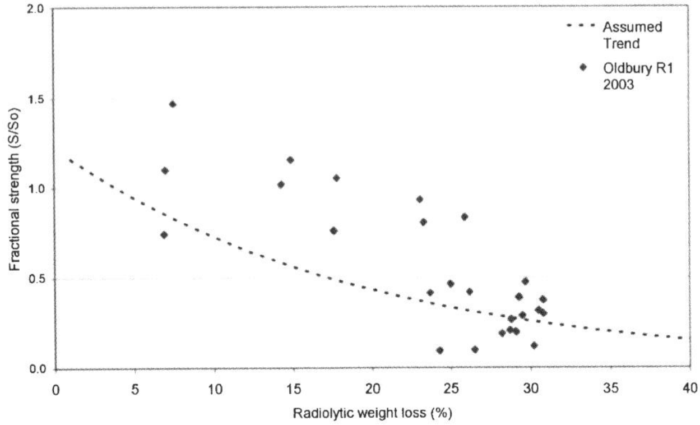

FIGURE 4: UTS values obtained using the ring burst technique on the Oldbury R1 2003 samples.

Extended Tensile Test

Tests on virgin cylindrical (ø5×10 mm) and rectangular (5×5×10 mm) specimens showed fracture in the graphite and not at the glue-graphite boundary with results in good agreement with reported values.

DISCUSSION

Most limitations associated with DYM measurements for the restricted reactor sample geometries have been overcome by the development of the Ultrasonic ToF and the Mk3 equipment. There remains a level of uncertainty over the test accuracy on irradiated and oxidised samples which is mainly due to the assumption that Poisson's ratio does not change significantly with irradiation and weight loss. As far as the UTS measurement is concerned, the sample size in combination with the increasingly non-uniform microstructure of the highly degraded samples reduces the confidence in the strength data. Although there is further work under consideration on the ring burst and extended tensile techniques, this development programme has demonstrated that the availability of a high accuracy technique for trepanned samples at high weight losses presents a significant challenge.

CONCLUSIONS

An extensive programme to develop and prove new measurement techniques for DYM and strength has been carried out. The conclusions are:

- There are several limitations associated with DYM and strength measurements of Magnox and AGR trepanned samples of high weight loss, the most important of which are the sample size and the microstructure.
- Of the four DYM techniques that were investigated, the Ultrasonic ToF was proven accurate to better than 4% on virgin PGA graphite wave speed and repeatable to better that 1%. The improved Mk3 rig has been fully validated and applied to the MTR samples with weight losses up to 61%.
- Of the five strength techniques that were investigated, the ring burst technique has been successfully applied on thermally & radiolytically oxidised samples with weight losses up to 30%. Since this method provides the strength value of the weakest part of the ring and involves machining difficulties, it is expected that application on higher weight loss samples will only provide significantly conservative values.
- The 3-point bend test on beams machined from discs is currently considered the best option provided trepanned specimens are of sufficient size.
- This development programme has highlighted the challenges faced in the measurement of strength and in particular the difficulties of high accuracy strength tests on highly degraded material.

Acknowledgements

The authors would like to acknowledge British Nuclear Group Ltd permission for publishing this work.

References

C747-93 (1993): Standard Test Method for Moduli of Elasticity and Fundamental Frequencies of Carbon and Graphite Materials by Sonic Resonance.

C769-89 (1989): Standard Test Method for Sonic Velocity in Manufactured Carbon and Graphite Materials for Use in Obtaining an Approximate Young's Modulus.

C1161-90 (1990): Standard Test Method for Flexural Strength of Advanced Ceramics at Ambient Temperature.

C749-92 (1992): Standard Test Method for Tensile Stress-Strain of Carbon and Graphite.

C651-91(1991): Standard Test Method for Flexural Strength of Manufactured Carbon and Graphite Articles Using Four-Point Loading at Room Temperature.

Techniques for Measuring the Properties of Unirradiated and Irradiated AGR Graphite

Stephen D Preston and William E Ellis[#]
Serco Assurance, Risley.
[#]Nexia Solutions, Sellafield.
Email: steve.preston@serco.com

Abstract
This paper presents some* of the measurement techniques that have been used to determine the physical properties of irradiated and unirradiated graphite over the last 30 years. It describes the methods used by the graphite manufacturers who routinely monitored the material as it was produced. The paper then surveys the techniques used to measure the effects of neutron dose and thermal irradiation on the AGR graphites during the MTR irradiations, before describing those currently used to characterise the trepanned cores taken from the operating AGRs. The constraints of the three work programmes described above has resulted in test samples of different sizes. This, and the need to handle irradiated graphite remotely, has to some extent forced changes in the testing procedure. The paper discusses the additional work used to qualify the new techniques and establish a relationship with previous methods.

Keywords
Material properties, irradiated graphite, measurement techniques

INTRODUCTION

The graphite in a reactor will have been characterised at various stages throughout its life:

1. The initial supply: Tests are performed by the manufacturer to ensure the material supplied meets the specification. The results of these tests are recorded on "heat certificates" for the material in the reactor.
2. Irradiation characterisation: The irradiation behaviour of the graphite is normally determined by irradiation in a high flux Materials Testing Reactor (MTR) which will allow the lifetime dose to be achieved on a reduced timescale.
3. Inspection during the operating period: Throughout the operating period of the reactor (especially towards the end of life) specimens are removed from the reactor (pre-installed specimens and trepanned cores) and characterised to ensure that the material is performing as predicted.
4. Decommissioning and waste: Properties such as the strength, density, radioactive inventory, oxidation, stored energy, *etc.* are needed for the decommissioning and waste disposal.

Because of the large coke particles traditionally used in nuclear graphites in the UK the properties of the material vary with the size of the specimen tested and this is particularly true of strength properties (Brocklehurst, 1977). This makes it difficult to compare properties measured by the manufacturer, with those obtained from the MTR programmes and the trepanned cores taken from the reactors since they are all measured on different size specimens (Table 1).

TABLE 1: Examples of specimen sizes use for material property measurements.

Property	Manufacturers data+		MTR irradiation		Trepanned cores	
	Technique	**Sample size**	**Technique**	**Sample size**	**Technique**	**Sample size**
Density	Mensuration	Whole brick	Mensuration	6 dia. x 12 long	Mensuration	19 dia. x 6 long
Open pore volume	Beckman Pycnometer	25 dia. x 38 long	Helium Pycnometer	6 dia. x 12 long	Helium Pycnometer	19 dia. x 6 long
Electrical resistivity	Kelvin Double Bridge	Whole brick	Electrical resistance meter	6 dia. x 12 long	-	-
Thermal diffusivity/ conductivity	-	-	Laser flash	6 dia. x 12 long	-	-
C.T.E.	Optical lever	19 dia. x 127 long	Silica dilatometer	6 dia. x 12 long	Silica dilatometer	19 dia. x 6 long
Young's modulus	Sonic velocity (ASTM C769)	19 dia. x 127 long	Electrostatic resonance	6 dia. x 12 long	Sonic velocity	19 dia. x 6 long
Bend strength*	4 Point (ASTM C651)	19 x 19 x 178	-	-	3 Point	6 x 6 x 18
Tensile strength*	UKAEA test Specification (ASTM C565)	Dog-bone 25 dia.(min) x 150 long	Glued end fittings	6 dia. x 12 long	-	-
Compressive strength*	UKAEA test Specification (ASTM C695)	45 x 45 x 57	UKAEA	6 dia. x 12 long	-	-

+Data taken from the AGL specification for Dungeness B graphite in 1966, *Destructive tests

The manufacturer has large blocks of material to sample and is able to test large specimens which better represent the properties of the bulk material. Space in the high flux positions in an MTR is limited so specimens tend to be small, which also helps to minimise temperature and dose variation across them. A standard size of specimen (usually 6 mm dia. by 12 mm long) was used for all the MTR property measurements. Trepanned cores are cut from reactor bricks and again they are usually small, typical sizes ranging from 12 to 19 mm diameter and lengths up to 30 or 40 mm.

TEST METHODS

The test methods used for the MTR irradiations in the 70's and 80's to support the AGR stations evolved from earlier irradiation experiments, new techniques being added as they were developed (*e.g.* the laser flash technique for measuring thermal diffusivity was added in the late 60's). In practice many of the MTR tests were a compromise between accurately measuring the material property and being able to perform the measurement on small specimens. The same specimens were measured before and after irradiation using the same measurement techniques and systematic errors were largely eliminated by expressing the irradiation effects as ratios. This cannot be done by the manufacturers or for the trepanned cores where absolute values are needed, making the correct choice of measurement technique and validation against National Standards essential.

MEASUREMENTS ON MTR SPECIMENS

The following tests were all measured on standard (6 mm diameter by 12 mm long) cylindrical specimens that had been accurately machined and fully characterised prior to irradiation using the same measurement techniques. Most of the MTR tests are non-destructive allowing the same specimens to be irradiated-measured-irradiated-measured over and over again. In this way the behaviour of individual specimens could be studied as a function of irradiation dose. A small number of specimens were used for destructive tests (*e.g.* strength) after each irradiation period and these were replaced by new ones in the next irradiation period, allowing the strength properties to be studied as a function of irradiation dose.

Bulk Density

Density was measured on the standard sized MTR specimens by measuring the length and diameter at a number of points, weighing the specimen and then simply dividing the weight by the calculated volume.

Open Pore Volume (OPV)

The Open Pore Volume was measured using a commercially available (Micromeritics) helium pycnometer.

Electrical Resistivity

Measured using a commercial (Keithley) four probe alternating current resistance meter.

Thermal Diffusivity / Conductivity

This was measured at room temperature using the laser flash technique, only a high power photographic flash lamp was used as the heat source in place of the laser and a thermocouple was used to measure the temperature rise on the rear of the specimen. Thermal conductivity was calculated as the product of the measured thermal diffusivity, the measured bulk density and the specific heat of graphite (695 $J.kg^{-1}.K^{-1}$ at 20°C).

Coefficient of Thermal Expansion (CTE)

This was measured in banks of silica dilatometers from room temperature to 120°C. The specimens were loaded into the dilatometers at room temperature, the dilatometer was set to zero and then it was immersed in an oil bath maintained at 120°C and the temperature left to stabilise. When the reading became constant the change in length of the specimen was divided by the temperature change and the expansion coefficient of silica added.

Dynamic Young's Modulus

This was measured at room temperature by electrostatic resonance. The cylindrical specimen was secured at the centre and high voltage electrodes place near (but not touching) the flat end faces. A high sinusoidal voltage was then applied to both electrodes, periodically attracting the ends of the specimen. The gap between the specimen and the electrodes also acted as a capacitive displacement transducer and by changing the excitation frequency and monitoring the displacement of the specimen ends, the fundamental resonance frequency could be determined. From this the dynamic Young's modulus (E) could be determined using $E=4L^2F^2d$ where L is the length of the specimen, F is the fundamental resonance frequency and d is the density.

Tensile Strength

This was measured at room temperature by bonding metal pieces to the end of the cylindrical specimens and then pulling them apart between two short lengths of chain to remove any bending moments.

Compressive Strength

This was measured at room temperature by crushing the specimen between two platens.

Flexural Strength

It was not possible to perform three or four point bend strength measurements on the size of specimen used in the MTR irradiations.

MEASUREMENTS ON TREPANNED CORES

Because of the high dose levels (up to 100 mSv/hr at contact) emanating from the cores the measurement techniques are designed to be performed in-cave.

Cutting

The Gilsocarbon cores trepanned from the channel bore of the CAGRs are typically 19 mm in diameter by some 40 mm long and for most of the measurements the cores are sliced into discs approximately 6 mm thick. For cutting a purpose built rig is used. The cylindrical cores are gripped in a three-jaw scroll chuck on the end of a swinging arm counterbalanced to lower the sample gently onto a circular saw blade rotating at a maximum speed of 400 rpm. The blade is 0.5 mm thick and 80 mm diameter with 128 teeth. Wherever possible, discs are cut successively to ensure parallel sides. The cutting rig is also used to produce flats on curved surfaces for thermal expansion measurements and to reduce the discs to beams for flexural strength measurements (see below). For these operations a pair of suitably spaced blades are ganged together to provide a parallel cut.

In all these operations a careful note is kept of the sample orientation and the sample is marked before cutting to ensure that this remains known. Sample preparation is vital to the quality of most of the measurements made, good quality cutting is vital and any residual abnormalities, such as "rags", must be removed before subsequent measurements.

Bulk Density

All metrology is performed using purpose built rigs. Measurements are made using Sony Magnescales accurate to 2μm with a repeatability of 1μm. The transducers are retracted by vacuum cylinder and a constant contact force is applied through light springs.

Diameter

Diameter measurements are made on the whole core. The core is held on a vacuum pad with the core axis vertical. Measurements are made using two horizontally opposed Magnescale transducers fitted with blade type anvils and mounted on Unislide track allowing different positions on the core axis to be measured. The support rotates allowing different diameters to be accessed. The process is fully automatic. The unit is calibrated against cylindrical reference standards and part of the support structure is measured on each diameter traverse to check for drift.

Length

Lengths are measured against a base plate using a single Magnescale with a hemispherical anvil. The process is semi-automated with process controlled measurements being made across a sample diameter and across a reference standard. For discs, measurements are made across four equi-spaced diameters on each side of the sample and suitable weighting is applied to yield a mean length for the specimen. The lower of the two values is assigned as the true mean length. This corrects for any doming of the sample during cutting and for the presence of any residual "rags".

Weight

All the samples were dried at 150°C for 2 hours in a vented fan oven which has been modified for in-cave use before being weighed and stored in a desiccator. Weighing is performed on a modified semi-microbalance. The balance has a readability of 0.01 mg up to a capacity of 80g and has automatic calibration using a built-in reference weight.

Correction for Surface Roughness

In order to check the validity of the bulk density calculation using the methods described above, commissioning trials were performed on graphite samples of varying thickness cut from a single block of graphite. Bulk densities obtained by immersion were compared with mensuration data. Figure 1 summarizes the results. The behaviour may be understood as systematic errors in volume (length) measurement. For immersion, water penetrates the surface giving a 5μm underestimate of the "true length" while the Magnescale tend to measure the roughness peaks yielding an 8μm overestimate of "true length". This correction has been carried forward into all our density calculations for irradiated cores. The validity of the correction is checked by comparing densities of whole cores with those of individual discs from the same core.

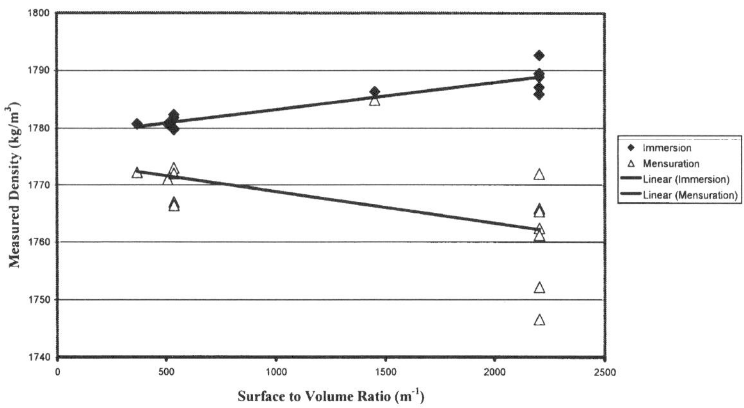

FIGURE 1: Variation of immersion and mensuration bulk density with surface to volume ratio.

Analysis of data from various campaigns indicates an overall accuracy for bulk density measurements of disks is 0.2% (1s.d.).

Helium Density

Helium density is measured using a standard commercial pycnometer modified for in-cave use. The instrument is simple in operation using pressure changes into a standard volume to measure the volume of a specimen chamber with and without a sample present. The instrument is calibrated against a standard volume sample. The instrument makes a number of estimates of volume, in our case thirty. The average is the best estimate of volume.

The instrument proved to be sensitive to ambient weather conditions, wind and changing air pressure. It was found that these could lead to systematic errors in volume well outside that expected from the scatter in measured volumes even though this value did increase somewhat in poor weather conditions. However, with careful choice of acceptable scatter and daily monitoring for drift, analysis indicates an overall accuracy in helium density of 0.1% has been achieved.

Dynamic Young's Modulus

In the technique used here, dynamic Young's modulus is determined from measurement of the velocity of ultrasound in bulk material. The method requires both the density and Poisson's ratio of the material to be known. In early testing time-of-flight was measured using a portable, off-the-shelf, tester. This instrument was subject to a large range of commissioning tests on graphite samples with a variety of lengths and diameter using a range of couplants. This study revealed that optimum performance could be obtained using 1 Mhz probes coupled to the test piece with distilled water. This pairing was particularly suitable as it gave a velocity which was independent of the length of the specimen.

The measurements made by this technique were compared to DYM values determined using the resonance technique. A direct comparison is difficult as the optimum sample shape for each method is very different. In this comparison specimens were taken from a small region of a single block and covered a range of geometries with differing acceptability to each technique. The two methods agreed within a few percent if Poisson's ratio measured using the resonance technique was used in the calculation of DYM from velocity. The velocity method was also compared with static Young's modulus measured by small strain bend tests. The moduli obtained agreed within 5%. This is considered perfectly satisfactory given the uncertainties in each measurement and the difficulties in making direct comparisons.

More recently the portable tester has been replaced by a less invasive and higher precision 1 Mhz tester developed at Berkeley Laboratories. Commissioning showed that time-of-flight measurements could be made with high reproducibility (±0.005 μsec) and with no evidence for any dependence on sample length. Retesting of irradiated samples measured with the portable tester revealed no systematic difference between the two methods.

The new tester is used to measure time-of-flight across the thickness of a 6mm disc which is typically 2μsec. Result from standards show that an accuracy of ±0.25% (1s.d.) is achieved and at this level of uncertainty errors in thickness proved to be a significant contributor to the overall error in velocity. For this reason it was essential to use a purpose designed sample holder to control the positioning of the measuring probes on the specimen.

In a single campaign on a reactor up to 72 DYM measurements are made covering twenty four cores. The measured velocities have been found to be well correlated with bulk density as shown in Figure 2. The residual scatter in velocity from this correlation shows the overall variability in DYM to be somewhat smaller than that expected from pre-irradiation measurements. This suggests that sufficient measurements are being made to characterize the overall variation in the reactor concerned.

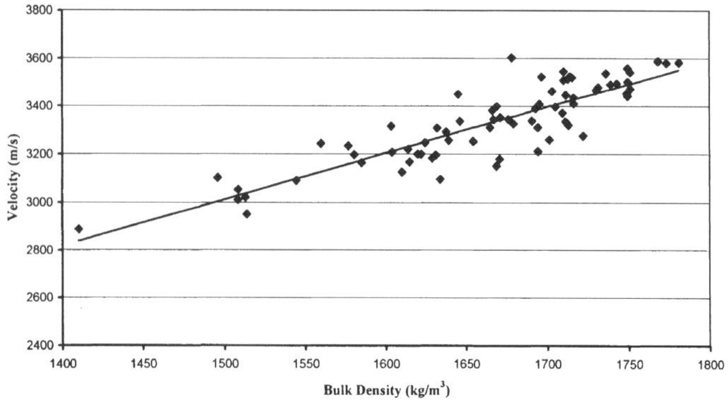

FIGURE 2: Variation of velocity with density in DYM measurement.

It is recognised from MTR data that the variation of DYM with irradiation is complex and the correlation with bulk density is somewhat of an autocorrelation. It is, however, a useful criterion for testing the internal self-consistency of measured data.

Fracture Strength

The small size of sample precludes the used of the standard four-point bend method to measure flexural strength and a three-point bend test was adopted instead. Beams nominally 6 mm square by 18 mm long are loaded over roller bearings with a 15 mm span. The method has been shown to give consistent results so long as a beam with a satisfactory surface finish is produced. An extensive test programme backed by theoretical assessments, both analytic and using finite element methods, has validated the stress assessment for this geometry.

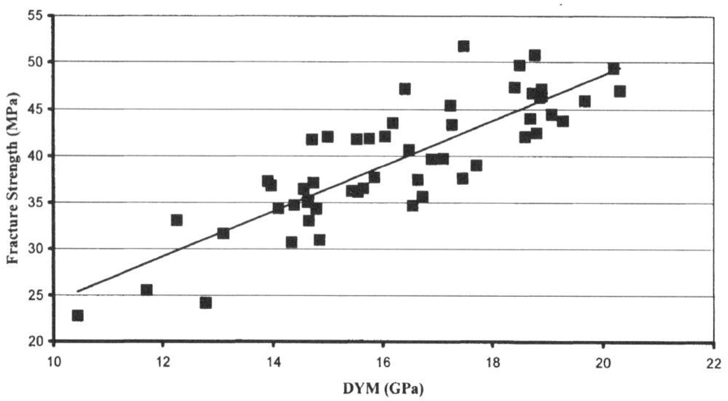

FIGURE 3: Comparison of fracture strength by 3-point bend with measured DYM

To-date all of the irradiated bend tests have been performed in a glove box. A jig for loading beams remotely has been designed but has yet to be tested in production. The results shown in Figure 3 are typically observed. The bend strength is proportional to DYM with a residual scatter of about 9% which is typical of that observed in pre-irradiation characterisation.

Thermal Expansion

Expansion is measured using a commercially available silica pushrod dilatometer modified for in-cave use with 19 mm width specimens. Measurements are made in static helium at 20 kPa to minimise thermal gradients. As the method is comparative with corrections being made for the relative expansion of the silica pushrod and the support tube, a number of unirradiated Gilsocarbon reference standards have been developed which mimic the geometry and thermal properties of the test pieces. This minimises the magnitude of the corrections. The standards were calibrated against silica and sapphire standards in the dilatometer and against a Fabry-Perot interferometer.

Each test uses two thermal cycles between ambient and 400 °C. A fast settling cycle is followed by a measurement cycle typically at 2 °C/min and an expansion coefficient over the range 100 °C to 300 °C is quoted. In a typical campaign measurements are made in three orientations on thirty specimens. Two measurements are made along the 17 mm length of the squared-up disc (see cutting) while the other measurement is along the 6 mm thickness.

The accuracy of the method is limited by the accuracy to which test pieces can be prepared. Re-measures are sometimes required on samples deemed to be of poor quality. In a campaign the value of CTE measured is typically 5.5 ± 0.2 μm/m/K (1s.d.).

CONCLUSIONS

Various techniques have been described for measuring the properties of unirradiated and irradiated graphite on a wide variety of specimen sizes. Checks are in place to ensure consistency between the different measurement techniques enforced by geometry and environmental considerations.

Acknowledgements

The authors would like to thank British Energy Generation Ltd for permission to present this paper.

References

ASTM C565; "Standard test method for tension testing of carbon and graphite mechanical materials", American Society for Testing and Materials, 100 Barr Harbor Drive, PO Box C700, West Conshocken, PA 19428-2959, USA.

ASTM C651; "Standard test method for flexural strength of manufactured carbon and graphite articles using four-point loading at room temperature", American Society for Testing and Materials, 100 Barr Harbor Drive, PO Box C700, West Conshocken, PA 19428-2959, USA.

ASTM C695; "Standard test method for compressive strength of carbon and graphite", American Society for Testing and Materials, 100 Barr Harbor Drive, PO Box C700, West Conshocken, PA 19428-2959, USA.

ASTM C769; "Standard test method for sonic velocity in manufactured carbon and graphite materials for use in obtaining an approximate Young's modulus", American Society for Testing and Materials, 100 Barr Harbor Drive, PO Box C700, West Conshocken, PA 19428-2959, USA.

Brocklehurst, J. E (1977), *Chemistry and Physics of Graphite*, 13, 145-279.

Modelling of Nuclear Graphite Behaviour Under Irradiation Conditions

Derek K L Tsang and Barry J Marsden
Nuclear Graphite Research Group, School of Mechanical Aerospace and Civil Engineering
The University of Manchester, PO Box 88, Manchester, UK.
Email: kwong-lai.tsang@manchester.ac.uk

Abstract
The graphite components in gas-cooled reactors provide channels for the fuel and control rods. During operation the dimensions and material properties of these graphite components are modified by fast neutron irradiation and, in the case of carbon dioxide cooled reactors, by radiolytic oxidation. Due to local gradients in the fast flux and temperature these dimensional and property changes result in the build up of component internal stresses. This paper describes a constitutive material model for the assessment of stresses and deformations in nuclear graphite components. Examples are given for typical stress analysis of nuclear graphite components under irradiation and radiolytic oxidation. The influence that the different irradiation induced strains have on resulting stress profile is discussed. These analyses provide an insight into the effect of irradiation in nuclear graphite components.

Keywords
Nuclear graphite, finite element analysis, stresses

INTRODUCTION

Nuclear graphite changes in dimensions and material properties due to fast neutron irradiation. Fast neutron damage to the graphite crystallite is mainly the result of a cascade of carbon atoms "knocked-on" caused by initial neutron collisions. These knock-on atom processes result in the creation of interstitial clusters of carbon atoms between the layer planes and produce graphite crystal "c-axis" expansion perpendicular to the layer planes. Furthermore, vacancies are produced in the layer planes; these vacancies are less mobile than the interstitials but either can collapse or diffuse to layer plane boundaries and produce contraction in the crystal "a-axis".

These large dimensional changes (up to ~20%) that can occur within the graphite crystallites eventually lead to microstructural changes which modify the nuclear graphite component bulk thermal and mechanical material properties as well as changes in bulk dimensions. The linear dimensional changes can exceed 3% during service life and this can significantly influence the geometry of the many graphite components that make up a typical graphite reactor core. Radiolytic oxidation is only significant in carbon dioxide cooled graphite moderated reactors. In a CO_2 system, ionising irradiation from gamma and neutron interactions leads to the dissociation of the CO_2 producing oxidising species which can react with carbon atoms mainly at the surface of the graphite internal open porosity. Peak component weight losses of up to ~40% may result due to this interaction, leading to exponential reductions in the graphite Young's modulus, strength and thermal conductivity.

These irradiation induced changes are complex functions of fast neutron dose (or fluence), radiolytic oxidation, and temperature. In addition some of these strains interact with each other. If these graphite material property and dimensional changes are not taken into account in the reactor design they could present a significant challenge to the reactor operation. Therefore assessment of the effects of graphite dimensional and material property changes on the integrity of graphite moderator structures is an important factor in graphite moderated reactor design, operation and component life prediction.

To model the irradiated graphite material behaviour is a complex task. Irradiation creep in graphite does not take place at constant volume as in most metals. Consequently, standard facilities provided in most finite element codes cannot be used directly for irradiated graphite assessments and therefore specially written coding is required to model the irradiation behaviour of nuclear graphite components. Recently Tsang and Marsden (2005) have developed a material modelling code for irradiated isotropic graphite. The model is based on data obtained on samples irradiated in Material Test Reactor (MTR) programme. Totally, there are seven different strains within the model and their interactions are also included.

In order to gain an understand the effect of each of these strains on the resulting stresses in a two-dimensional hypothetical graphite brick, several assessments have been systematically carried out and the resulting stress profiles determined by including a different number of strains in each assessment.

CONSTITUTIVE EQUATIONS

In this section, a brief introduction to the material constitutive model for nuclear graphite is given. For more detail descriptions about the material model can be found in Tsang and Marsden (2005). The total strain within the irradiated graphite model is the sum of seven different strain components:

$$\boldsymbol{\varepsilon}^{\text{total}} = \boldsymbol{\varepsilon}^{e} + \boldsymbol{\varepsilon}^{th} + \boldsymbol{\varepsilon}^{dc} + \boldsymbol{\varepsilon}^{pc} + \boldsymbol{\varepsilon}^{sc} + \boldsymbol{\varepsilon}^{ith} + \boldsymbol{\varepsilon}^{idc} \quad [1]$$

where $\boldsymbol{\varepsilon}^{e}$ is the elastic strain, $\boldsymbol{\varepsilon}^{th}$ is the thermal strain, $\boldsymbol{\varepsilon}^{dc}$ is the dimensional change strain, $\boldsymbol{\varepsilon}^{pc}$ is the primary creep strain, $\boldsymbol{\varepsilon}^{sc}$ is the secondary creep strain, $\boldsymbol{\varepsilon}^{ith}$ is the interaction thermal strain and $\boldsymbol{\varepsilon}^{idc}$ is the interaction dimensional change strain. However, the interaction strains will not be considered in this paper.

The elastic strain $\boldsymbol{\varepsilon}^{e}$ is related to the stress $\boldsymbol{\sigma}$ by means of the usual Hooke's law of linear elasticity, *i.e.*

$$\boldsymbol{\sigma} = \mathbf{D}\boldsymbol{\varepsilon}^{e} \quad [2]$$

where $\mathbf{D} = \mathbf{D}(E, \nu)$ is the material matrix and is a function of material properties, Young's modulus, E, and the Poisson's ratio, ν. The Poisson's ratio ν is assumed to be invariant to irradiation at 0.2.

The thermal strain is defined as

$$\boldsymbol{\varepsilon}^{th} = \boldsymbol{\alpha}_{(T_{ref} - T)} \left(T - T_{ref}\right) \quad [3]$$

where $\boldsymbol{\alpha}$ is the mean coefficient of thermal expansion (CTE) between temperature $\left(T_{ref}-T\right)$ °C and T_{ref} is a reference temperature. The CTE is a function of irradiation dose, irradiation temperature and real temperature.

The dimensional change strain $\boldsymbol{\varepsilon}^{dc}$ is based on curves fitted to experimental results (Kelly 1985) obtained in a MTR research programme, which are a function of irradiation dose, irradiation temperature and graphite oxidation rate. The dimensional change strain value is calculated by interpolation of the experimental data, although it is possible to use the fitted equation directly.

The primary creep strain, which is often referred to as transient creep (Kelly and Brocklehurst 1977), is a recoverable under irradiation when the load is removed is defined as

$$\boldsymbol{\varepsilon}^{pc} = 4\varphi e^{-4\gamma} \int_0^{\gamma} \mathbf{D}_c \boldsymbol{\sigma} e^{4\gamma'} d\gamma' \qquad [4]$$

where $\varphi(T)$ is a parameter depended on temperature T, γ is the irradiation dose and $\mathbf{D}_c$ is the creep material matrix which is a function of the irradiated modulus and is also assumed to be modified by structural changes. Poisson's ratio for creep is taken as being the same as the elastic Poisson's ratio and to be invariant to irradiation. Primary creep is characterised by a rapid deformation which soon decrease in rate with increasing irradiation dose.

The secondary creep strain, which also called steady state creep, is irreversible strain and is defined as

$$\boldsymbol{\varepsilon}^{sc} = \xi(T) \int_0^{\gamma} \mathbf{D}_c \boldsymbol{\sigma} d\gamma' \qquad [5]$$

where $\xi(T)$ is a temperature dependent parameter. Secondary creep is characterised by a linear dependence on the creep strain. The total creep within the model is the sum of the primary creep and the secondary creep.

TWO DIMENSIONAL MODEL

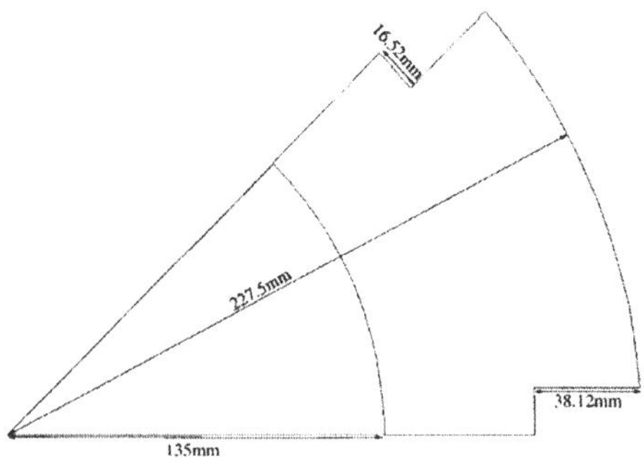

FIGURE 1: Geometry of hypothetical graphite brick.

A two-dimensional hypothetical graphite brick model is depicted in Figure 1. Only one eighth of the brick was used in the calculation by applying symmetric condition. The internal radius and external radius are 135 mm and 227.5 mm, respectively. The brick model is analysed for a period of 30 full power years (fpy). The irradiated graphite material data used in the analysis are based on that given by Brocklehurst and Kelly (1993).

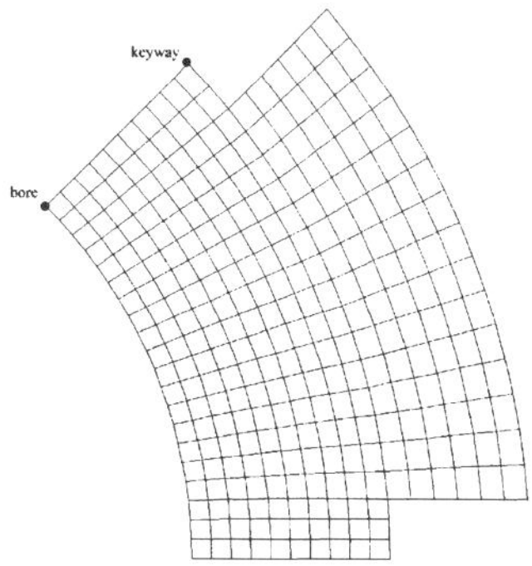

FIGURE 2: Finite element mesh.

The mesh configuration for the above model is given in Figure 2. Generalised plane strain elements (ABAQUS 2005) are used in the model. There are in total 327 nodes and 290 four-noded elements in the model. The reference node for the generalised plane strain element is set at the origin. It is assumed that symmetry condition apply on both radial boundaries. Also it is assumed that the plane is free to move axially, but without any rotation about the reference point. The cylindrical coordinate system is used in the analysis.

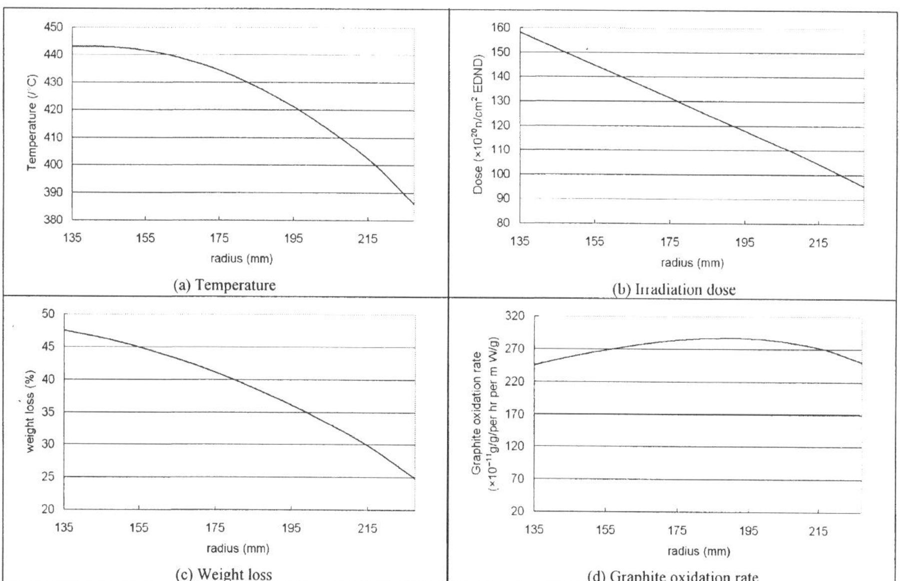

FIGURE 3: Plots of predefined variables at the end of 30 fpy through the brick thickness.

Predefined time-dependent reactor field variables are required by the analysis. They are irradiation dose, irradiation temperature, radiolytic weight loss and graphite oxidation rate. Simple profiles have been created to model the reactor condition. Figures 3a, b, c and d show the predefined field variables along the brick thickness. Figure 3a shows the temperature profile at the end of 30 fpy. The pre-start-up and post-shutdown temperatures are 20 °C. Figure 3b presents the dose profile. For simplification, it is assumed that the dose rate (irradiation flux) does not vary with time during at-power operation. Figure 3c shows the weight loss profile at the end of 30 fpy. The maximum and minimum weight loss occurs at the inner radius and outer radius, respectively. Figure 3d shows the graphite oxidation rate. The maximum rate occurs at about two-third of the way through the brick thickness.

NUMERICAL RESULTS

Four sets of numerical results will be presented in this section. Different strains are considered in each set of result. Table 1 shows which strains are included in each analysis. In the first analysis only the thermal and the elastic strains are considered. The results from this analysis are considered as the baseline solution. Then the dimensional change strain, primary creep strain and secondary creep strain will be added into the model one by one. Consequently, the influence of each strain on resulting stress can be ascertained. In particular, the solutions time histories are presented at two locations. These are at the keyway and at the bore (see Figure 2). For ease of presentation only variables in hoop direction will be depicted here.

TABLE 1: Strain option considered in each set of results.

Set	Strain calculated
1	Elastic strain and thermal strain
2	as Set 1 plus dimensional change strain
3	as Set 2 plus primary creep strain
4	as Set 3 plus secondary creep strain

Elastic Strain and Thermal Strain

In this subsection only elastic strain and thermal strain are considered in the analysis. The solutions are given in Figure 4. The stress profiles are shown in Figure 4a. The stress is due to the thermal gradient across the brick radius. With increasing dose, the keyway stress changes from tensile early in life to compressive later in life, whereas the bore stress changes from compressive early in life to tensile late in life. The stress change during life is caused by the irradiation induced changes in Young's modulus and the changes in CTE. The stresses reduce back to zero after the shut down. The elastic strains are given in Figure 4b. As expected, the elastic strain curves are similar in shape to the stress profiles. The thermal strains are given in Figure 4c. At the beginning of life, the thermal strain is higher at the bore. This is due to higher temperature at the bore. However, the bore thermal strain is lower later in life due to of the higher dose received at the bore. The profiles of Young's modulus are given in Figure 4d. The Young's modulus curves indicate a sudden rise just after the initial start-up, due to the so called "pinning effect" (Kelly 1985). The Young's modulus continues to fall after initial rise, due to the effect of radiolytic weight loss.

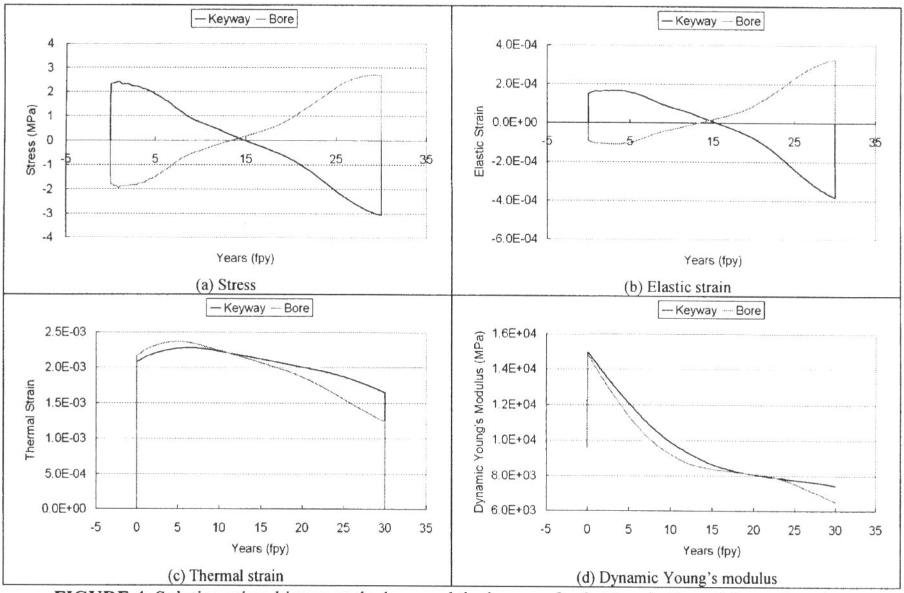

(a) Stress (b) Elastic strain (c) Thermal strain (d) Dynamic Young's modulus

FIGURE 4: Solutions time history at the bore and the keyway for Set 1 - elastic and thermal strains.

Dimensional Change

In this section dimensional change strain has also been included, which is a function of dose, temperature and radiolytic oxidation. The stresses at the keyway and the bore are given in Figure 5a. The keyway and bore stress profiles show opposite trends. The bore's stress continues to rise after a negative jump after the initial start-up whereas the keyway's stress continues to drop after a positive jump. The maximum tensile stress at the bore and compressive stress at the keyway are 60 MPa and 65 MPa, respectively. The jump after the initial start-up and shut-down are caused by change of temperature. The elastic strains are given in Figure 5b.

The dimensional change strain is given in Figure 5c. The larger shrinkage at the bore is due to by the higher temperature and irradiation dose. The curve is starting to "turn around" at the bore near the end of life. It is important to note that the dimensional change strain is one order higher than the thermal strain and elastic strain. The corresponding dimensional change rate profiles are shown in Figure 5d.

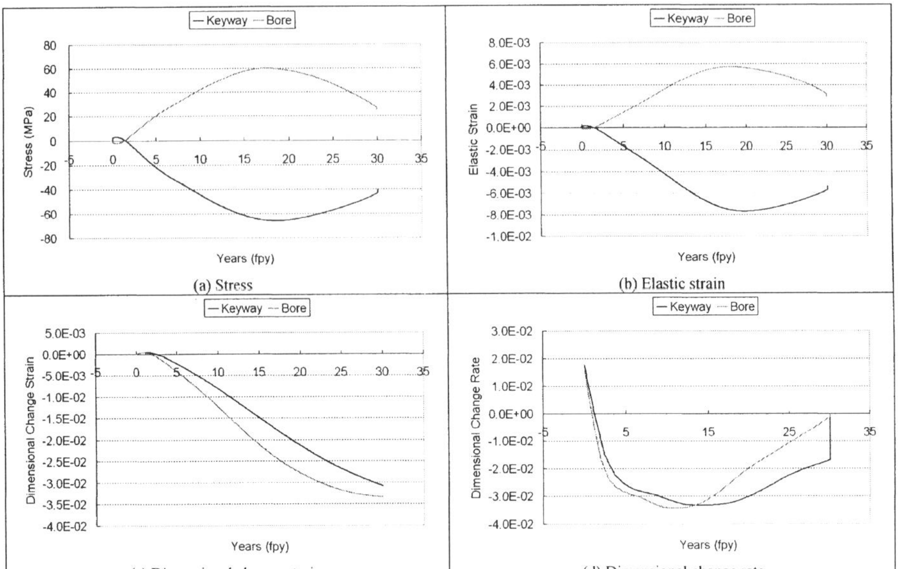

(a) Stress (b) Elastic strain (c) Dimensional change strain (d) Dimensional change rate

FIGURE 5: Solutions time history at the bore and the keyway for Set 2 – elastic, thermal and dimensional change strains.

Primary Creep

In this subsection the primary creep strain has also been included. The results with primary creep strain included in the model are given in Figure 6. The stress curves are given in Figure 6a. The stress profiles are very similar to the previous results. However, the maximum magnitudes of stress have been reduced by about a third. Also there are spikes in the stress profile after the initial start-up. These spikes are caused by the primary creep due to its short time scale. The jump at the shut-down is caused by change of temperature. The primary creep strains are given in Figure 6b. The shapes of the primary creep strains are similar to the stress. The primary creep strain and elastic strain are in similar order, but the primary creep strain is larger than the elastic strain. At the final shutdown there is no irradiation dose thus no jump occurs in the primary creep strain.

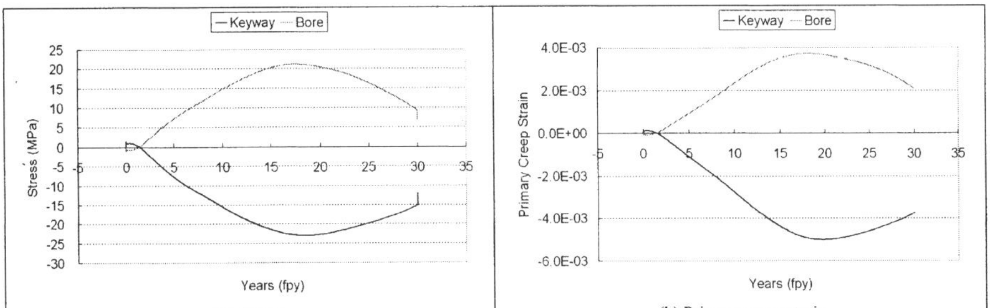

(a) Stress (b) Primary creep strain

FIGURE 6: Solutions time history at the bore and the keyway for Set 3 – elastic, thermal, dimensional change and primary creep strains.

Secondary Creep

In this subsection, the secondary creep strain has also been included, which is a function of stress, dose and radiolytic oxidation. The stress curves are given in Figure 7a. The shape of these curves change significantly from previous section. The maximum tensile stress at the bore is reduced to about 2.5 MPa whereas the maximum compressive stress at the keyway is about 3.0 MPa. Another important feature is that the stress returns to early stress status later in life, *i.e.* the tensile bore stress becomes compressive again around 20 fpy and the keyway compressive stress becomes tensile again around 22 fpy. The spikes due to the primary creep become more noticeable here. The elastic strain and the primary creep strain are given in Figures 7b and 7c. The orders of magnitude for both strains are reduced by a factor of one when compared with previous results and again primary creep strain is larger than the elastic strain. However the secondary creep strain in Figure 7d is the second largest strain within the model, the dimensional change strain being larger. The secondary creep strain is positive at the bore and negative at the keyway.

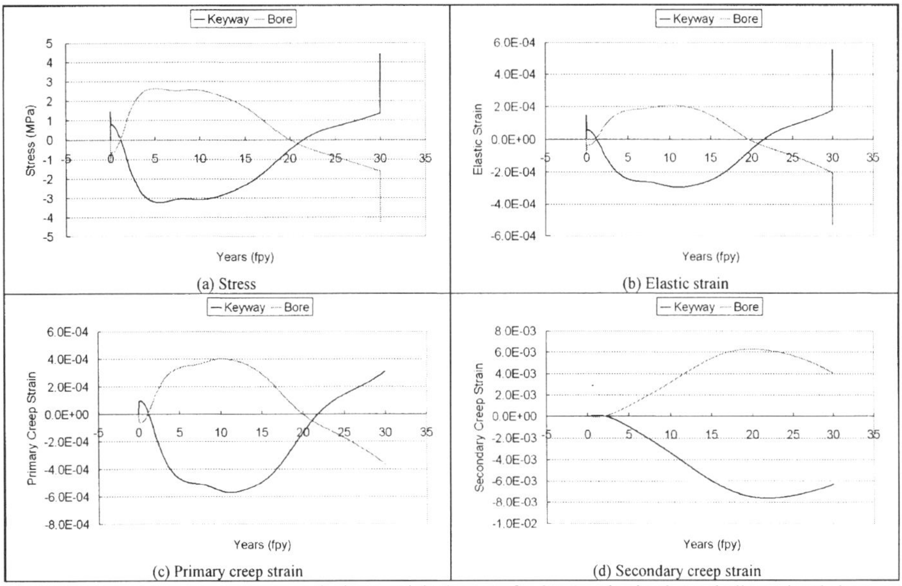

FIGURE 7: Solutions time history at the bore and the keyway for Set 4 – elastic, thermal, dimensional change, primary creep and secondary creep strains.

CONCLUSIONS

In this paper the constitutive equation for nuclear graphite under irradiation conditions has been described. A series of finite element analyses for a two dimensional hypothetical nuclear graphite brick are presented, by adding strain onto the model one by one. The following conclusions can be drawn from these results:

- The main source of stress within the graphite brick is irradiation induced dimensional change. Since the dimensional change is temperature, dose and radiolytic weight loss dependent, and there are temperature and dose gradient across the brick, this causes significant self-induced "shrinkage" stresses.

- Radiolytic oxidation significantly elevates stresses through changes to Young's modulus, dimensional change rate and irradiation creep, although it should be noted that it also reduces strength.
- Irradiation creep strain is particularly important in reactor design. Without creep strain self-induced shrinkage stresses would build up to levels exceeding the bricks failure strength.
- Temperature change at start-up and shut-down cause a rise in stress, due to thermal gradients these thermal gradients creep out during irradiation, but return in the opposite sense.

Acknowledgements
The authors would like to thank HSE for financial support. The views expressed in this paper are those of the authors and do not necessarily represent those of the HSE.

References

Abaqus (2005), Hibbit, Karlsonn and Sorensen Inc.

Brocklehurst J. E. and Kelly B. T. (1993). Analysis of the dimensional changes and structural changes in polycrystalline graphite under fast neutron irradiation. *Carbon*, 31, 155-178.

Kelly B. T. (1985). The radiolytic corrosion of advanced gas-cooled reactor graphite. *Progress in Nuclear Energy*, 16, 73-96.

Kelly B. T. and Brocklehurst J. E. (1977). UKAEA reactor group studies of irradiation-induced creep in graphite. *Journal of Nuclear Materials*, 65, 79-85.

Tsang D. K. L. and Marsden B. J. (2005). The Development of a Stress Analysis Code for Nuclear Graphite Components in Gas-Cooled Reactors, Submitted to *Journal of Nuclear Materials*.

Predicting the Stresses and Deformations of Irradiated Graphite Moderator Bricks

Chris J Jones
Serco Assurance, Walton House, 404 Faraday Street, Risley, Warrington, Cheshire, UK.
Email: chris.jones@sercoassurance.com

Abstract
During the lifetime of a station, the graphite moderator bricks are subject to irradiation induced deformations which in turn lead to a build up of internal stresses in the brick. Excessive brick and column bowing and shrinkage could impede refuelling operations or control rod movements. The build-up of internal stresses may eventually lead to crack initiation at the bore or keyway roots of the bricks which then has implications for the integrity of the core. Hence, it is important to have reliable predictions of the brick deformations and the time at which the initiation of brick cracking may occur. This paper describes the methods used to predict stresses and deformations in graphite bricks with an emphasis on the AGR cores. The paper describes the requirements of such predictions, the data needed for the assessments and the important features and properties of the bricks and fuel that need to be considered. The methods of validating the predictions are discussed, as are the limitations of the analyses and likely future developments.

Keywords
Irradiated graphite, stress analyses, brick deformation

INTRODUCTION

During the lifetime of a graphite moderated reactor the graphite bricks are subject to irradiation induced deformations which in turn lead to internally generated stresses. An understanding of these deformations and stresses is necessary as both may have safety implications for the core. Excessive brick and column bowing and shrinkage could impede refuelling operations or control rod movements. A build-up of internal stresses, or external stresses arising from interactions of deformed components, may eventually lead to crack initiation at the bore or keyway roots of the bricks which then has implications for the integrity of the core. An understanding of the behaviour of the stresses and deformation of a brick following partial, or through-wall, crack formation, is also required to aid assessments used in safety cases aimed at extending the operating life of the reactor cores.

Calculations to predict the deformations and stresses of graphite moderator bricks have developed over the years from simplified two-dimensional analyses to detailed three-dimensional studies using finite element codes such as FEAT and ABAQUS. These codes require modules that incorporate the models of irradiated graphite behaviour and these models have themselves developed over the years to become more complex and detailed as understanding of that behaviour has developed. Serco Assurance has extensive experience with the analysis of irradiated graphite moderator bricks and has carried out work on AGR, Magnox, RBMK and PBMR core designs. It is the AGR cores, however, that have been studied in most detail and it is the experience of the analyses of the AGR moderator bricks that is primarily reflected in this paper.

METHODOLOGY

Geometry

Finite element analyses of graphite moderator bricks now use large three-dimensional models that aim to provide good estimates of brick deformations and stresses. A typical finite element mesh is shown in Figure 1.

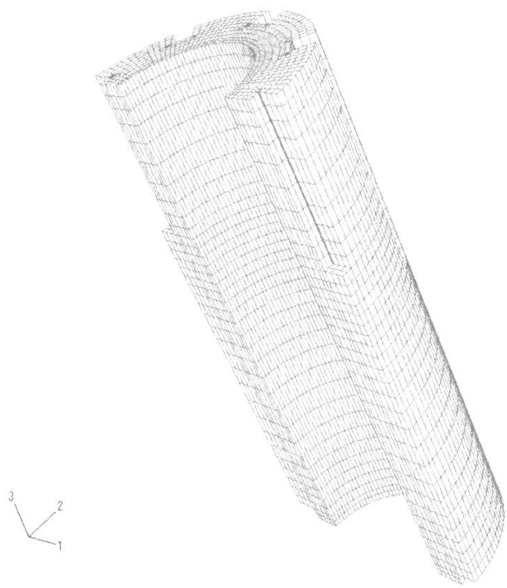

FIGURE 1: Typical finite element mesh of a graphite moderator brick.

The geometry of the brick is often simplified to keep the model at a manageable size. Unless specifically required by a particular analysis the end features are generally omitted, as these do not significantly contribute to the stresses and deformations of a stand-alone brick. The methane holes are also generally omitted although calculations including the holes have been carried out to assess the stress raising influence that they may have at the bore of the brick. The small radii of the keyway roots require a high level of mesh refinement which is not required for the rest of the brick model. Hence generally a "global" model, such as that in Figure 1, will only have enough mesh refinement at the keyway roots to indicate the relative magnitudes of the root stresses. When more precise keyway root stresses are required then a sub-modelling technique is used in which a more highly refined model of the region just around the keyway root is used, with the boundaries of that model driven by the displacements calculated from the global model.

Field Variables

The deformation of the bricks is driven by the irradiation damage dose and influenced by the irradiation temperatures and the weight loss arising from radiolytic oxidation. Distribution of these three field variables through the brick must be provided and are utilised by the modules that calculate the irradiated graphite behaviour. The distribution of the damage dose, temperature and weight loss through a brick is complex and is influenced by such factors as neighbouring channels, the position of fuel element interfaces, coolant flow, the location of methane holes and the position within the brick. The field variables are calculated by a suite of codes that provide the data in a form that can be utilised by the stress analyses.

Material Properties

The stress analyses require many material properties to be defined, both unirradiated and irradiated. Some values are used directly, *e.g.* unirradiated Young's modulus or Coefficient of Thermal Expansion (CTE), whilst others are coefficients that are used within the specific graphite module of the finite element code to specify some behaviour of the brick, *e.g.* various coefficients used by the weight loss model or the creep strain model. Other data such as the damage dose and temperature dependencies of properties such as Young's modulus, CTE, creep coefficients must also be specified.

User Material Subroutines

As previously noted, the finite element analyses use specially written subroutines that model the irradiated behaviour of the graphite. The subroutines must define the constitutive equations relating the stress and strain in the graphite and must account for elastic, thermal and irradiation effects. Furthermore, to be useful in analyses of actual graphite bricks, the equations must be implemented in a code that is stable, robust and does not incur large time penalties when running. The User Material Subroutine (UMAT) developed by Serco Assurance for AGR graphite core analyses includes the latest endorsed models of graphite behaviour as set down by British Energy Generation Ltd. (BEGL). The UMAT also includes a number of variants on the approved models to allow sensitivity studies to be carried out to assist in the ongoing development of material models by BEGL. The UMAT has proven to be stable and allows large analyses to be completed in acceptable timescales.

STRESS ANALYSES

Typical stress analyses for graphite moderator bricks are carried out to model a full lifetime of operation, nominally 30 full power years, with periodic shutdowns included. Typical time history hoop stress plots for the bore and keyway root of a brick are shown in Figure 2.

Figure 2 shows that the tensile stress at the bore of the brick peaks early in life and then tends to reduce, eventually changing from a tensile to a compressive stress. Conversely, near the perimeter of the brick, such as at a keyway root, the stress is compressive early in life, becoming tensile towards the end of life.

For cracks to initiate and grow, a tensile stress field is required. With this consideration there are a number of important conclusions that can be drawn from results such as Figure 2. The tensile stresses at the bore early in life imply that initially it is the brick bore that may be more susceptible to crack initiation. Later in life the stress raising effects of the keyway roots become important with the potential for crack initiation at that location. The magnitude and timing of the stress peaks are also important as is a consideration of the irradiation effects on reducing the graphite strength. A further important point on Figure 2 is the time at which the keyway root stress "turns around" from compressive to tensile stress. Whilst the periphery of the brick is in compression it is unlikely that a through wall crack could occur. However, once the outer part of the brick is in tension through wall cracking becomes possible. In particular, at the keyway root the stress tends to rise rapidly after stress turn around to a level at which crack initiation may be predicted. Hence the time of stress turn-around may be a good predictor of when through wall cracks may become possible.

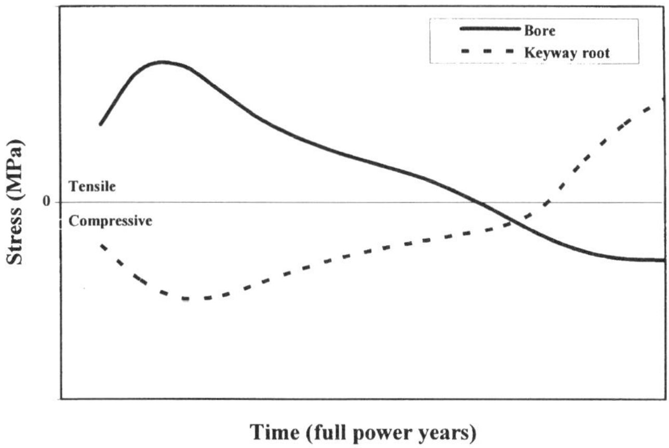

FIGURE 2: Typical time history of stress at the brick bore and keyway root.

The magnitude of the stresses and the timings of the stress peaks clearly depend on the input data used in the analyses. Some of these data vary from analysis to analysis due to known variations in material properties or field variables. For example, there are known differences in the unirradiated Young's modulus of the graphite used at the different AGR stations and the field variable distributions vary according to the location of the brick in the channel and the position of the channel in the core. However, there may be other differences deliberately introduced to allow for uncertainties in the data on which the material models are based or to ensure the analyses are conservative in their estimates of time to crack initiation.

Conservative and Best Estimate Analyses

Historically, stress analyses that have been carried out in support of safety cases have been deliberately conservative. These conservatisms have been present in input data, in the manner in which the material models have been defined and in the calculational route used to solve the graphite constitutive equations. The use of conservative analyses is intended to give confidence that tensile stress magnitudes are over-predicted and that actual times to crack initiation are therefore longer than implied by the analyses. When making conservative assumptions care must be taken in interpreting the results, for example an analysis that may be conservative for keyway root stresses may not be conservative for bore stresses. Nonetheless the conservative approach is also often simpler, as some of the finer details of an analysis or a solution route can be omitted, as long as it has been shown to be conservative to do so. However, the conservative approach becomes problematic when seeking to validate the calculational route with comparisons against measured data.

If predictions from the stress analyses are to be compared with measurements from the reactors then clearly a conservative approach is unsuitable as by the nature of the calculations good agreement between prediction and measurement would not be expected. For comparisons with measurements "best estimate" analyses are required where the data and material models are all as close to the actual reactor conditions as believed practically possible and the solution procedure does not deliberately include conservative approximations. Recently, there has been a move towards performing more "best estimate" analyses. The material models and the data used by those models have been revised by BEGL with the aim of making them more representative of the actual reactor graphite behaviour. This has become increasingly possible as more data from the AGR cores has become available, the original models having been based on Materials Test Reactor data. The field variable data have also been revisited to take account of the latest understanding of weight loss distribution and the effects of the fuel element interfaces at which there is a "dose depression", *i.e.* a local reduction in dose near the fuel ends and a corresponding local reduction in weight loss. Additionally, the Serco Assurance graphite UMAT has been updated to include the latest material models and to remove some of the inherent conservatisms that were present in the previous solution procedure. Best estimate type analyses then offer the opportunity to provide validation of the calculated brick deformations.

VALIDATION

Validating the finite element stress analyses by comparing calculated and measured brick stresses has not proven possible as there are no practical methods in use to measure the in-reactor brick stress state. Whilst in theory it may be possible to infer these stresses from strains measured in trepanned samples this technique has not been pursued. However, there are good measurements of the brick bore diameters which are obtained from the Channel Bore Measuring Unit (CBMU), used to investigate the shape of the bores of a number of channels at each reactor outage. Comparison of the measured and calculated bore diameters, ovalities and channel bow provide an opportunity to validate the calculated strains in the irradiated bricks. If it can be shown that the existing models provide good agreement between the predicted and measured brick deformations, then this would give confidence that the models are giving a reasonable prediction of brick stresses.

Ideally, a highly detailed analysis should be undertaken for a specific brick in a specific channel for which good quality CBMU data are available, preferably from more than one outage. This detailed analysis would include brick specific field variables that account for neighbouring channels, power history of the channel, any fuel shuffling and brick specific material properties using data derived from trepanned samples from the brick. However, obtaining such detailed, specific, data would be difficult and time consuming and so appropriately averaged field variables, and brick data are used but with the aim of obtaining the "best estimate" of the actual data.

A true best estimate analysis would also include such details as the weight of the column of bricks above the brick being studied and interactions between the keys and keyways that may occur through the later stages of the core life. However, for simplicity these factors have been omitted from current analyses although their effects have been studied previously.

Figure 3 shows a typical comparison of predicted and measured bore radial displacements. The figure compares measured bore profiles up the height of an entire channel as obtained at two separate outages, three years apart, and as predicted by a best estimate finite element analysis. The predicted profile corresponded to a time slightly earlier than that at which the 2004 measurements were made.

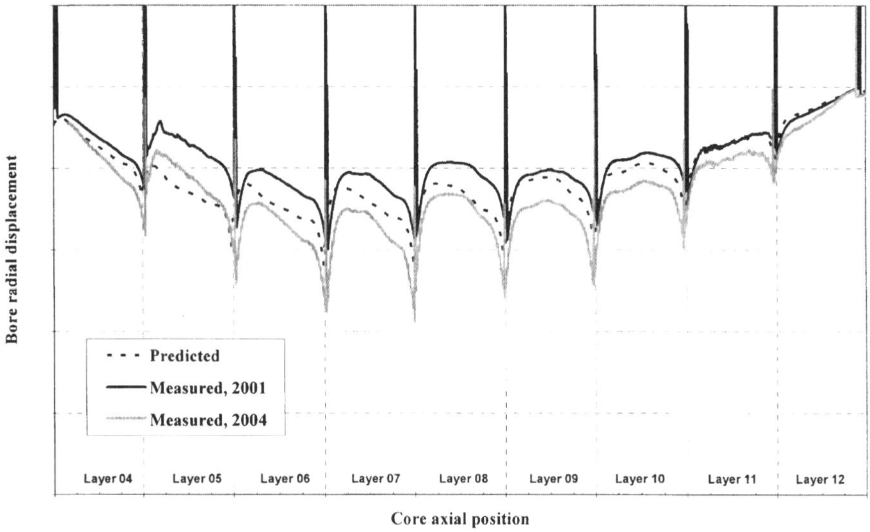

FIGURE 3: Comparison of calculated and measured bore profiles.

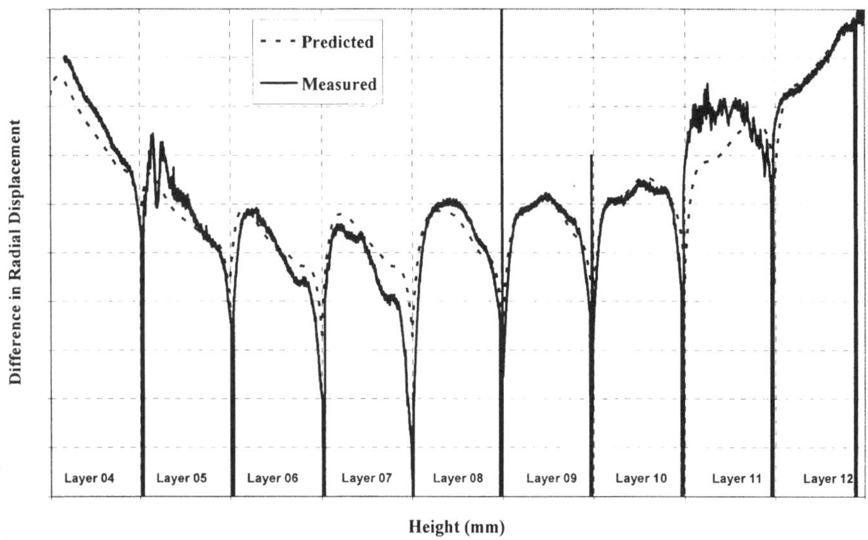

FIGURE 4: Comparison of the calculated and measured difference in radial bore shrinkage over a 3 full power year interval.

The observed agreement between measurement and prediction is considered to be good. Although there is a time offset between the prediction and measurement, the trends in the profiles are closely matched. Note that the large spikes in the measured data correspond to the brick interfaces. A time offset is equivalent to an offset in the applied damage dose which may be accounted for by the analysis not using a channel specific power history.

The shape of the predicted bore displacements over the length of each brick show good agreement with the measurements and show the same asymmetries both along the length of a brick and up the height of the channel. The apparent time offset in the analyses can be effectively removed by considering the difference in bore displacements during the interval between measurement sets. A comparison of the change in bore shape as measured between the two outages and as calculated over an equivalent time period is shown in Figure 4.

The agreement between prediction and measurement seen in Figure 4 is good and shows that for these analyses the predicted change in bore shapes were a good representation of what was actually observed in the core. This in turn provides a measure of confidence in the calculation of stresses in these bricks.

Whilst agreement between calculated and measured bore profiles, such as shown in Figures 3 and 4, is encouraging, there are still other factors to consider in the analyses and good agreement must be shown over a range of core positions. Additionally, the sensitivity of the results must be considered and in particular the range of stress results that may arise from analyses that give similar displacement results.

UNCERTAINTIES IN THE ANALYSES

Specifying the precise reactor conditions and the full interplay of the material properties provide a limitation to the accuracy of the analyses. Determining the specific conditions that each brick has experienced in core would be difficult and therefore sensitivity studies are required to investigate the effects on brick behaviour of variability in dose, temperature and weight loss.

A further source of uncertainty in the analyses arises from the variability in material properties that occur between graphite batches, within graphite batches and within a single brick. Again, sensitivity studies are necessary to scope these uncertainties.

The graphite material properties depend on the reactor conditions and are not themselves independent from each other. The interactions between irradiation induced dimensional change, creep strain and CTE for example, are complex and work by BEGL to refine the material models is ongoing. Sensitivity studies and comparisons with measurement provide an input to this process of model refinement.

FUTURE DEVELOPMENTS

As noted in the previous section more detailed analyses could be undertaken for a specific brick for which material properties were known from trepanned samples. However, there are also other factors that could be investigated both to improve understanding of the sensitivity of the calculated brick stresses and deformations and to provide further validation of the analyses.

The three-dimensional analyses include brick shrinkage in the axial direction, but currently assume a fixed position for the fuel element interfaces relative to the bricks. Including the shifting axial dose distribution that follows as the column of bricks shrink relative to the fuel elements would be expected to add to the fine detail of the comparisons between prediction and measurement.

Most analyses currently carried out consider a "stand-alone" brick, *i.e.* interactions with neighbouring bricks are not included. However, it is known from previous analyses that the weight of bricks in the column above the brick under consideration can, in some circumstances, have a significant influence on the calculated ovality of the brick. Additionally interactions between the brick keyways and adjacent keys can influence the brick stresses later in life.

The analyses of bricks with either partial through-wall or full through-wall cracks have also been carried out previously but with conservative models and material properties. Repeating such analyses, predicting crack development over a fixed period of time and comparing that with crack development observed to have occurred between outages would also potentially provide either further validation or assist in the development and refinement of the material models.

CONCLUSIONS

The finite element analysis of graphite core bricks has advanced considerably over recent years to reach a position where large three-dimensional calculations can be routinely carried out. The type of analyses carried out has shifted from conservative, from which results are difficult to validate, to best estimate, from which results can be partially compared to measurements.

Uncertainties in the detailed reactor conditions experienced by the bricks, in the material property variations and in the interaction of the irradiated material properties require sensitivity studies to scope the uncertainties and provide data to help refine the material models.

Intended future developments hold the promise of improving the predictive ability of the analyses and increasing confidence in the calculated brick stresses. Finite element stress analyses can assist in understanding the current condition of the graphite cores and in determining the consequences of single or multiple-cracked bricks that may occur in ageing reactors.

Acknowledgements
The author would like to thank British Energy Generation Ltd for permission to present this paper. The views expressed in this paper are those of the author and do not necessarily represent those of the British Energy Generation Ltd.

Explicit Modelling of Graphite Core Components during a Reactor Thermal Transient

Colin D Elcoate, Gareth M Davis and John F B Payne[#]
Frazer-Nash Consultancy Limited, 1 Trinity Street, College Green, Bristol, UK.
Email: c.elcoate@fnc.co.uk and g.davis@fnc.co.uk
#Nexia Solutions Limited, Berkeley Centre, Berkeley, Gloucestershire, UK.
Email: john.b.payne@nexiasolutions.com

Abstract
An understanding of the behaviour of radially-keyed graphite cores and the internal stress and deformation of individual core components under normal operation and transient conditions is essential in assessing core geometry, component integrity, inspection strategy and consequently core safety. Stresses in graphite moderator bricks at reactor normal operating conditions are governed by irradiation induced dimensional change whilst during a reactor thermal transient and when the reactor is off-load additional thermal stresses are present. During a reactor thermal transient the gas coolant temperature can change very rapidly in conjunction with changes to the gas flow rate in the fuel channel, nuclear heat deposition in the graphite and thermal radiation from the fuel. All of these effects need to be considered to accurately predict the temporal and spatial distribution (both within a brick and from brick to brick) of stress in graphite moderator bricks during a thermal transient. This paper describes recent advances in the modelling approach developed to predict stresses in Pile Grade 'A' moderator bricks in Magnox gas cooled reactors during a typical reactor shutdown transient. The predictions from the approach described in this paper support the subsequent overall structural integrity calculations and ultimately the graphite core safety case.

Keywords
Shutdown transient, PGA graphite, FEAT

INTRODUCTION

An understanding of the behaviour of radially-keyed graphite cores and the internal stress and deformation of individual core components under normal operation and transient conditions is essential in assessing core geometry, component integrity, inspection strategy and consequently core safety. Current graphite core safety cases supporting continued operation of the four operating gas cooled Magnox reactors (Wylfa, Oldbury, Dungeness 'A' and Sizewell 'A') are based on four legs:

- Structural integrity.
- Monitoring.
- Inspection.
- Consequences.

The structural integrity leg adopts a deterministic methodology which includes the calculation of stresses in individual core components at various reactor operating conditions. The operating conditions for which explicit calculations are carried out include:

- Full power steady state normal operation.
- Shutdown transient.
- Steady state shutdown off-load.

Irradiation induced stresses, thermal stresses and stresses induced from loads acting on individual core components are considered in the assessment. The stresses are compared to the brick failure stress (or strength) to provide an indication of the likelihood of failure of any core component at any time during the core's operation.

Stresses in graphite moderator bricks at reactor normal operating conditions are governed by irradiation induced dimensional change whilst during a reactor thermal transient and when the reactor is off-load additional thermal stresses are present. During a reactor thermal transient, the gas coolant temperature can change very rapidly in conjunction with changes to the gas flow rate in the fuel channel, nuclear heat deposition in the graphite and thermal radiation from the fuel. All of these effects need to be considered to accurately predict the temporal and spatial distribution (both within a brick and from brick to brick) of stress in graphite moderator bricks during a thermal transient.

This paper describes recent advances in the modelling approach developed to predict stresses in Pile Grade 'A' (PGA) graphite moderator bricks in Magnox gas cooled reactors during a typical reactor shutdown transient. Before the methodology is described in detail, it is important to understand the governing factors behind the generation of thermal stresses in graphite bricks during and after a shutdown transient. These governing factors are described in the next section.

FACTORS GOVERNING THERMAL STRESS

During a reactor thermal transient the gas coolant temperature can change very rapidly inducing a thermal shock at the fuel channel wall. This is most noticeable at the top of the core where the coolant temperature can fall of the order 100 °C in less than 2 minutes. The thermal shock reduces with distance down the core. Towards the bottom of the core the internal stresses change gradually from the normal operation steady state value to the off-load steady state value. Towards the top of the core, the stresses peak early in the transient (<10 minutes). After the peak, the stresses reduce as the temperature gradients in the brick reduce and then in the later part of the transient the stresses change gradually to the steady state off-load value. The time scale of the full shutdown transient (of the order of days) relative to the initial sharp drop in bulk gas temperature (of the order of minutes) means that the two regimes (thermal shock and off-load steady state) are independent of each other.

The thermal strain at any point in the brick is given by $\alpha\Delta T$, where α is the coefficient of thermal expansion (CTE) and ΔT is the change in temperature from a reference temperature which is chosen as the normal operating temperature. During a shutdown transient and in the off-load steady state condition the elastic strains in the brick are changed by the thermal strains. The change in elastic strain at any given point in the brick from the normal operating value (pre shutdown) can be approximated by calculating the difference in thermal strain at the position of interest relative to a weighted average thermal strain in the brick. At a section where there is constant total strain across the brick (*i.e.* away from the brick ends) the difference in elastic strain change between any two points on that section is equal to the difference in thermal strains at the same points. Hence, at a section with constant total strain the difference in the thermal strain between the point of interest (the fuel channel wall for example) and a weighted average in the brick governs the thermal stresses.

During a shutdown transient it is the product of α and the difference in ΔT between the fuel channel wall and a weighted average that governs the thermal stresses (to a first approximation). This understanding allows the effect of changes in temperature on brick thermal stresses during a reactor shutdown to be assessed as follows:

- If there is a change in the difference in ΔT between the point of interest and the weighted mean by a factor *f*, the difference in elastic strains between the point of interest and the weighted mean, and hence the thermal stress, changes by the factor *f*.
- To a first approximation, if all the coefficients of thermal expansion, α, increase by a constant *c*, the difference in elastic strains between the point of interest and the weighted mean changes by the product of *c* and the difference in ΔT between the point of interest and a weighted average. Hence the thermal stress increases by the product of Young's modulus, the constant *c* and the difference in ΔT between the point of interest and a weighted average.

After a shutdown, when the reactor is off-load, it is the product of ΔT and the difference in α between any given point and a weighted average, that governs the thermal stresses (to a first approximation) at the given point. This understanding allows the effect of changes in temperature on brick thermal stresses after a reactor shutdown to be assessed as follows:

- If there is a change in the difference in ΔT between the point of interest and the weighted mean by a factor *f*, the difference in elastic strains between the point of interest and the weighted mean, and hence the thermal stress, changes by the factor *f*.
- To a first approximation, if all the coefficients of thermal expansion, α, increase by a constant *c*, the difference in α between any two points does not change hence the thermal stresses do not change.

With an appreciation of the factors that govern thermal stress in a graphite brick during and after a reactor shutdown transient, it is timely to consider the most appropriate method to predict them. Previous and current modelling approaches are described in the next section.

MODELLING APPROACH

Prior to the safety case developed in support of the Wylfa Periodic Safety Review (PSR) in 2004, a simplified and conservative approach was used to predict the thermal stresses during a reactor shutdown transient. The approach used steady state two dimensional finite element models. The differential temperature between the bulk of the brick and the fuel channel wall during the transient was simulated by reducing the temperature of the fuel channel wall elements and carrying out a steady state calculation. The differential temperature in the brick was typically set equal to the change in coolant temperature which it has been shown can overestimate the thermal stresses by a factor of roughly 2 at the top of the core.

To provide more realistic, and less conservative, predictions of peak stress during a shutdown transient an alternative modelling approach has been developed. The current methodology utilizes station data and a number of computational codes to provide a full spatial and temporal distribution of stress during and after a shutdown transient. The high level methodology can be summarized as follows:

- Station operating data are extracted before, during and after a typical station shutdown transient. These data include bulk gas inlet and outlet temperatures, gas pressures, gas circulator speeds and reactor thermal power.
- Calculations are carried out using the whole plant thermal hydraulic modelling code MACE (Cooper 1997). MACE takes relevant station data and calculates the temporal variation of channel inlet flow and temperature for the channel of interest.
- The temporal variation of channel inlet flow and channel inlet temperature form part of the inputs to subsequent PANTHER calculations. The neutron diffusion and thermal hydraulic code PANTHER (2003) is used to predict the temporal variation of gas temperature and effective heat transfer coefficient at the fuel channel wall throughout a single channel. The PANTHER model includes the effects of convection from the surfaces of the brick, conduction within the brick and thermal radiation from the fuel element in the channel.
- Predictions of the spatial and temporal variation of temperature in a brick is carried out using two dimensional FEAT (2005) finite element calculations. Thermal properties are derived from the full power normal operating condition of the brick (radiolytic oxidation, fast neutron fluence, irradiation temperature). The initial brick temperature distribution is taken from full power normal operating condition FEAT calculations. The thermal boundary conditions are derived from the previous PANTHER predictions.
- Predictions of the spatial and temporal variation of thermal stress in a brick are carried out using three dimensional FEAT (Elcoate, 2001) finite element calculations. Mechanical properties are derived from the full power normal operating condition of the brick (radiolytic oxidation, fast neutron fluence, irradiation temperature). The initial stress state of the brick is taken from full power normal operating condition FEAT calculations.
- Partial validation of the finite element predictions is possible by comparing temperatures at discrete locations against data from thermocouples embedded in graphite moderator bricks. Validation of the approach is discussed later in the paper.

The complete predictive route presents a revised and improved methodology in support of graphite core safety cases. However, the paper concentrates on the finite element predictions of temperature and stress which present the largest single improvement over previous calculations. Temperature predictions are discussed in the next section.

TEMPERATURE PREDICTIONS

Before considering the temperature predictions through a reactor thermal transient (*e.g.* shutdown) it is worth summarizing the approach to predicting brick temperatures at full power normal operation. Two dimensional, steady state, finite element calculations are carried out at discrete times through the reactors operating history (approximately every 5 full power years). The thermal material properties are taken from the PGA graphite materials database and where appropriate include a temporal and spatial variation dependent on radiolytic oxidation, fast neutron fluence and irradiation temperature. The thermal models include nuclear heating, conduction, thermal radiation, convection from each of the brick surfaces and convective heat transfer from porous flow through the brick. The inclusion of convective heat transfer from porous flow through the brick is only important when radiolytic oxidation (weight loss) is significant. Generally, this approach is adopted when considering

Oldbury moderator bricks (~30-40% weight loss in a peak rated position at the current mean core irradiation), but not the other operating Magnox reactors that exhibit much less radiolytic oxidation (~20-25% weight loss in a peak rated position at the end of generation). A typical through wall temperature distribution is shown in Figure 1. Note that the figure also contains thermocouple data which are used to validate the predictions. This aspect of the methodology is discussed later in the paper.

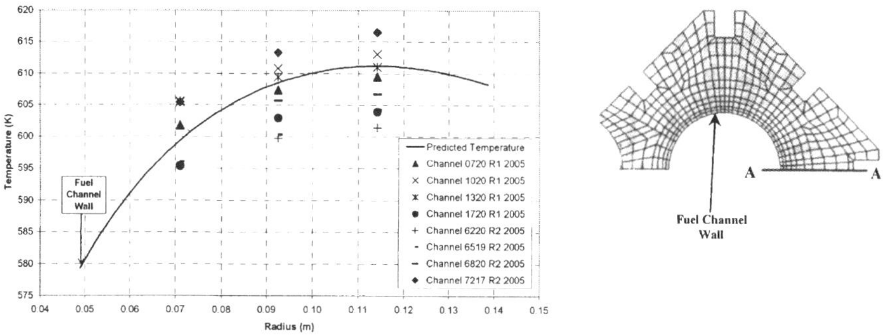

FIGURE 1: Predicted variation of temperature along a radius (A-A) of an Oldbury octagonal moderator brick at core mid-height in the flattened region (showing thermocouple data).

Two dimensional, transient, finite element calculations are carried out to predict temperatures throughout and after a reactor thermal transient (*e.g.* a reactor shutdown). The thermal material properties are taken from the PGA graphite materials database and where appropriate include a temporal and spatial variation dependent on radiolytic oxidation, fast neutron fluence and irradiation temperature. The thermal models include conduction, convection from each of the brick surfaces and thermal radiation from the fuel element. Nuclear heating (which decreases rapidly during a reactor shutdown transient) and convective heat transfer from porous flow through the brick are not included in the transient models as the effects are judged to be of second order when considering the driving force behind the transient thermal stresses, namely, the difference in temperature between the fuel channel wall and the bulk of the brick as discussed earlier in the paper. The thermal boundary conditions are taken directly from the preceding PANTHER analyses as outlined in the previous section. The boundary conditions are generally taken from a channel with an above average gas outlet temperature while the final steady state off-load temperature of the bricks is generally taken to be 20°C. Both of these modelling assumptions make for conservative calculations of thermal stress. As discussed the governing factor behind the generation of thermal stresses during a shutdown transient is the difference between the temperature at the fuel channel wall and the bulk of the brick. This difference is typically ~50 °C at the top of the core and occurs less than 10 minutes into the transient. As discussed earlier this difference is approximately half of that assumed by previous methods and is more representative of actual reactor conditions. Having described the approach to predicting temperatures in graphite bricks at full power normal operation and during a reactor thermal transient, the next section describes how these temperatures are fed into the subsequent stress predictions.

STRESS PREDICTIONS

Before considering the stress predictions through a reactor thermal transient (*e.g.* shutdown) it is worth summarizing the approach to predicting brick stresses at full power normal operation. Three dimensional, transient, finite element calculations are carried out over the reactor's operating history. Boundary conditions are applied to simulate an unconstrained brick and to prevent rigid body motion. The mechanical material properties are taken from the PGA graphite materials database and where appropriate include a temporal and spatial variation dependent on radiolytic oxidation, fast neutron fluence and irradiation temperature (predicted as described in the previous section). To a first approximation, the total strain rate of a brick is equal to its average rate of dimensional change and its average elastic strain is zero. However, fast neutron dose and temperature vary across the brick, so the dimensional change rate at any given point generally differs from the average. Thus, the elastic strain at any given point is generally not zero, and its magnitude depends on the deviation of the dimensional change rate at the point from the average and on the deviation of the total strain rate at the point from the average. The rate of dimensional change of PGA graphite is approximately constant for fast neutron fluences from $\sim 20\times10^{20}$ n cm^{-2} to $\sim 50\times10^{20}$ n cm^{-2}. Near the middle of the brick, the total strain rate is approximately constant across a section (this approximation is very good for axial strain, but not so good for hoop strain). Consequently, in this range of fast neutron doses, and at any given point near the middle of the brick, a steady elastic strain is established which produces a constant rate of creep strain that balances the deviation of the dimensional change rate at the point from a weighted average dimensional change rate.

Three dimensional, transient, finite element calculations are carried out to predict stresses throughout and after a reactor thermal transient (*e.g.* a reactor shutdown). Boundary conditions are applied to simulate an unconstrained brick and to prevent rigid body motion. The mechanical material properties are taken from the PGA graphite materials database and where appropriate include a temporal and spatial variation dependent on radiolytic oxidation, fast neutron fluence and irradiation temperature. The mechanical properties are assumed constant during and after the thermal transient. The only loads on the finite element models during the shutdown transient are the thermal loads defined by the temperature predictions discussed in the previous section. Peak tensile and compressive stresses are extracted from the finite element models to support subsequent structural integrity calculations. Towards the top of the core the stresses peak in less than 10 minutes after which they change gradually to the off-load stress state.

In the final structural integrity assessment the likelihood of failure of any core component is calculated by considering all potential stress sources. These include normal operation and thermal stresses (as discussed above) plus stresses induced from keyway loads (as a result of gas pressure differentials for example) and stresses induced from the weight of the column of bricks above. The combination of stresses takes cognizance of the spatial and temporal variance of each stress source and is not discussed further in this paper.

The preceding sections of this paper discuss the factors that govern thermal stresses during a reactor thermal transient and the current methodology of predicting them using a series of computational codes. The next section gives a brief overview of the level of validation that supports the predictions.

VALIDATION

Several computational codes are used in the current methodology for predicting thermal stresses during a reactor transient. MACE and PANTHER (Cooper, 1997 and PANTHER 2003) are standard UK tools, the former for whole plant thermal hydraulic modelling (primary and secondary circuit) and the latter for whole reactor neutron diffusion and thermal hydraulic modelling (primary circuit). Both have a substantial body of validation evidence supporting their application in Magnox reactors. FEAT (2005) is a powerful finite element tool for simulating a variety of physical problems, incorporating non-linear stress analysis including plasticity and graphite creep (Elcoate, 2001). The code has been developed within the nuclear industry and is particularly well suited to dealing with some of the industries more challenging problems such as graphite analysis. The entire thermal and stress prediction route has been verified by comparison with analytical solutions and comparison with other independent finite element codes.

Station data and material property measurements support the predictive route at all levels. The data ranges from bulk gas inlet and outlet temperatures, gas pressures and gas circulator speeds that support the MACE calculations through to measurements of coefficient of thermal expansion and Young's modulus from trepanned samples that support the definition of material properties for the FEAT stress calculations. Validation of the final temperature predictions from FEAT is achieved by comparison with data from thermocouples embedded in the graphite moderator bricks. Typically, three thermocouples are located towards the top of certain bricks on three different radii. This is shown diagrammatically for an Oldbury octagonal brick in Figure 2. There is considerable scatter in the data from the embedded thermocouples. The scatter is due to three effects, channel to channel variation, brick to brick variation and measurement uncertainty. The comparison between the predictions and the measurements is judged to be reasonable as demonstrated in Figure 1. If anything, the predictions are generally slightly conservative (in terms of predicting tensile stresses at the fuel channel wall) due to slightly steeper temperature gradient at the fuel channel wall than the data suggests.

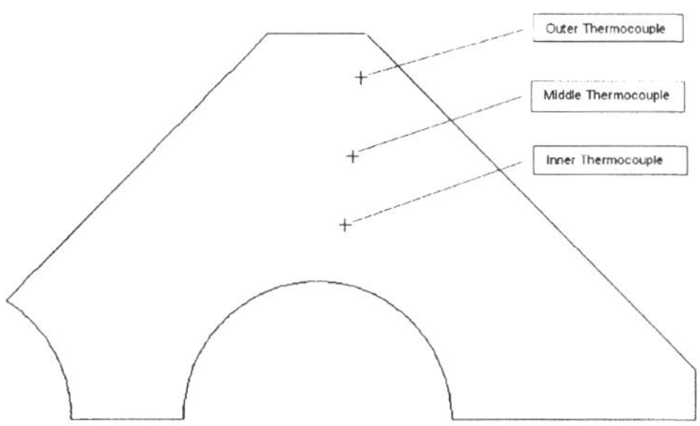

FIGURE 2: Position of embedded thermocouples in an Oldbury octagonal moderator brick.

At this time, there is no direct validation of brick stresses available although the UK nuclear industry is continually reviewing the possibility of taking existing methods to measure residual stresses and developing them to take in core measurements when the reactor is off-load. However, it must be recognised that making measurements in a reactor presents a significant challenge given the environment and limited core access. In the absence of direct validation of brick stresses sensitivity studies are carried out on key parameters such as the coefficient of thermal expansion, Young's modulus and reactor operating conditions. These sensitivity studies, in isolation and in combination, are used to support safety case assessments.

CONCLUDING COMMENTS

An understanding of the behaviour of radially-keyed graphite cores and the internal stress and deformation of individual core components under normal operation and transient conditions is essential in assessing core geometry, component integrity, inspection strategy and consequently core safety. Stresses in graphite moderator bricks at reactor normal operating conditions are governed by irradiation induced dimensional change whilst during a reactor thermal transient and when the reactor is off-load additional thermal stresses are present. This paper describes recent advances in the modelling approach developed to predict stresses in Pile Grade 'A' moderator bricks in Magnox gas cooled reactors during a typical reactor shutdown transient. The main conclusions can be summarized thus:

- An improved methodology has been developed to predict the spatial and temporal distribution of temperature and stress in individual graphite core components during a reactor thermal transient.
- The methodology links available station data and a number of well established computational codes to give realistic predictions during one of the most onerous load cases the core experiences under normal operating conditions.
- The methodology is underpinned by validation of the predictions throughout the majority of the calculation route. However, direct validation of brick stresses is not possible at the time of publication so safety case assessments are supported by a range of sensitivity studies.

Acknowledgements
The authors would like to thank British Nuclear Group who gave their permission to publish this paper and funded all development activities as part of their graphite safety case programme. The views expressed in this paper are those of the authors and do not necessarily represent those of the British Nuclear Group.

References

Cooper, S. and Hart, A. C. D., (1997), MACE Program for Modelling Thermal-hydraulic Transients in Nuclear Systems. *Nuclear Electric Limited,* EPD/GEN/REP/0181/97, Issue 1.

Elcoate, C. D. and Payne, J. F. B., (2001), Pile Grade A Graphite – Constitutive Modelling, *16th International Conference on Structural Mechanics in Reactor Technology.*

FEAT (2005), Finite Element Analysis Toolbox User Guide, *FeatPlus Limited,* Version 3.7.0.

PANTHER (2003), AGR and PWR Reactors PANTHER User Guide: Volume 1: General Description Section., *British Energy Generation Limited,* REP/BBDB/0015/GEN/03, ED/PANTHER/UG/5.5, Release 5.5 Revision 000.

PART D

The Consequences of Graphite Deterioration: Whole-Core Behaviour

PBMR Graphite Core Structures
Fitness for Purpose Management Strategy

Mark Davies, Mark Mitchell and Pieter Venter
Pebble Bed Modular Reactor (Pty) Ltd; P.O.Box 9396, Centurion 0046, Rep. of South Africa
Email: mark.davies@pbmr.co.za

Abstract
The Pebble Bed Modular Reactor (PBMR) is a high temperature, helium-cooled, graphite moderated, direct cycle plant generating 168 MW_e with a design lifetime of 40 years. The reactor core consists of an annular pebble bed (450,000 fuel spheres) that is formed and supported by graphite reflectors which in turn are supported by restraint straps and tie rods and the core barrel. Reactivity control is achieved using control rods that move in channels in the side reflector and small absorber spheres introduced into channels in the central reflector. PBMR is well aware of the challenges facing graphite moderated reactors and has implemented an integrated management strategy at the design phase to address the long term fitness for purpose issue. This paper will present the specific challenges facing PBMR and the process PBMR has embarked on to demonstrate fitness for purpose throughout the planned lifetime, including: life assessment and qualification during the design phase; design of a suitable Surveillance, Test and Inspection (STI) programme; and, design of the maintenance of the CSC including replacement.

Keywords
PBMR, NBG-18 Graphite, life assessment, STI, maintenance

INTRODUCTION

The Pebble Bed Modular Reactor (PBMR) is a high temperature, helium-cooled, graphite moderated, direct cycle plant generating 168 MW_e with a design lifetime of 40 years. PBMR is a development of the high-temperature gas-cooled reactor technology originally developed in Germany. The design utilises all ceramic, spherical fuel elements containing TRISO-coated fuel particles with Low Enriched Uranium (LEU) dioxide. The basic safety goal of PBMR is the retention of fission products within the TRISO coated particles under all reactor conditions, normal operation, anticipated operational occurrences and accidents. PBMR is able to meet these requirements in a natural and passive way by relying only on gravity, conduction, convection and thermal radiation. The graphite core structures play an important role in ensuring the passive safety of the PBMR.

PBMR is well aware of the fitness for purpose issues facing graphite moderated reactors subject to high fluence and temperature. To address these issues PBMR has implemented an integrated management strategy at the outset of the design. This paper will present the challenges facing PBMR and the process PBMR has embarked upon to demonstrate fitness for purpose throughout the lifetime of the plant.

REACTOR DESIGN

The Main Power System of a PBMR module consists of a graphite-moderated, helium-cooled Reactor Unit (RU) in which the working fluid is heated by nuclear fission, and a direct, Brayton cycle Power Conversion Unit (PCU) in which the heat within the working fluid is converted into electrical energy by means of a turbine-driven generator (Venter *et al.*, 2005). The RU consists of the fuel, the Core Structures (CS), the Reactivity Control System (RCS) and the Reserve Shutdown System (RSS), all contained inside the Reactor Pressure Vessel (RPV). The CS is further subdivided into the Core Structure Ceramics (CSC) and Core Barrel Assembly (CBA) as shown in Figure 1 which also shows a horizontal cross section of the reactor unit. The CSC has four sub-assemblies, the Bottom Reflector (BR), the Side Reflector (SR), Top Reflector (TR) and the Central Reflector (CR). These reflectors are manufactured from blocks that are arranged and interconnected according to design principles developed for the German HTR programmes (HRB, 1985).

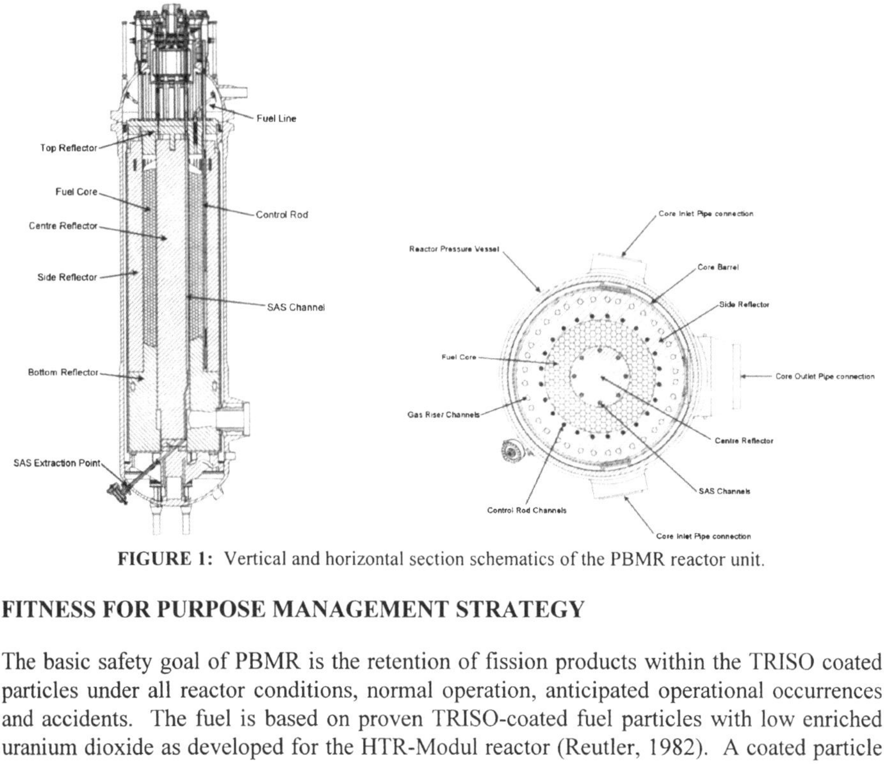

FIGURE 1: Vertical and horizontal section schematics of the PBMR reactor unit.

FITNESS FOR PURPOSE MANAGEMENT STRATEGY

The basic safety goal of PBMR is the retention of fission products within the TRISO coated particles under all reactor conditions, normal operation, anticipated operational occurrences and accidents. The fuel is based on proven TRISO-coated fuel particles with low enriched uranium dioxide as developed for the HTR-Modul reactor (Reutler, 1982). A coated particle consists of a kernel of uranium dioxide surrounded by four coatings. The first layer deposited on the UO_2 kernel is porous carbon followed by a layer of dense pyrocarbon, a layer of silicon carbide and a final layer of pyrocarbon. The outside diameter of the finished coated particle is approximately 1 mm. The porous carbon layer is designed to collect the fission products without over pressurising the coated particle and to accommodate any mechanical deformation of the UO_2 kernel. The silicon carbide and pyrolytic carbon layers are designed to provide a barrier that will contain the fuel and fission products up to very high temperatures, approximately 1600 °C. Approximately 15000 coated particles are mixed with graphite powder and a phenolic resin and pressed into a 50 mm diameter ball. A 5 mm thick layer of carbon is added to form an outer zone without fuel and the resulting spheres are sintered and annealed to make them hard and durable. The features of the reactor are low power density (~5 MW/m^3), high thermal inertia, strong negative temperature coefficient and low excess reactivity. These characteristics are combined with a high surface area to volume ratio to ensure that decay heat can be removed from the core under all circumstances without threatening the integrity of the fuel.

The CSC plays an important role in ensuring the passive safety of the PBMR in that it will:

- Control of heat generation via core geometry and access for reactivity control system.
- Maintain decay heat removal capacity via core geometry and material properties.

As a result there are several unique design challenges that face the CSC and some are illustrated below:

- The pebble bed could generate significant lateral loading during temperature transients.
- The CSC must assure that the shape of the pebble bed cavity is such that the residence time for fuel remains within defined limits.
- Differential thermal expansion of the pebble bed and the CSC/CBA.
- The graphite will be subject to significant irradiation exposure.
- Large number of parts to be designed.

An integrated system engineering approach illustrated in Figure 2 was adopted to manage the challenges posed by the demonstration of continued fitness for purpose of the CSC throughout the design life. This integrated approach includes functional analysis, hardware breakdown, material selection and qualification, lifetime assessment, examination (surveillance, test and inspection) and maintenance. These are discussed in the following sections. The entire system engineering structure for the RU, including specifications, requirements, plans *etc* is maintained in a relational database called DOORS (www.telelogic.com). The CS management strategy is a subset of this database that ensures full compliance tractability.

CSC Functional Requirements

The CSC functional requirements are derived from the passive safety features via the RU functional requirements and have been defined as:

- Form and support the annular pebble bed and transmit mechanical loads due to dead weight, lateral loading due to the pebble bed, seismic loadings and pressure differentials to the CBA.
- Provide a passive heat transfer path between the fuel and the core barrel.
- Allow the insertion of the neutron absorbers from the Reactivity Control System (RCS) and the Reactor Shutdown System (RSS) into the core by providing channels in the Side Reflector (SR) and Central Reflector (CR), respectively.
- Provide neutron reflection while protecting the metallic components from exposure to high neutron fluence levels with the BR, SR and TR.
- Permit the circulation of fuel through the core by providing sphere inlets in the TR, the geometry of the top surface of the BR and the provision of three de-fuel chutes.

Hardware Breakdown

The CSC consists of assemblies with sub-assemblies and parts. Materials are assigned to each part. The complete geometry of the PBMR module is maintained in a 3D CAD model. This defines the geometry for all subsequent assessments, *e.g.*, neutronic, thermal hydraulic, structural. There are approximately 400 unique parts within the CSC that must be designed. A number of these parts have been selected as master parts that envelope a slave set of parts in terms of geometric features within a sub-assembly. A range of loading is applied to the master part. Design reports are produced for the master part, which summarise all of the assessments that have been completed to demonstrate fitness for purpose.

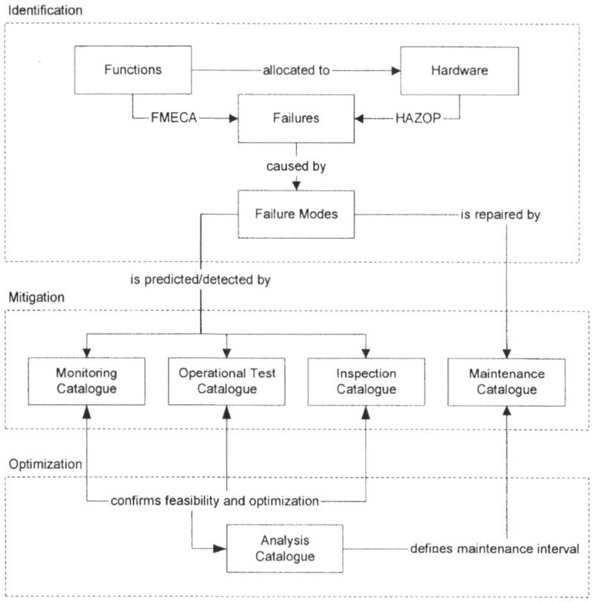

FIGURE 2: Graphite core structures integrated system engineering approach.

Material Selection and Qualification

The functionality of the CSC must be maintained in a harsh irradiation environment for the full 40-year design life of the PBMR (the peak fluence is 280 x 10^{20} EDN at 800 °C after 40 years of operation). The fluence-temperature locus of the core facing surfaces of the inner side reflector and outer central reflector are shown in Figure 3. The range of data available from irradiation programmes on similar graphites is also shown in Figure 3. The majority of this data is available on ATR-2E, an extruded pitch-coke nuclear graphite, that was developed as a reference material for High-Temperature Reactors (HTR) in Germany. The SGL Carbon Group, the original developer and manufacturer of ATR-2E has been chosen as the supplier of the PBMR reflector graphite. The SGL grade NBG-18 graphite has been chosen for the CSC. This is a vibration moulded, medium-grain size graphite produced from NSCC coke. NSCC coke is also used by SGL for the extrusion of graphite fuel sleeves for the AGR programme. A significant amount of work has already been done to qualify this material and the manufacturer has initiated an Integrated Manufacturing and Quality Control Plan to ensure compliance with the PBMR material specification. ATR-2E forms the basis for the PBMR graphite database. The currently available range of medium grain, nuclear graphites are expected to have similar, but not identical irradiation induced behaviour to that of ATR-

2E. The ATR-2E irradiation campaigns were carried out at Petten and the experimental data was supplied to PBMR by FZJ (G Haag, 1999). The ATR-2E data has been supplemented with another, similar graphite, VQMB, which was irradiated at Petten and DFR. AEA Technology supplied the experimental data (BJ Marsden, 2002). VQMB data generated in the Dounreay Fast Reactor (DFR) was not incorporated into the PBMR graphite database due to uncertainty regarding the irradiation temperatures. PBMR has also not adopted the Equivalent Dido Temperature correction in the development of the material database.

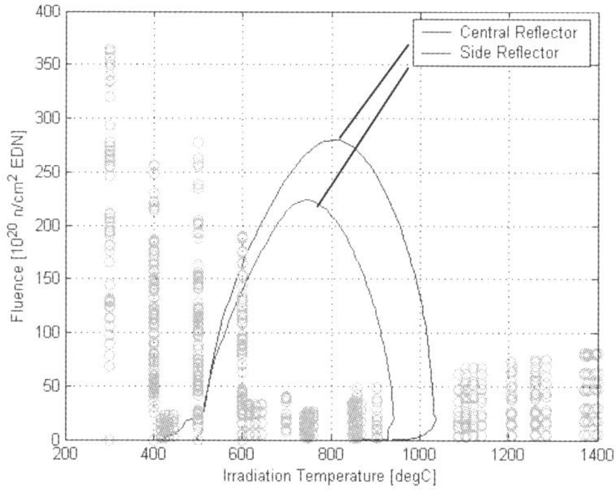

FIGURE 3: Fluence-temperature requirements for PBMR and available irradiation data.

The PBMR graphite database covers irradiation-induced changes for the following properties: dimensional change; Young's modulus; Coefficient of Thermal Expansion (CTE); thermal conductivity; and irradiation induced creep. The PBMR graphite database was originally developed by AEA Technology, modified by PBMR, and peer reviewed by FZJ and Westinghouse. Whilst there is experimental data available across the complete PBMR temperature range, there is a lack of information at high fluence. PBMR will carry out a specific MTR programme using production material up to lifetime fluences at temperatures between 600 and 1000 °C.

Life Assessment

The pebble bed facing graphite blocks are subject to high flux and temperatures as well as significant gradients in flux and temperature. Due to the irradiation induced material response, the blocks will be subjected to significant internal stresses. The blocks will also be subjected to external loading because of the PBMR design and the nature of the pebble bed, *e.g.* differential fluid pressure. For the purposes of design, the point of crack initiation within a block is defined as the end of the useful life of that block. Failure evaluation of parts is completed in accordance with the methods developed in support of the German HTR programme and presented in the draft KTA rule for the design of the ceramic internals of a high temperature reactor (KTA, 1992 390 /id). The graphite reflector part with the minimum lifetime will, for the purposes of design, determine the service life of the core structures and the time at which the components should be replaced. At a later stage, PBMR may carry out damage tolerance assessments to assess the effect of multiple block failure on safety functionality.

PBMR graphite lifetime assessments are carried out as best estimate with perturbations. One such assessment on a side reflector block (Mitchell, 2005) included the variability of the following parameters:

- Variation in the geometry of the part, representing design changes to the CSC that may be feasible should the study indicate suitable benefit.
- Variation in the graphite material properties, to account for the natural variability in the graphite material.
- Variation in the irradiation behaviour of the material to account for uncertainties in this behaviour.
- Variation in the loading of the parts, specifically in terms of the fast neutron flux and temperature fields.

Only internal stresses were considered assuming either uniform or normal distributions of design parameters. A fully automated analysis route was developed to minimize the manpower required. For the side reflector blocks the predicted lifetime of a part is defined as at the point where the probability of crack initiation reaches 10^{-2}, as prescribed in the KTA rule. The evolution of the probability of crack initiation, as a function of the operating time, for the best estimate calculation is shown in Figure 4 and the life of the part is estimated as approximately 40 Full Power Years (FPY). The graphite block lifetime analyses will be revisited once the PBMR MTR programme has been completed.

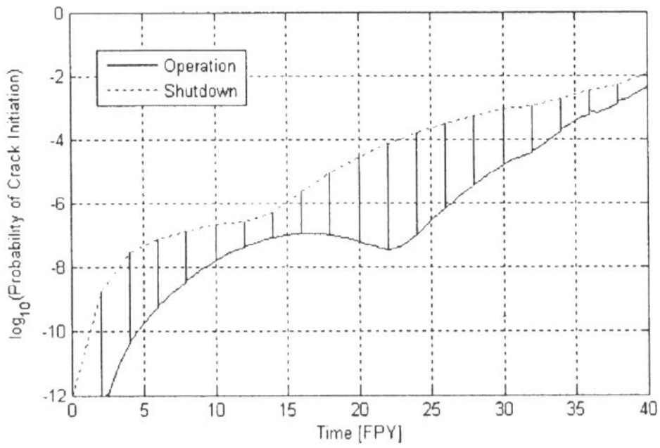

FIGURE 4: Evolution of the probability of crack initiation with operating life.

Examination (Surveillance, Testing and Inspection)

The objective of examination is the demonstration of continued fitness for purpose of the CSC with regard to both nuclear safety and economic risk. To achieve this, the following will need to be defined: 1) What to examine? 2) How to examine? 3) When to examine?

As part of a project to update their Safety Assessment Principles, the Nuclear Installation Inspectorate has recently published ten principles specifically applicable to address Maintenance, Inspection and Testing requirements. The PBMR examination strategy has been developed with the objective of compliance with these principles. The PBMR examination approach will investigate the following possibilities:

- Direct measurement of an individual parameter at a specific location.
- Indirect evaluation of an individual parameter at a specific location via direct measurement of the parameter at a remote location.

- Indirect evaluation of an individual parameter at a specific location via indirect measurement of another parameter.

The PBMR examination approach includes surveillance, testing and inspection and each may be carried out on-line or off-line as the case may be to predict or detect deterioration in CSC functionality. Surveillance is the monitoring of plant parameters, testing is specific operational procedures (distinct from commissioning tests) to obtain plant parameters and inspection is the direct measurement of parameters. To produce a robust and auditable examination plan PBMR has adopted a three-stage process: i) identify all risks; ii) identify all mitigation strategies; and iii) optimise mitigation strategies.

Figure 2 shows the risk identification process. An Failure Modes, Effects and Criticality Analysis (FMECA) is carried out to determine the nuclear safety or economic risk based on qualitative risk ranking criteria and to identify failure modes where more detailed quantitative assessment or design change may be required. A structural hazard and operability study (HAZOP) is also carried out at various levels of hardware breakdown, sub-assembly, master part, *etc* to identify structural nuclear safety or economic risk based on qualitative risk ranking criteria and to define areas where more detailed structural assessment or design changes are required. These are combined with a review of international experience to compile an initial risk of failure catalogue, which contains the failure effect and the evidence that may manifest itself within the reactor. At this stage the catalogue is subjected to an internal PBMR peer review and baselined.

Potential mitigation strategies are then identified for each initial risk. PBMR mitigation strategies include analysis and examination (surveillance, testing and inspection). The analyses could include quantitative probability of failure assessment, consequence assessment and quantification of the failure effect on reactor operating parameters. Potential examinations must either demonstrate that failure will not happen prior to next examination or detect that the failure has already happened or will happen prior to the next examination. For each initial risk analysis, surveillance, operational testing and inspection catalogues are identified. At this stage the examination catalogue is subjected to an internal PBMR peer review and baselined. An extract from the examination catalogue is shown in Figure 5. A hierarchy of examination options has been implemented by PBMR as surveillance > testing > inspection in descending order of preference. The various examination options for risk mitigation are then optimised. Each option is assessed for versatility, feasibility and cost. The number of risks to which the mitigation strategy can be applied is the versatility. The cost is defined as the total incremental cost including development, demonstration, capital expenditure and O&M. A number of CSC relevant reactor parameters will already be captured by the MPS C&I system. The cost is then defined as the additional cost required over and above the MPS C&I system weighted by the versatility (no of applicable risks). At this stage the examination strategies are optimised for versatility and feasibility. For each risk a number of prediction and detection options are required dependent upon the perceived severity of the risk.

Maintenance

The CSC design is intended to be maintenance free - there are no moving parts, ignoring differential thermal expansion and fuel sphere movement. However, given the significant uncertainty in graphite irradiation-induced material behaviour, a maintenance catalogue is also developed as the final risk mitigation strategy. With a few exceptions, the identified maintenance options tend to be replacement and PBMR has developed a CSC replacement concept. All the pebble bed facing components can be replaced remotely. The concept consists of a manipulator that removes the blocks one by one and loads them onto a skip. The skip is moved from the reactor cavity to a containment location above the RU where the blocks are transferred to transport casks. Two independent studies commissioned by PBMR confirmed the feasibility of the reflector replacement concept concluding that the replacement concept is within the capabilities of present day technology and can be completed within a timescale that will be acceptable with respect to the required overall availability of the plant.

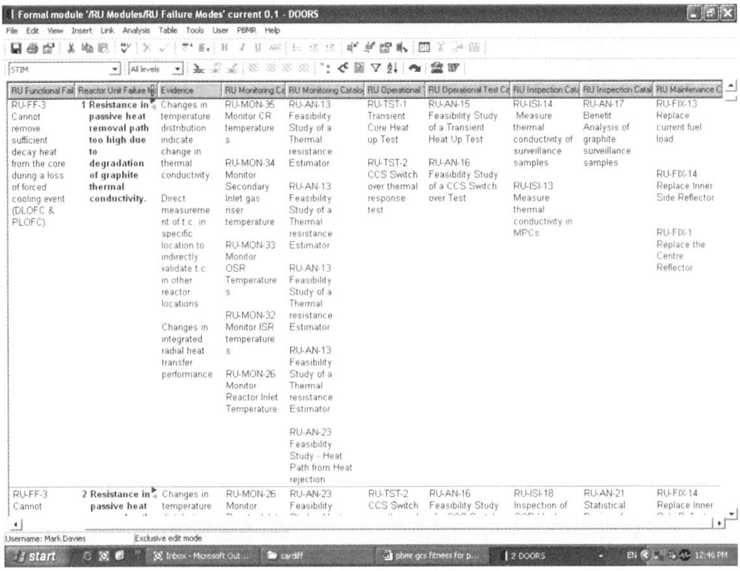

FIGURE 5: PBMR examination catalogue.

CONCLUSION

PBMR faces some serious challenges to demonstrate the continued fitness for purpose of the CSC for the complete lifetime of the plant. PBMR has adopted an integrated system engineering approach at the design stage to mitigate any risks posed by these challenges.

References

Bäumer, R., et al., 1990, AVR - Experimental high-temperature reactor, 21 years of successful operation for a future energy technology, Association of German Engineers (VDI), ISBN 3-18-401015-5.

Haag G (1999), private communication.

HRB, 1985, HKG 300 MWe Nuclear Power Plant Hamm-Uentrop with Thorium High-Temperature Reactor, Description of Site and Power Plant with Reference to Safety-Related Aspects, Druckschrift nr. D HRB 1429 84 E.

KTA, (1992), "KTA-3232 Keramische Einbauten in HTR-Reacktordruckbehälten, Sicherheitstechnische Regel des KTA", KTA-3232.

Marsden BJ, Wignall C (2002), Graphite Design Database – AEAT 6289 Issue 2.

Mitchell MN (2005), Identification of key parameters defining the life of graphite core components, SMIRT18-C05-1

Reutler, H. and Lohnert, G.H., 1982, Der modulare HTR - Ein neues Konzept für den Kugelhaufenreaktor, Atomwirtschaft, pp. 18-21.

Venter PJ, Mitchell MN, Fortier, F (2005), PBMR reactor design and development, SMIRT18-S02-2.

Seismic Modelling of an AGR Nuclear Reactor Core

B Kralj, S J Humphreys and B G J Duncan
Atkins Nuclear, 240 Aztec West, Almondsbury, Bristol, UK.
Email: berislav.kralj@atikinsglobal.com

Abstract
The work concerns the seismic response of an Advanced Gas-cooled Reactor (AGR) core late in its life. An LS-DYNA finite element model has been developed that represents each graphite brick. The contact between the bricks and their keys has been modelled with non-linear coaxial spring/dampers allowing for limited free motion. The model is a three dimensional representation of all the core layers and a neutron shield, making use of a symmetry by only modelling half of each layer. Different layers and groups of bricks have been allocated different spring/damper properties in order to accommodate differing initial geometry and levels of irradiation flux exposure. The core's restraint system has also been modelled. Bespoke programs have been developed in Visual Basic for Applications to automatically generate the key file used by LS-DNYA and deal with the post-processing of results files. Sensitivity studies were done to assess the effects of changing the magnitude of the event and the effects of irradiation on the bricks.

Keywords
AGR Core, 3D FE model, seismic analysis

INTRODUCTION

With the risk from earthquakes being low in Britain, the design of earlier AGR (Advanced Gas-cooled Reactor) nuclear power stations, designed during the late 1960's, did not consider the consequences of seismic loads. More recently, however, it was accepted that the stations could (even though with a very low probability) experience a seismic event of significant severity. Consequently, it became a requirement to demonstrate that these earlier nuclear power stations, of which Hinkley Point B and Hunterston B were the first to start operating, can be safely shut down in the case of a seismic event. In order to satisfy this requirement, British Energy asked Atkins to develop a 3-dimensional computer model capable of predicting the seismic responses of typical AGR reactor cores and of assessing their behaviour during an earthquake event, with particular emphasis on evaluating:

- Loads on individual keys.
- Loads on the elements of the restraint structure.
- Patterns of peak brick displacement through each layer.
- Double crack separation/shear.
- Fuel stringer squeezing force.
- Retention of key engagement.
- Control rod insertion path alignment during and following the event.

CORE MODEL

Core description

The AGR core comprises of stacks of graphite bricks providing vertical channels for fuel assemblies, control rods and coolant flow. The core is supported vertically on a diagrid plate, supported on a number of pin jointed columns. Lateral stability of the core is provided by support structure consisting of restraint rods, restraint beams, centralising brackets and Warwick links tied into the restraint ring, which is attached to the boiler shield wall, Figure 1. The stacked bricks are interconnected on plan by a system of keys and keyways allowing the core to expand and contract thermally whilst retaining its cylindrical shape, Figure 2.

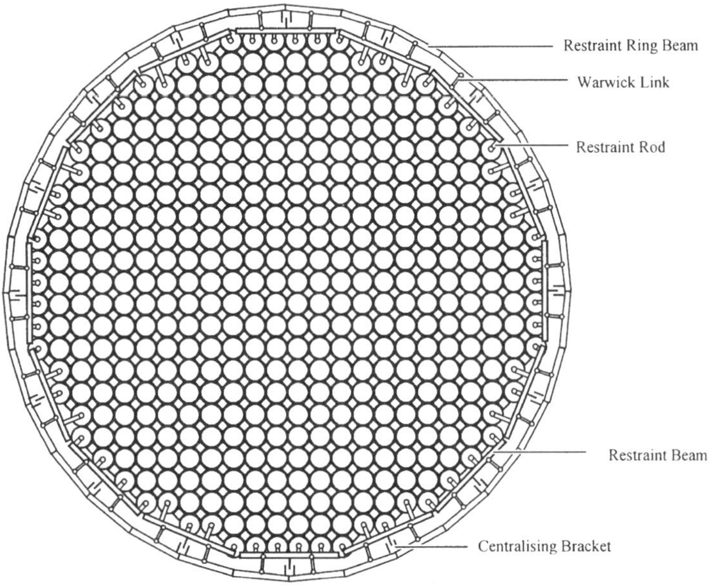

FIGURE 1: Schematic representation of a graphite core layer.

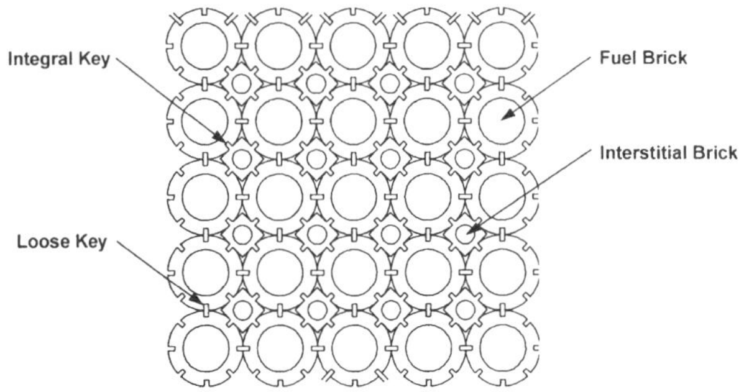

FIGURE 2: Schematic representation of the keying system on plan.

The graphite bricks are also keyed together vertically by their end features, while offsets in the arrangement of fuel and interstitial bricks would prevent any gross relative shear displacements (Figure 3). The finite element model of the core has been developed so that it simulates all of the significant interactions between adjacent bricks, and between the bricks and the core restraint structure.

FE Model - Bricks

The aim of this work is to investigate the dynamics of the core, but it is not to consider stresses caused in the bricks in detail. Because of that, and in order to keep the size of the finite element model manageable, a decision was made to not adopt a 3D model of each brick as a solid body (with 'brick' type 3D finite elements) with contact surfaces. Instead, each brick in the core is modelled as a rigid, vertical body, with inertial and other properties attached to it. In order to represent possible interactions with the adjacent bricks along its height, each rigid body was attributed seven planes in which rigidly connected nodes were to be defined in order to attach spring/dampers to the brick for modelling contact between adjacent bricks, Figure 4.

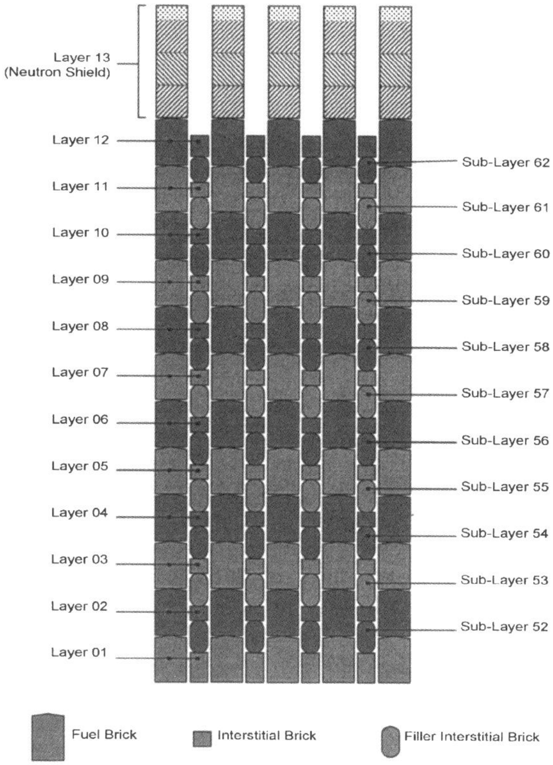

FIGURE 3: Schematic representation of a keying system in vertical plane.

Brick Interactions

Bricks can interact with adjacent bricks through a number of mechanisms as summarized in Table 1 below. Since the main aim of the work is to investigate dynamics of the core when subject to seismic loads, contacts between the bricks are predominantly short duration impacts. Impact stiffness of the contact was based on values measured in experiments, while a coefficient of restitution of 0.65 was used. In order to simplify the model and allow more control over its behaviour, all of these interactions were modelled through a number of springs and dampers connecting relevant nodes.

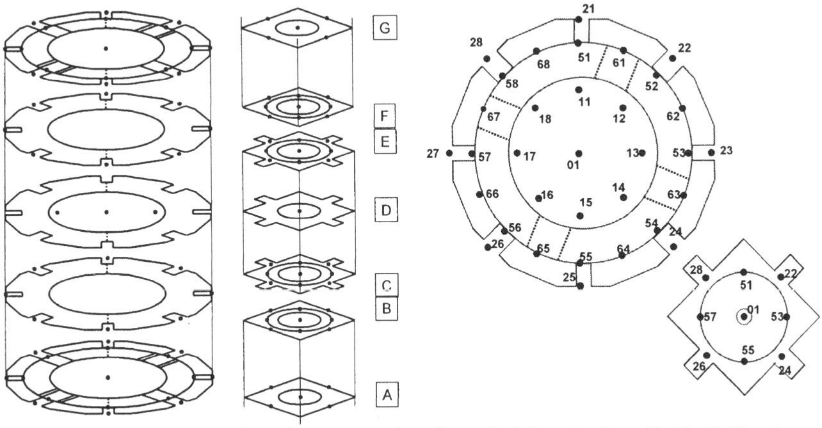

FIGURE 4: Schematic representation of a vertical discretisation of bricks (left) and all the possible nodes in a single plane (right).

TABLE 1: Summary of interactions in the keying system and their locations.

Nodal Plane	Vertical Position in Fuel Brick Column	Vertical Position in Interstitial Brick Column	Interactions
A	FB bottom surface	FIB centre of mass	FB – FB lateral direct contact FB – FIB lateral direct contact FB connection with FB below
B	N/A	FIB top surface	FB – FIB lateral direct contact FIB connection with IB above
C	Bottom of loose bearing key	IB bottom surface	FB – IB lateral direct contact FB – FB loose key FB – IB integral key IB connection with FIB below
D	FB centre of mass	IB centre of mass	None
E	Top of loose bearing key	IB top surface	FB – IB lateral direct contact FB – FB loose key FB – IB integral key IB connection with FIB above
F	N/A	FIB bottom surface	FB – FIB lateral direct contact FIB connection with IB below
G	FB top surface	N/A	FB – FB lateral direct contact FB – FIB lateral direct contact FB connection with FB above

Key: FB – Fuel Brick IB – Interstitial Brick FIB – Filler Interstitial Brick

The keying system was designed to allow some free motion before becoming engaged by specifying clearances in the system. In order to model these clearances, the springs modelling the impact behaviour have force displacement curves as depicted in Figure 5. Energy dissipation during the impact, resulting in a restitution coefficient less than 1, is modelled as an equivalent viscous damping using dashpot elements in parallel with the springs. LS-DYNA modelling allows for this by defining the viscous damping force to be proportional to the velocity, with a proportionality ratio equal to the dashpot damping. Each damper is defined by both a displacement graph and a velocity graph, with their ordinates multiplied together, so that they only come into play whilst contact is actually made between the components.

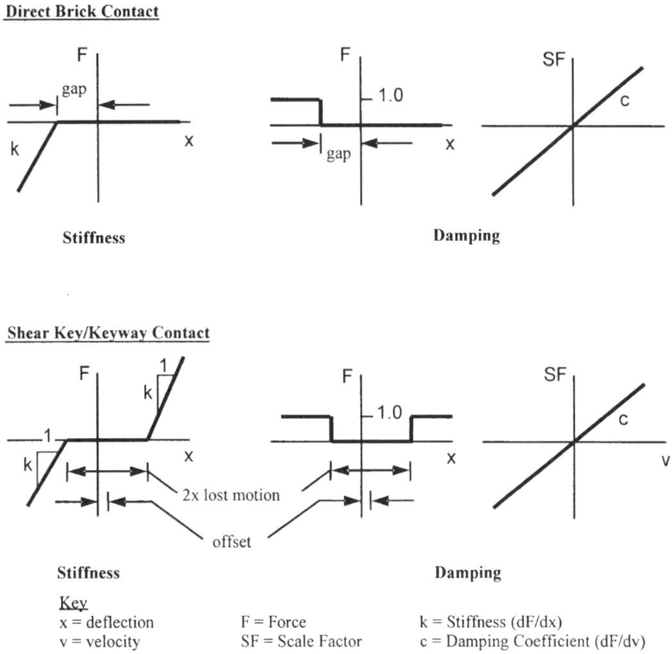

FIGURE 5: Definition of contact spring and dashpot properties.

Dimensions of the bricks and keys, as well as clearances in the keying system were calculated based on the geometry of the bricks after a prescribed number of full power years, *i.e.* the effects of irradiation on dimensions, stiffness and shape of the bricks, keys and keyways were taken into account. The developed model has the capability to account for potential damage induced in bricks due to irradiation, thermal or other defects. For example, development of a single vertical through-the-wall crack would cause opening of the brick's cross section into a C shaped profile. This would lead to changes in the apparent radius of the brick, as well as offsetting and changed clearances in the shear keyways. If a secondary crack develops, the brick would split vertically into two half-bricks. The developed model was designed to be able to simulate these; positions of the cracked bricks and crack plane orientations can be randomly distributed within the core's more irradiated zones with prescribed probabilities of primary and secondary cracks occurring.

Connections of the bricks within a stack is twofold; horizontally, the keying system at the ends of the bricks only allows limited relative horizontal motion of the bricks; vertically, the ends of the bricks are designed so that the moderator brick above rocks on the brick below (*i.e.* there is no full area contact). This rocking feature was designed to help evenly distribute stresses due to thermal expansion. For the rocking motion, two dominant rocking directions were designed; higher stiffness for rocking in a core's tangential direction, and lower stiffness for rocking in the core's radial direction. These end features were also modelled by nonlinear springs connecting relevant nodes on brick ends, Figure 6.

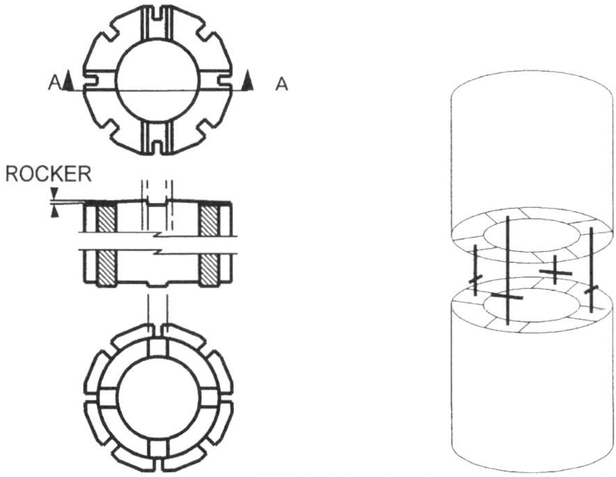

FIGURE 6: Modelling of axial (vertical) features.

RESTRAINT STRUCTURE MODEL

Description

The main components of the restraint structure are depicted in Figure 1, and comprise of:

- Restraint links. One end of the restraint links is spigotted into peripheral bricks at a mid-layer plane whilst the other end is fixed into a restraint beam.
- Restraint beams. There are 16 beams per horizontal mid-layer plane, each beam supporting four restraint links.
- Two Warwick links per beam, creating a trapeze on plan. The Warwick links are pin jointed at both ends.
- One centralizing bracket per restraint beam designed to locate the restraint beam circumferentially in relation to the restraint ring beam. The brackets were designed to allow free relative motions in radial and vertical directions.
- The Warwick links are connected into the restraint ring beams, which are connected into the boiler shield wall.

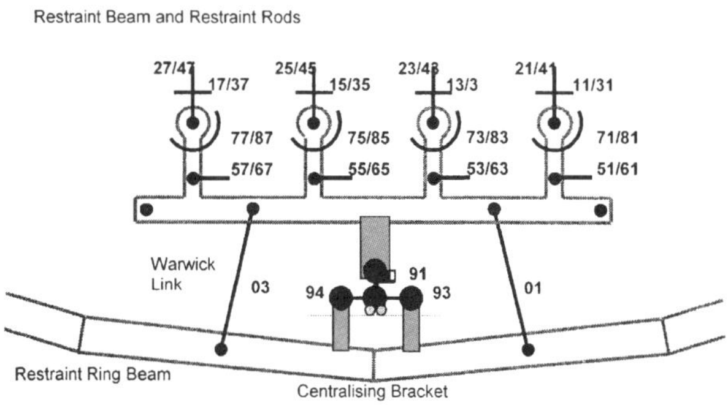

FIGURE 7: Diagram showing numbering of restraint structure elements.

FE Model

The FE model of the core models the restraint structure up to the boiler shield wall, which is assumed to be rigid and is where the input motion due to the seismic event is applied to the model. Figure 7 shows a finite element model of the restraint structure in more detail. Each restraint beam with its associated restraint links was modelled as a single rigid body. Flexibility of the components and clearances in the system were modelled using direct springs and dampers. The centralising bracket spring properties were developed by considering a separate, more detailed component finite element model incorporating material and geometrical non-linearity. Results from this component model were than transformed into a load-displacement curve for the respective spring.

KEYFILE GENERATOR

It was recognised that general purpose finite element mesh generators are not suitable for developing this type of finite element model, and so a decision was made to develop a purpose made program able to generate a complete keyfile for inputting data to LS-DYNA. The keyfile generator program was written in Visual Basic for Applications, in combination with Excel. While it is generic in certain aspects, it is applicable without modifications only to a single core design. The generator reads results of preliminary calculations defining the geometry of the core and the keying system, restraint structure properties, probability of damage to the core, and automatically creates a LS-DYNA keyfile. The total number of nodes used in the model was 283202, whilst the number of elements totalled 264296. The topology, *i.e.* numbering system, of the FE model's constituents has been developed such that it is quickly apparent to an experienced user the exact whereabouts of any particular node or element just by looking at the number, and similarly any given component's number can quickly be ascertained by following the topology calculation's conventions.

POSTPROCESSOR

The main aim of the work is to demonstrate that, during and following the seismic event, the core can be safely shut down and held down. Based on this requirement, the following aspects of the core performance need to be assessed for the seismic analysis:

- Integrity of the keying system.
- Integrity of the core support structure.
- Fuel and control channel alignment, *i.e.* interaction of the channels with the fuel stringers and control / sensor rods to allow free movement of these components and free flow of coolant.
- General level of displacement of graphite bricks.

Again, the general purpose finite element post-processor is not suitable for dealing with models like this and displaying results in an appropriate format for assessment. Because of that, the post-processing tool was developed which automatically extracts all the required results from LS-DYNA output files, carries out the required analyses and assessment of the components, and displays results in user friendly format.

SAMPLE RESULTS

The analyses are run for an acceleration time history extracted at relevant locations from the calculated responses of an earlier and simplified finite element model of the reactor. A full time history analysis was carried out with a time step set sufficiently short to prevent any instability in results. Sensitivity studies were carried out to investigate the effects of various time-step sizes on the model behaviour. At the beginning of the analyses, the model is left for a certain period of time to settle down into an equilibrium position, before the seismic accelerations are applied. The analysis is run for the whole duration of the assumed seismic event. Outputting results of the analyses at each time step would lead to a large results file, making them un-manageable. Because of that, the results output frequency is set to be variable during the analyses, with the highest frequency during the strong motion period of the seismic event. Sensitivity studies were carried out to assess the effects of different output sampling rates on the accuracy of the presented results.

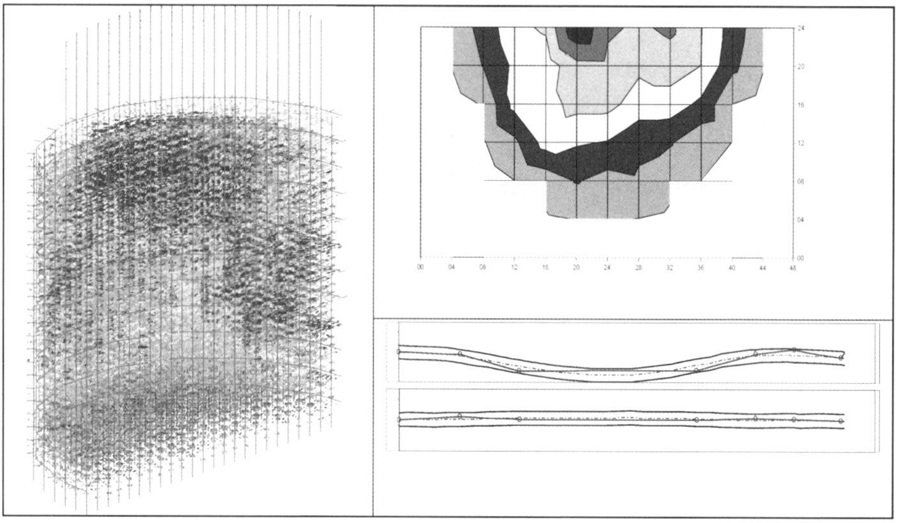

FIGURE 8: Example results; contour plot of velocities (left), displacement contour plot at one layer (top right), deformed channel profile with sensor rod (bottom right).

CONCLUSION

A finite element model was developed to evaluate effects of seismic event on a reactor core at various times in life. Effects of potentially damaged bricks on the core response were modelled. Purpose made programs were developed for generating a finite element mesh for LS-DYNA and to post-process analyses results.

Acknowledgements

The authors would like to thank Mr. AG Steer for his advices in developing the model of the core. The authors would also like to thank British Energy Generation Ltd for permission to present this paper. The views expressed in this paper are those of the authors and do not necessarily represent those of the British Energy Generation Ltd.

References

LSTC, LS-DYNA Theoretical Manual and Keyword User's Manual, Livermore Software Technology Corporation, 2003.

Measurement of AGR Graphite Fuel Brick Shrinkage and Channel Distortion

Aiden Cole-Baker and James Reed
British Energy, Barnett Way, Barnwood, Gloucester, UK.
Email: aidan.cole-baker@british-energy.com and jim.reed@british-energy.com

Abstract
When subject to fast neutron irradiation, the AGR graphite undergoes dimensional change. Amongst other factors, this change is controlled by the cumulative fast neutron irradiation dose. As part of the ageing management of the reactor cores, the effect of the dimensional change in the graphite can be directly measured using the Channel Bore Measurement Unit (CBMU). This inspection technique is deployed in a number of fuel channels at each reactor periodic shutdown. The data from the CBMU is used to assess fuel brick diameters, brick ovality and channel distortion (channel bow and channel tilt). The CBMU data collected can be used to monitor the ageing of the cores through development of the brick shapes and levels of channel distortion with time.

Keywords
Inspection, ageing management, AGR

INTRODUCTION

The AGR reactor cores consist of many hundreds of fuel and interstitial channels, each made up of a number of graphite bricks. Each power station consists of two reactors and fuel channels are referenced using a coordinate system of a letter and a number, or two numbers. Channel references can be preceded with the reactor number.

When subject to fast neutron irradiation, the graphite undergoes dimensional change. In addition to their moderating function, the geometry of the AGR reactor cores must continue to allow refuelling, shutdown and cooling activities during their working lives, thus the structural integrity of the cores must be assured and any distortions limited. Core inspection with a Channel Bore Measurement Unit (CBMU) has been used on stations to examine the effects of fast neutron irradiation on the dimensional change of the fuel channels. The dimensional change of the graphite manifests itself as changes in the diameters of the fuel bricks. This change is not uniform along the length of the brick or indeed around its circumference. Thus diameter is a function of height within a brick as well as height within the core, and also angular orientation, which gives rise to an ovality (or cross-diameter difference). By characterising the brick shapes (by defining lambda parameters or barrelling as described below) and trending the data, the onset of shrinkage stress reversal can be predicted. In addition, individual brick tilts are measured which allows the channel shape to be determined. Significant deviation from channel straightness could interfere with fuel movements. Assessment of the channel shapes from the CBMU data allows the level of core distortion to be inferred.

The CBMU can also detect significant cracks and provide information on the influence of cracking on brick and channel distortion. Correlating the CBMU information with that observed from TV inspections also allows the type of cracking to be more fully characterised.

THE CHANNEL BORE MEASUREMENT UNIT

The CBMU consists of two sets of four wheels, one set at the top, the other at the bottom. These are geared to keep the CBMU centred in the channel. Feelers arranged every 90° and measure the distance to the channel wall; combining these measurements give a measure of the diameter. Measurements are taken in one-millimetre height intervals. The CBMU also contains two tilt transducers, arranged 90° apart. When the CBMU is entirely within a brick, the tilt of that brick can be determined. Scanning the brick in diametrically opposite directions allows the tilt zero offsets to be eliminated from the calculations of channel bow and tilt. These parameters give a measure of the channel distortion and ultimately imply levels of core distortion.

DIMENSIONAL CHANGE

Diameters and Shrinkage

The average diameter from readings taken in orthogonal directions is used to measure the shrinkage of the graphite fuel bricks. These are monitored at periodic shutdowns against perceived limits that could potentially inhibit fuel movement and coolant flow. In general, all channels are expected to exhibit the same rate of shrinkage, although there is scatter in the data due to dose variations across the core and differences in virgin properties between bricks. Monitoring of repeat channels gives the opportunity to monitor changes more closely. As an example, channel 24:70 at Hunterston B (HNB) Reactor 4 has been inspected five times (Figure 1) and channel H15 at Heysham Stage 1 (HYA) Reactor 1 was measured prior to start-up and four times since (Figure 2). Following the peak rated layer mid-brick shrinkage of repeat channels highlights the scatter in all the data but a broad shrinkage trend can be observed.

Monitoring against Shrinkage Stress-Reversal

Although CBMU analysis monitors against a minimum diameter for fuel movements, the data can also be used to monitor against the onset of shrinkage turnaround. A cylinder that is subjected to greater shrinkage stresses at the bore compared to the outside will tend to "barrel in" at the ends, as seen in Figure 1 and Figure 2. The extent of the barrelling will increase while the rate of shrinkage at the bore remains above that at the outside. This differential shrinkage results in a tensile stress at the bore. Once shrinkage turnaround begins, the rate of increase of barrelling will fall until a peak is reached at the point when shrinkage rates are equivalent across the bore. Several factors complicate the analysis, including:

- The geometry is not a pure cylinder as the bricks contain full- or part-length keyways, depending on the station;
- The bricks have fuel end dose depressions within their length;
- There will be differences in cross-brick shrinkage rates resulting from power differences in the subject and neighbouring channels.

The values calculated are known as lambda parameters at Hinkley Point B and Hunterston B, or brick-end barrelling at Heysham Stage 1 and Hartlepool. Monitoring of these parameters remove the uncertainty in the absolute CBMU calibration and these values are expected to be more sensitive to the onset of shrinkage stress reversal.

Analysis of the barrelling at Heysham Stage 1 and Hartlepool (HRA) has shown that the rate of barrelling continues to increase linearly with neutron irradiation (Figure 3). This linear increase is also confirmed by repeat channel data. Thus it is concluded that Heysham Stage 1 and Hartlepool are not exhibiting signs of reaching the onset of shrinkage stress reversal. At subsequent inspections new data will be added to the plot and examined for deviation from the linear trend exhibited by the historical data. In the mean time, the inspection data feeds into stress analyses for comparison of the material property and stress analysis models.

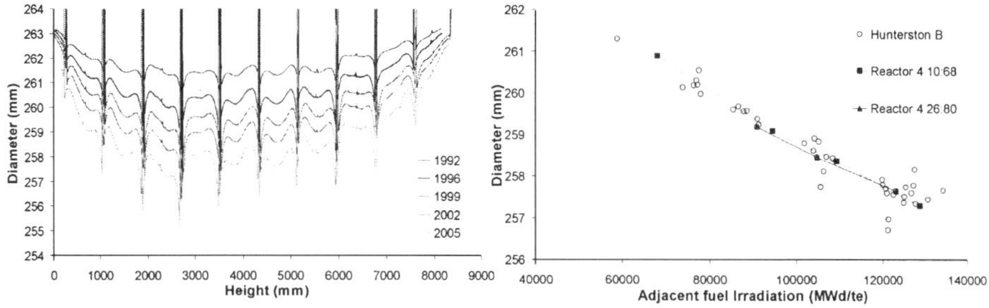

FIGURE 1: Change in diameter profiles with inspection year of channel 24:70 at Hunterston B Reactor 4 and the layer 7 shrinkage.

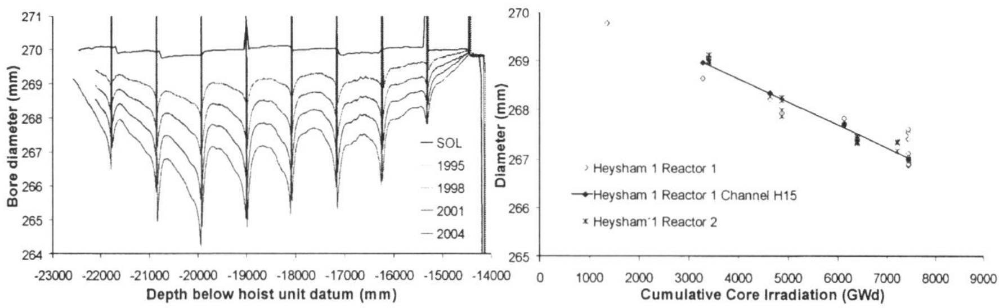

FIGURE 2: Change in diameter profiles with inspection year, including start of life (SOL) data of channel H15 at Heysham 1 Reactor 1 and the layer 7 shrinkage.

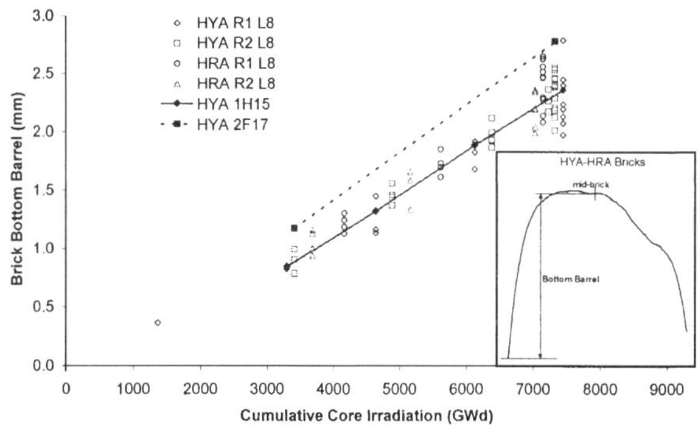

FIGURE 3: Extent of 'Barrelling' at Heysham Stage 1 (HYA) and Hartlepool (HRA) showing a continuing increase.

At Hinkley Point B (HPB) and Hunterston B power stations, the bore shapes develop differently due to different fuel end dose depression heights. It has been calculated that the data for layers 4 and 6 are likely to show the onset of shrinkage stress reversal earliest. Figure 4 displays the data for the layer 6 $\lambda_1^{(bottom)}$. Prior to the 2005 inspections, finite element predictions of bore shapes were parametrically scaled to give the best fits to the Hunterston B lambda data. These suggested that the peak in lambda behaviour had been reached at Hunterston B and was relatively sharp. Therefore, repeat channel inspection data should have shown clear evidence of peaking. Examination of all the 2005 data, including repeat channels from the HNB periodic shutdown, Figure 4, suggests that the lambda values are in fact reaching a plateau, suggesting a delay in the shrinkage stress reversal.

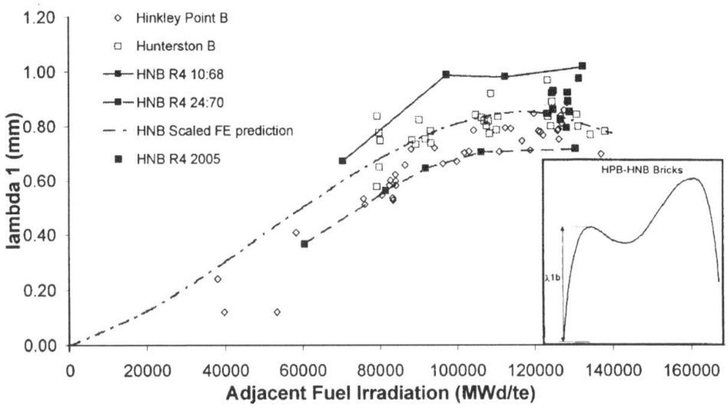

FIGURE 4: Hinkley Point B (HPB) and Hunterston B (HNB) Lambda 1 layer 6 (bottom) compared with the FE calculated data which has been scaled to give the best fit to the reactor data. Two repeat channels from HNB Reactor 4 (R4) are shown.

In summary, at Heysham Stage 1 and Hartlepool, the rate of barrelling continues to increase suggesting that the onset of shrinkage stress reversal has not been reached. The shape parameter observations at Hinkley Point B and Hunterston B made to date are consistent with the understanding that has developed from previous inspections with the exception that the rate of peaking appears to be slower than predicted. Monitoring of these trends against predictions will continue at subsequent inspections.

Ovality

In general, ovality (or cross-diameter difference) is driven by cross-channel flux gradients, so the ovality of non-peripheral channels should be low. However, analysis of historical ovality data identified a link between the shape and magnitude of the ovality and the orientation of the rocking features in the Hinkley Point B - Hunterston B brick design.

Predictions based on the mechanical properties models and stress analyses have been made as to the way in which ovality should develop with irradiation. By monitoring ovality during the inspections, a comparison to predictions can be made (Figure 5) and, in general, the data supports the predictions.

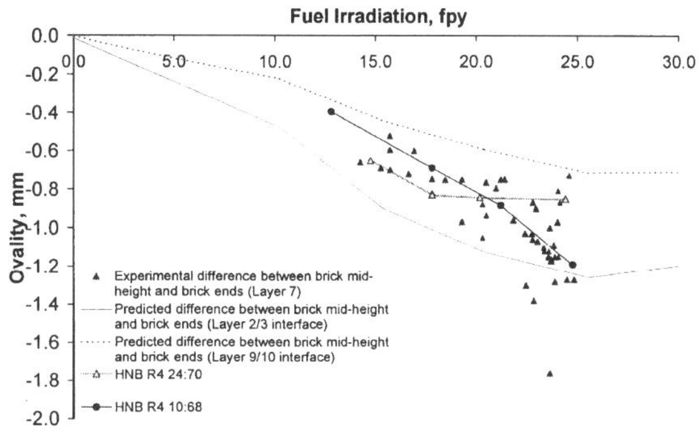

FIGURE 5: Comparison of predicted and measured ovality at Hinkley Point B and -Hunterston B.

In summary, the CBMU is used to monitor the irradiation-induced shrinkage of the graphite core and the data provides an important assessment of the onset of shrinkage stress reversal. The data is used to confirm predictions and, if necessary, provide a feedback to allow development of the underlying understanding of graphite behaviour.

CORE DISTORTION

From the orthogonal tilt transducers in the CBMU, the overall channel profile (defined by the channel bow and channel tilt) can be determined. It has been suggested that the measurement of four channels equispaced around a reactor is sufficient to enable global core distortions to be bounded.

Channel Bow

Channel bow is the measure of the distance between the axis traced from the top of the bottom brick to the top of the top brick and the top of each brick relative to this axis. If sufficient channels evenly spaced across the core are measured, the levels of distortion in the remaining channels can be estimated. Again, measurement of repeat channels provides the greatest insight to channel bow development, for example, the channel bow for 1H15 from Heysham Stage 1 is shown in Figure 6(a) and shows that, within the measurement uncertainties, the shape and magnitude has not changed. Trending the data with time, Figure 6(b), suggests that after an initial period of increase, the channel bows have reached a plateau. During inspections, the observation of a large channel bow outside of the observed range to date would have implications for the levels of core distortion.

Channel Tilt

Channel tilt alone has no implication for fuel movement. Channel tilts have only recently been calculated and require more information to eliminate zero-offsets. The available data again suggests that channel tilts are not changing significantly with increasing irradiation (Figure 7), however there is a significantly larger scatter in the data and increased uncertainty in the measurements (indicated by the variation in the repeat channel) compared with channel bows. The channel tilt data collected from inspections is currently being used as inputs in whole core modelling.

In summary, channel bow and tilt data is used to characterise the core distortion. The levels of distortion across a core can be bounded from an evenly spaced inspection campaign.

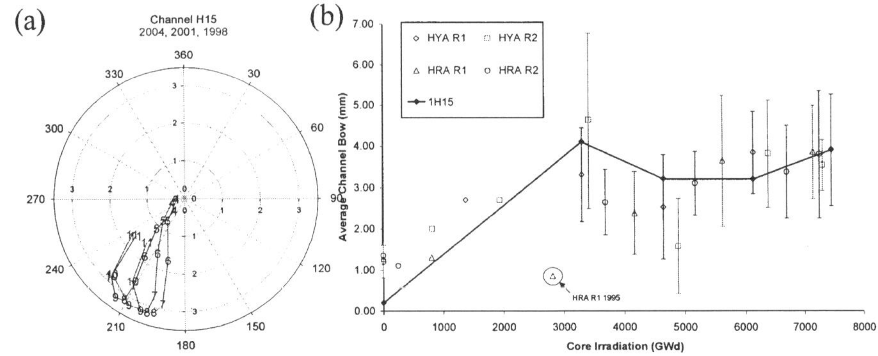

FIGURE 6: (a) Repeat inspections of channel H15 at Heysham Stage 1 Reactor 1 showing that the channel bow has not changed; and (b) a plot of average channel bow (for a campaign) with increasing irradiation suggesting that the channel bows have reached a plateau (repeat measurements of HYA 1H15 channel bow shown).

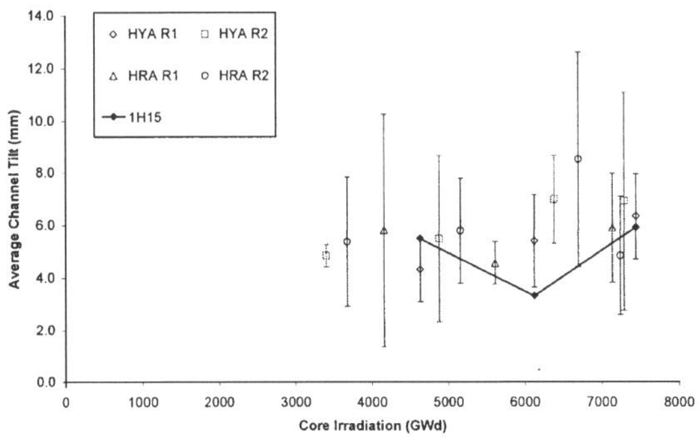

FIGURE 7: Average campaign channel tilt plotted against Irradiation (repeat measurements of 1H15 at HYA also shown).

DETECTION AND ANALYSIS OF DEFECTS

The CBMU measurements can also indicate the presence of cracks within bricks and provide further information beyond that of a visual image. The additional information provided by the CBMU allows cracks to be fully characterised to include their influence on bore distortion. The response of the CBMU differs depending on the type of crack within a brick. For circumferential cracks, the individual feelers will cross over approximately perpendicularly to the direction of travel of a crack. This results in a feeler response that can be readily recognised and interpreted. Figure 8 compares the image constructed from the TV footage with that from the CBMU. One of the CBMU feelers traversed the trepanned hole allowing orientations to be confirmed. Measuring the divergence of a feeler profile (or dishing) from that from a similar but uncracked brick, as shown in Figure 9(a) is providing some information on the distortion caused by circumferential cracking. Measuring the dishing in the feeler profiles for repeat crack inspections shows that the dishing is increasing with time, as such, an estimate of the ages of the cracks can be made, Figure 9(b). Assuming a simple linear relationship, such estimates of Hinkley Point B and Hunterston B cracking suggest that these cracks initiated many years ago.

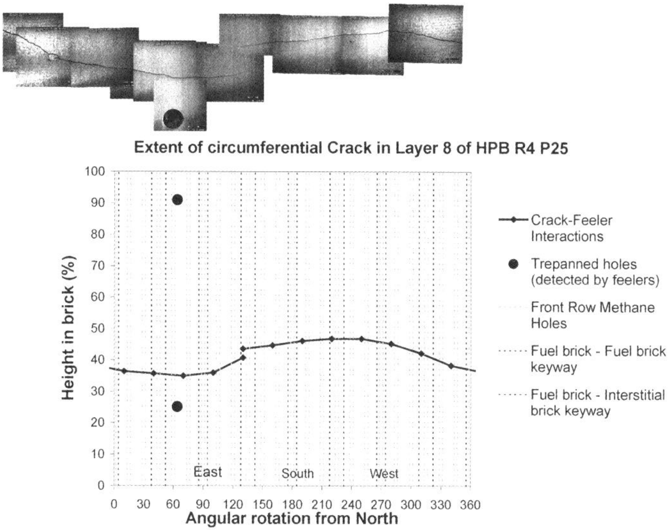

FIGURE 8: Comparison of TV image to that developed from the CBMU information.

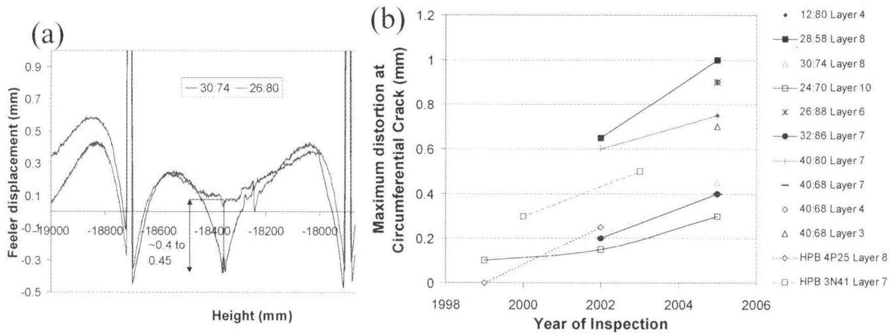

FIGURE 9: (a) Comparison of feeler traces between a cracked and uncracked brick in Hunterston B illustrating the method used to determine the "crack face dishing"; and (b) the development of crack face dishing over time for the cracked layer in the given fuel channel.

Axial cracks may initiate either at the inner bore during early life, or from the keyways in late life as illustrated in Figure 10. From visual inspections, the two types of crack may appear similar, but stress analysis has shown that the two types of cracking can be distinguished with the CBMU. To date, only bore-initiated cracks have been detected and these show an increase in the ovality, but no change in the mean diameter. Bricks have been observed with two full-length axial cracks. From an analysis of the CBMU data, the magnitude of the ovality changed significantly with only a relatively small increase in the diameter, as shown in Figure 11(a). A detailed investigation of the feeler displacements around the bore of the cracked brick also provided evidence to support the findings that the cracks were bore initiated. The "pinching in" of the profile indicates the orientation of the cracks. This behaviour is typical of all axial cracks observed to date, all of which have been classified as bore-initiated.

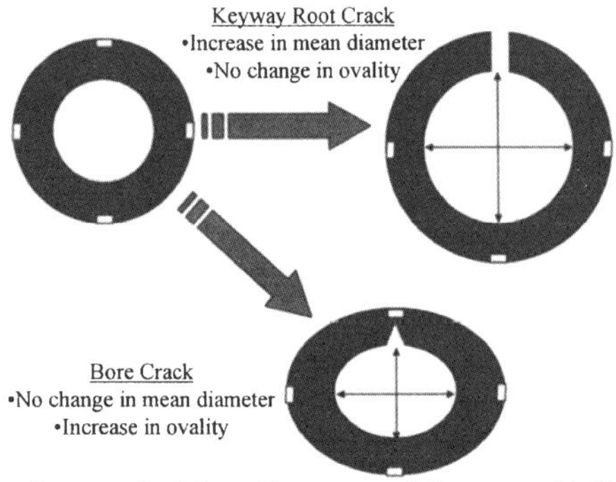

FIGURE 10: Schematic diagram of axial cracking types and the expected influence on brick geometry.

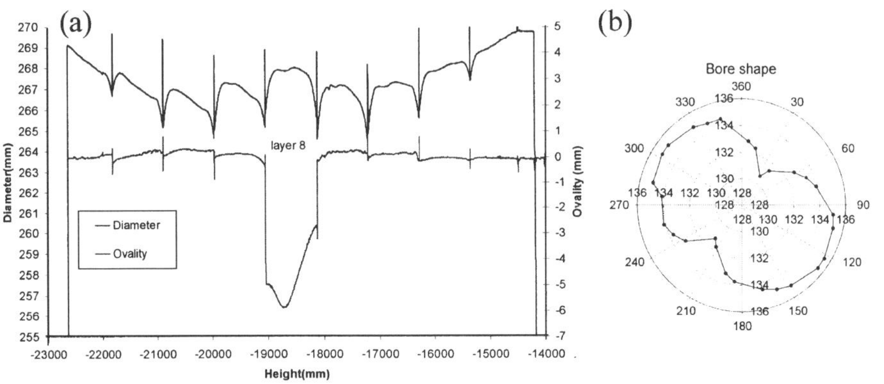

FIGURE 11: (a) CBMU data for twice axially cracked brick showing a large change in the ovality with minimal change to the diameter; and (b) interpretation of CBMU feeler data to give a bore shape showing the orientation of the cracks.

CONCLUSIONS

- The diameter, ovality, channel bow and channel tilt are measured at selected channels during reactor periodic shutdowns.
- The data is used to monitor the ageing of the graphite cores to add confidence that the core can continue to provide fuel movement and cooling functionality.
- Diameter and ovality data can be used to monitor against the onset of shrinkage stress reversal.
- Channel distortion data is used to bound the overall core distortion.
- The inspection data is also used as inputs to numerous models, with a feedback loop generating a greater understanding of the ageing behaviour of the graphite.
- Analysis of CBMU data can provide further information on brick distortions caused by cracking enabling crack types to be determined.

Acknowledgements

The authors would like to thank British Energy Generation Ltd for permission to present this paper.

Investigation of Degradation of AGR Graphite Core Geometry using a Whole Core Scale Model

Elisabeth Castro
AMEC NNC Limited, 601 Faraday Street, Birchwood Park, Warrington, UK
Email: elisa.castro@amecnnc.com

Abstract
The ability to operate the Advanced Gas-cooled Reactors (AGRs) depends on securing safety cases for the graphite cores. This requires ongoing research to support assessments of the functionality of the AGRs through life. The ageing effects in an AGR graphite core manifest themselves in a variety of ways including physical changes corresponding to dimensional changes and ultimately distortion of the individual components and initial geometrical configuration of the core. In addition, the ageing process of the graphite core may lead to the inducement of through thickness full length axial cracking of the fuel bricks and perhaps the possibility of double cracking of the fuel bricks. Whilst a single axial crack may not be detrimental to the functionality of the reactor, a pair of full length axial cracks could lead to significant changes to the fuel, and possibly adjacent control rod, channel geometry. These, in turn, may affect the functionality of the core in respect of fuel and control rod movement and other criteria. A number of experimental test programmes have been carried out to investigate the behaviour of cores with doubly axially cracked fuel bricks, using components increasingly representative of the actual AGRs. The test work investigated the tolerability of doubly cracked bricks directly in some rigs, but was always designed to provide data to assist in the validation of computational models of doubly cracked brick behaviour. Of these, a whole core array model was first developed at $1/8^{th}$ scale to determine potential fuel channel displacements of Hinkley Point B (HPB) / Hunterston B (HNB) using highly simplified components. A quarter scale model of the Hartlepool (HRA) / Heysham 1 (HY1) AGR core is currently in production. All of the keying components are represented and properly scaled. It is intended to use this rig to demonstrate the tolerability of large numbers of doubly cracked bricks more directly, and provide validation of the conservatism of more representative computational models. By tilting the rig, it will be possible to simulate a uniformly distributed load on the core, and thus generate extreme core movements. The model will determine the largest gross displacements of the fuel channels as well as the shear and separation of selected cracked fuel bricks. The measurements of the individual brick displacements allow the effects of cracked bricks on channel geometry to be assessed. In order for the measurements not to be influenced by the measurement system, optical techniques are being developed. Images of the whole core array and selected fuel channels pre- and post-displacement will be analysed. A machine vision system has been developed to aid this process. It allows the movement of large numbers of targets to be observed simultaneously. Whole core rig models enable a number of accurate simulations to be tested, and the results can be used to validate the equivalent computer models. Uniting physical modelling with finite element analysis provides the most effective basis for continued operation with increasing graphite degradation.

Keywords
Whole core, doubly cracked bricks

INTRODUCTION

Whole core modelling is part of British Energy's graphite core safety case strategy. It is intended to determine the extent of tolerable cracking in the graphite core. The whole core experimental rig work was originally based on a $1/8^{th}$ scale model of a HPB/HNB reactor cores being the lead stations. The $1/8^{th}$ scale core rig was developed in stages, firstly as a single layer rig, then a full array, testing channel distortions, and more recently as a core with a limited number of doubly cracked bricks. The rig was made up of simplified fuel and interstitial brick components scaled by a factor of 1:8, however, the clearances were maintained at full scale to facilitate displacement measurements (note: the results obtained were purely for geometric validation of computational models).

FIGURE 1: Photo of the 1/8th scale core rig with a number of small doubly cracked brick clusters.

$^1/_8{}^{TH}$ SCALE CORE RIG

The 1/8th scale core rig design simplified the HPB/HNB components by incorporating keys into the bricks: the fuel brick included two integral bearing keys and the interstitial brick included four integral keys. These simplified components represented the primary keying system and allowed it to transmit the loads in a representative manner. The secondary keying system, comprising filler keys and spacer keys, were omitted. An assessment of the effect of the primary and secondary keying systems was made using the computational models. The outcome suggested that the spacer and filler keys could be omitted, as the rigidity of the AGR core is only slightly reduced.

The full scale key / keyway clearances were maintained in order that the maximum displacements observed were equal to those possible in the reactor core. Unfortunately, this meant that other factors, *i.e.* rotations and key engagement were not representative. Had the 1/8th scale core rig been completely scaled, *i.e.* having clearances in the order of 0.13 – 0.19 mm, it would not have been practical to observe the displacement effects.

Doubly Cracked Bricks

The 1/8th scale core rig was not intended to model the effects of doubly cracked bricks. As the core rig is tested with increasing fuel brick cracking, the full scale key / keyway clearances cause the displacements observed to become overly pessimistic. Single layer tests to determine the upper bounds of possible displacements with varying percentages of doubly cracked bricks showed that key disengagement occurred much earlier than would be expected in the real AGR geometry.

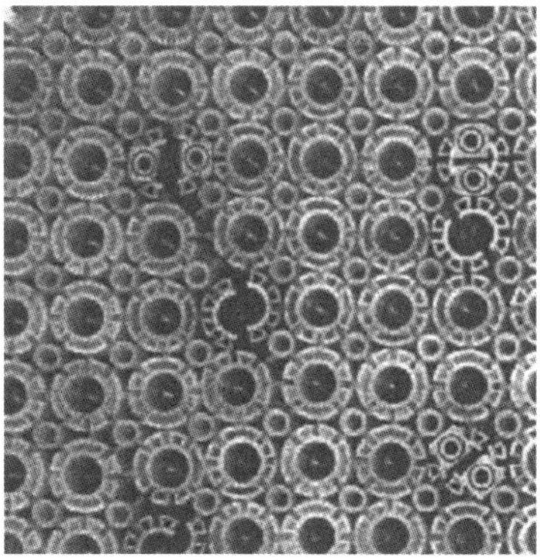

FIGURE 2: Photo of the single layer $1/8^{th}$ scale core rig showing key/keyway disengagement.

The results represent a very onerous worst case scenario – the bricks are on a smooth base with no end face keying to constrain them. The rig is no longer valid as the full scale clearances allow unrepresentative disengagement.

NEW SCALE CORE RIG

A new scale core rig was designed to replace the $1/8^{th}$ scale core rig, as the inherent pessimisms were no longer able to represent the proposed cracking scenarios accurately. The real key / keyway clearances used in the $1/8^{th}$ scale core rig facilitated the measurement of displacement, at the expense of the depth of key engagement. A true scale rig which replicates the salient features seen in the full scale reactor would be less pessimistic, as the key disengagement and rotation would be in proportion, thus reducing erroneous effects, such as the possibility of keys falling out of place.

Core Model

In order to model increasing crack tolerability, it was necessary to produce a rig with equally scaled components and clearances, incorporating the keying system components accurately. The most frequently occurring fuel brick design was selected from HPB/HNB and HRA/HY1. The fuel brick component dimensions were arbitrarily scaled to establish a feel for the proportions of the core. The scaled fuel brick had to meet two criteria, firstly it had to be easy to manipulate and secondly it had to be easy to monitor. A full array full height core takes time to build and the easier it is to handle the scale fuel bricks the faster it is to assemble. Making the scaled fuel bricks too small however, would reduce the clearances beyond measurable means. Extracting accurate data from tests would have become more difficult with smaller components, as the effects of instrumentation error are greater. The possible scale factors were reviewed for a fully scaled core rig and the smallest practical scale derived was approximately a quarter scale. Although tolerances of 0.02 mm could theoretically be achieved to produce a fifth scale core rig, in reality the experimental techniques available to analyse the results were not sufficiently accurate.

Reactor Core Components
HPB/HNB reactor core design was considered as the prime candidate for the new scale model being the lead stations. However, the complexity of the core geometry was very difficult to model accurately whereas the HRA/HY1 reactor core design has a much simpler core geometry. Still, the core geometry was simplified further for the scale rig due to economic constraints. The fuel brick, interstitial brick and loose key designs were taken from the central layers of the reactor core and these were replicated to fill the core. The rocking features were omitted from the scaled fuel brick – the effect of these was not deemed to be significant for the initial test programme. The selected components were accurately scaled down, minimising potential pessimisms in rig results.

Materials
No attempt to scale density or friction was carried out as the model is principally for extreme displacements under static loading. Therefore, the most important material property was dimensional stability. It was necessary to select a material which would maintain the tight tolerances set. A range of potential polymer options were researched, however the available manufacturing techniques could not meet the tolerances required to produce an accurate quarter scale rig. Previous experience with aluminium extrusions for doubly cracked bricks in the 1/8th scale model had been promising and this avenue was pursued. This method allowed the bricks to be machined in the final stages to meet the requisite tolerances.

¼ SCALE CORE RIG

The following design constraint was placed on the new quarter scale core rig frame:

- The deflection of the frame must stay within tolerable limits (approximately 0.5 mm) to reduce the need for later corrections to the measurements.

The quarter scale core rig was designed to model a full height full array core, *i.e.* 24 brick diameter and 12 layers high. A full array core was estimated to weigh 35 tonnes. The rig frame design went through several iterations in an attempt to reduce the size of the frame whilst maintaining the necessary stiffness. A structural analysis package (STAAD) was used to optimise the design. The quarter scale core was restrained by the outer ring of fuel bricks in the rig in much the same way as the reflector bricks restrain the real core. Additional stiffeners in the form of 15 mm thick steel plates were also used to prevent the core periphery from distorting during tilting. The mass of the complete rig was calculated to be approximately 45 tonnes. Although the frame could have been made from 150 mm deep beams, in terms of load capacity, the strict deflection limits meant that the rig frame was built from 533 mm deep beams. The rig frame analysis showed that the beams were only being subjected to 2% of their strength.

As with the 1/8th scale core rig, the 1/4th scale core rig is being developed in stages. The first stage was to carry out a test programme using a single layer of bricks on a smooth surface, thereby defining the boundaries of the core movement in the rig prior to introducing doubly cracked bricks. A small fraction of the core components (2 layers) has been procured so far, and single layer commissioning tests are currently underway. The next stage is to increase the understanding of the behaviour of the core by adding doubly cracked bricks and increasing the number of layers until a full height full array core is created.

Rig Testing

The quarter scale core rig does not model the transient thermal expansion and contraction of the core. It is purely a geometric model used for static testing, modelling the quasi-static end conditions. The tests were designed to induce geometric channel distortions by tilting the rig, applying a uniform load to the core, to generate extreme displacements. The applied load was sufficient to displace the bricks, using up all the available clearances and allowing the greatest channel distortion to be observed. The octagonal shape was chosen to allow the rig to be tilted in any of the eight directions. This enabled both cardinal and oblique loading of the rig to be assessed. Cardinal loading of the rig transmits the loads via the interstitial bricks whilst oblique loading transmits the loads via the bearing keys.

FIGURE 3: Photo of the ¼ scale core rig tilted to 40° with a single layer of doubly cracked bricks.

Measurement System

An optical measurement system was developed to determine the displacements of the bricks during testing. It consisted of a high resolution digital still camera located approximately 5 m above the single layer of bricks on a pyramidal frame mounted on the 1/4th scale core rig. Images of the core array were captured before, during and after testing. The stills were then analysed using a machine vision system. The system was able to identify each brick by pattern matching and map the respective centre positions. The system has successfully identified 99% of the bricks in a layer. Using the derived coordinates, it was possible to determine the vector displacement of each brick position by subtracting the start co-ordinates from the end co-ordinates.

FIGURE 4: Photo of the 1/4th scale core array taken from the digital still camera suspended above the rig, prior to tilting towards the northwest (oblique tilt).

Analysis

The first test programme was to determine the basis against which introduction of doubly cracked bricks can be compared to. The rig was filled with intact bricks and the results of tilting in various directions were assessed. Some comparison against the similar 1/8th scale test programmes were made, however the 1/8th scale rig results were based on 20 targeted channels and therefore the 1/4th scale rig results had better statistical quality being able to monitor the majority of the fuel bricks in the layer.

The data was extracted from the captured images, and after processing to remove anomalies caused by camera misalignment, the results were plotted. An example of which is given in Figure 5. There were some gaps in the initial data; these were due to the pattern matching system not identifying the brick position. This is a result of shadows cast in the rig and the lighting system is under review to determine how to minimise this effect.

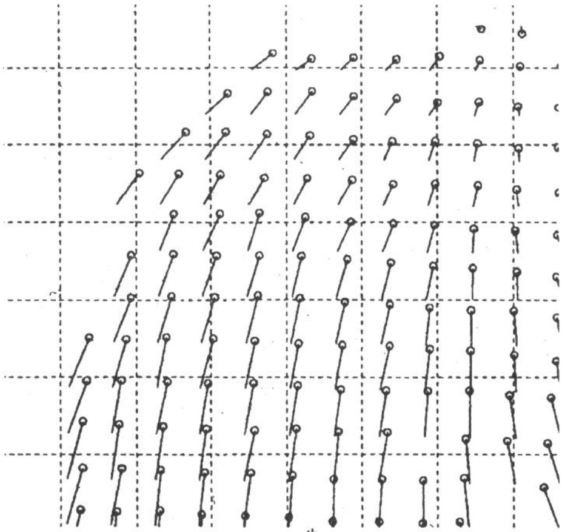

FIGURE 5: Example graphical representation of the 1/4th scale core rig movement. The data has been taken from the third test and shows the start and end position (represented by a circle) vector after tilting the rig up to 40° and lowering it back down. The vector displacements have been amplified by a factor of 50 to enable the movement to be easily observed.

CONCLUSIONS

Preliminary test work suggests that the new 1/4th scale core rig will be able to provide much more representative data than its predecessor. Previous test programmes using the 1/8th scale core rig were only able to monitor 20 channels with ease. The data obtained was limited and it was not possible to make significant advances in the understanding of the core behaviour. The development of the machine vision system has facilitated monitoring of large arrays and the possibility to greatly improve understanding of core behaviour. Furthermore, the coverage and diversity of data generated is now in line with that produced using computational models thus improving confidence in the combined results.

Acknowledgements

The author would like to thank Dr S J Pountney and the Development & Test Services Team in assisting with the experimental programme. The authors would like to thank British Energy Generation Ltd for permission to present this paper. The views expressed in this paper are those of the author and do not necessarily represent those of the British Energy Generation Ltd.

References

Shaw, D. J., Salih, M. H. S., and McLachlan, N. (2006). Development and Validation of Whole Core Modelling Methodology for AGR Graphite Cores. Presented at the Ageing Management of Graphite Reactor Cores Conference, 28 to 30 November 2005, Cardiff, UK (see also this book).

The Feasibility of Introducing Core Stiffening Devices and Diverse Shutdown Systems at Hinkley Point B / Hunterston B and Hartlepool / Heysham 1

Michael W Davies
AMEC NNC Limited, Booths Park, Chelford Road, Knutsford, UK.
Email: mike.davies@amecnnc.co.uk

Abstract
Two studies have been carried out by AMEC NNC with the aim of making the cores at Hinkley Point B / Hunterston B (HPB/HNB) and Hartlepool / Heysham 1 (HRA/HYA) more tolerant to widespread brick cracking. The first of these looked at the feasibility of introducing stiffening devices into specific core channels to limit the potential for brick displacements / column bowing. The second of these looked at the existing diverse shutdown systems with regard to their ability to safely shut down both reactors simultaneously and to provide long-term hold-down, even with widespread brick cracking. This would be required if the black control rods, *i.e.* the primary shutdown system, failed to enter the core in sufficient numbers. Possible enhancements to the existing systems were investigated, together with totally new systems. This paper shows that the requirement for both can be met using a single (dual-purpose) plant modification, although other options for each are possible.

Keywords
AGR, core stiffening devices, diverse shutdown systems

INTRODUCTION

It has long been predicted that keyway root initiated cracking of fuel bricks will arise in some AGR cores in late life. However, a number of bricks have already been observed where the cracks have originated at or near the bores. In both cases the cracks could be full-height and full-depth. Most affected bricks are expected to have one such crack (referred to as singly-cracked bricks), but some may develop a second such crack (referred to as doubly-cracked bricks). It has already been demonstrated that widespread singly-cracked bricks can be tolerated, with up to 10% doubly-cracked bricks. However, with an increasing number of doubly-cracked bricks the potential for core movement will increase and this could significantly impede the movement of control rods and fuel stringers. A point could be reached when an insufficient number of the black (shutdown) rods could enter the core to achieve shutdown and provide long term hold-down.

It must be stressed that it is not the intention of British Energy to operate reactors with severely distorted / degraded cores that leads to any challenge to the safety function of the AGRs. Nevertheless, as part of an assessment to increase the safety envelope of an AGR, several options have been assessed in terms of prevention, protection and mitigation. Two of the options are considered here. One option would be to introduce stiffening devices into the core that would effectively 'pin' selected channels in space and thus limit the maximum brick displacements that could arise anywhere in the core to an acceptable level. Another would be to ensure that one or more diverse shutdown (DSD) systems are available, *i.e.* systems that are independent of the primary shutdown (PSD) system, the control rods. These would achieve shutdown and long-term hold-down even if control rods were unable to enter the core in sufficient numbers. In either case a full assessment of the safety benefits and disbenefits must be fully explored before further investment in the plant infrastructure is considered. This paper gives details of the two feasibility studies carried out by AMEC NNC to investigate these options.

CORE STIFFENING

Objective

The minimum potential brick movements will be for a core with an intact keying system. This was found to increase if the bricks became singly cracked, but it has been demonstrated that even with very large numbers of singly cracked bricks the limit on acceptable core movement is not reached. However, as the number of doubly cracked bricks increases it is likely that the limit will be exceeded at some point (the percentage of doubly cracked bricks at which this will occur has not yet been determined, but is the aim of on-going work). One way of preventing the core reaching this limit is to install stiffening devices in a number of channels. The number of channels required will depend on the effective stiffness of each device.

Assessment Criteria

In order to assess the different options a list of criteria was drawn up. The device:

- Must limit column deflections to less than the specified value.
- Must not alter the existing brick positions for the worse.
- Must allow for (differential) thermal expansion.
- Must not unduly affect the moderator performance (material compatibility).
- Must not impinge on safety.
- Must not increase the risk of brick cracking during implementation and in service.
- Must be capable of being deployed, given the geometric and access constraints of the core structure and the guide tubes.
- Should ideally be reversible.
- Should allow the channel to be available for continued use (*e.g.* as a fuel or control rod channel).
- Should be simple to decommission.
- Should maintain adequate channel flow and not cause undesirable acoustic tones.
- Should have acceptable long-term integrity.

Comparison of Available Channels

There are two types of channel available – interstitial channels and fuel channels. These are compared below, and based on the comparison it was decided that using fuel channels was the better option overall.

Interstitial channels	**Fuel channels**
Only around one quarter are accessible from the pile cap. They only have a small diameter (~125 mm) and thus only a low stiffness is achievable. Most are control rod channels and so sacrificing some of these could adversely affect reactivity control and shutdown / hold-down capability.	All channels are accessible from the pile cap. They have a larger diameter (~ 265 mm) and thus a higher stiffness is achievable. Sacrificing fuel channels will reduce the power output of the core, although it may be possible to compensate for this.

Materials

It is essential that the material used is compatible. Two material types were identified as candidate materials, namely nuclear grade graphite and carbon composites. The benefits and disbenefits of each are summarised below and on the basis of the comparison it was decided that graphite was the best material to use.

Nuclear grade graphite	Carbon composite
Relatively low stiffness (low YM)	Relatively high stiffness (high YM)
Relatively low strength	Relatively high strength
Fast neutron irradiation tolerant	Fast neutron irradiation intolerant (?)
Known oxidation rate/effect on properties	Unknown oxidation rate/effect on properties
Pre-characterised for use in reactor	Not pre-characterised for use in reactor

Preferred Option/Channel Locations

The different possibilities for the design and materials were ranked against one another using the criteria given earlier. The preferred option was found to be beam of graphite installed in sacrificed fuel channels with stabilising brushes at particular locations, probably at two locations per brick, one a distance below, and the other a distance above, the brick-to-brick interfaces (see Figure 1). Due to the current limitations for graphite extrusions (~3 m), the beam may have to be made up of three sections. The design of the interface will have to be such that it maximises the stiffness of the overall beam. The ends of the beam will be such that they will interface with the fuel plug unit and the fuel support at the bottom of the channel in the same way as the stringers. This means that little alteration will be required to the plug unit and they will be installed / removed in the same way as a fuel assembly.

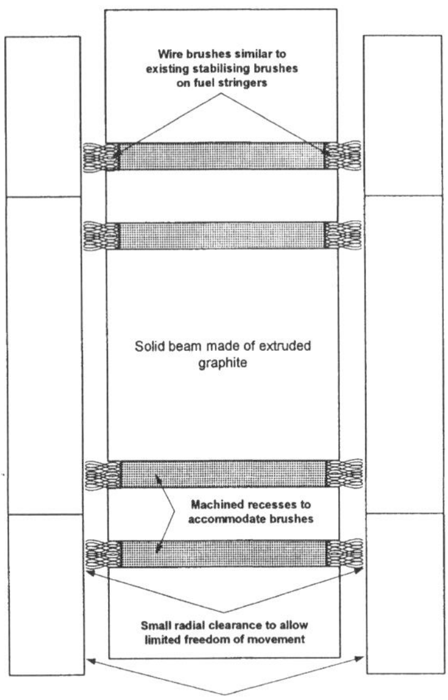

FIGURE 1: Possible locations of stabilising brushes.

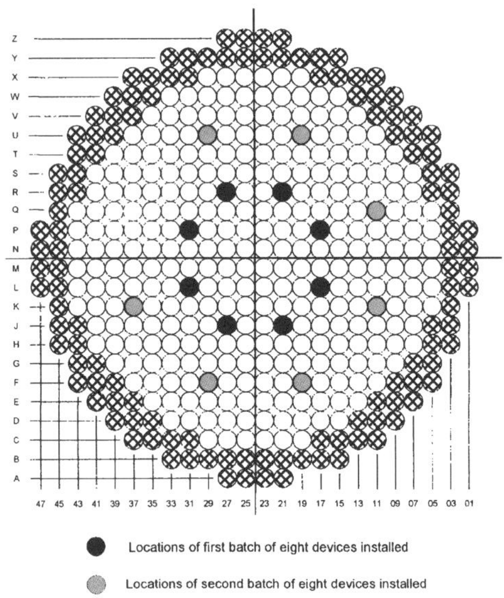

FIGURE 2: Possible locations for stiffening devices.

It is envisaged that eight beams may be sufficient to provide the required stiffening for the core. However, it may be necessary to increase this to, say sixteen. Possible locations for the first eight devices and if necessary the second eight devices are shown in Figure 2. However, there is a possibility of using the stiffening devices as a DSD system. This possibility is covered below.

DIVERSE SHUTDOWN SYSTEMS

Objective

The PSD system in the AGRs comprises the black control rods, although some negative reactivity can be inserted by the grey rods. During the design of the AGRs there was a need to include a diverse shutdown and hold-down capability in case of top dome failure, which although a very low probability event, could nevertheless prevent control rods dropping freely into the core. There is now considered to be an additional requirement for a diverse shutdown / hold-down system due to brick cracking which could potentially result in a core distortion that would prevent a sufficient number of rods entering the core to effect shutdown and hold-down.

Proposed Additional Safety Guidelines

As a result of the need to cater for severely distorted cores, some additional Safety Guidelines (SGs) for a DSD system have been proposed. These are that:

1. It should be able to shut down both reactors simultaneously (in the short term) and hold them down (in the long term) independent of the PSD system.
2. It should be operable with significant core distortion / degradation.
3. It should be activated automatically if control rods have failed to enter the core.
4. It must have a seismic withstand capability for a 10^{-4} per annum event.
5. It must be readily recoverable in the event of spurious initiation.

Current DSD Systems

Hinkley Point B / Hunterston B

At HPB/HNB there is the Enhanced Shutdown (ESD) system and the nitrogen injection system. The ESD relies on the grey control rods and (black) sensor rods being motored into the core in the event of the failure of the PSD. It is not therefore a rapid shutdown system, and would be equally affected by core distortion / degradation as the black rods. The nitrogen injection system is manually operated and requires at least half an hour from manual initiation before it can inject nitrogen of the required quality. It is not therefore a shutdown system, but can be used for medium to long-term hold-down in combination with the ESD system, which may be the more significant issue.

Hartlepool/Heysham 1

At HRA/HYA there is a nitrogen injection system, which automatically injects nitrogen gas into 165 interstitial channels in the reactor core on detection of either low top dome differential pressure or high channel gas outlet temperature to provide rapid reactor shutdown. It is referred to as the Secondary Shutdown (SSD) system. An emergency button is provided on the desk in the Central Control Room to allow manual initiation of the system. The system is designed to be capable of holding the pressurised reactor in the shutdown state for at least 12 hours in the absence of insertion of any of the control rods.

There is also a boron ball injection system that is designed to fill 32 of the SSD channels in the core with 3 mm diameter boronated glass balls. The boron balls are stored in 32 steel hoppers arranged in four groups around the outside of each pressure vessel and are blown into a reactor using the station CO_2 supply. The boron ball system is operated by manually opening valves and it takes a few hours to discharge all the balls. It provides a means of inserting additional neutron absorber into the core to prevent any risk of reactor re-criticality following a shutdown by the SSD system. At least 12 hours would be available after tripping of the SSD system before there is a requirement to initiate the boron system. Although it is referred to as the Tertiary Shutdown (TSD) system, it clearly does not provide a rapid shutdown capability.

Assessment against Additional Safety Guidelines
Each of the above systems was assessed against the additional SG's to identify any shortfalls. The results are summarised below.

Proposed additional Safety Guideline	HPB/HNB		HRA/HYA	
	ESD	Nitrogen	Nitrogen	Boron ball
Shutdown both reactors simultaneously	No	No	No	No
Operable with significant core distortion	No	Yes	Yes	Yes
Activated automatically if control rods have failed to enter the core	No	No	No	No
Seismic withstand capability for a 10^{-4} per annum event	Yes	No[1]	No[1]	No[1]
Readily recoverable	Yes	Yes	Yes	Yes/(No)[2]

[1] Currently being reviewed [2] Might not be possible to extract all the balls

Possible Enhancements to the Current Systems
From the table above, it can be seen that none of the current systems satisfies all of the proposed additional SG's. The study looked at what could be done to enhance the systems in order to try and meet them if considered necessary.

Hinkley Point B/Hunterston B
The ESD system is judged not to be sufficiently diverse from the PSD system when it comes to distorted cores, as it would be equally affected, and so was not considered further. The nitrogen injection system needed to be enhanced to operate on both reactors simultaneously to satisfy the first SG above. This could be achieved in one of three ways – build a replica plant, add additional plant and segregate, or demonstrate that there is sufficient time post trip to allow nitrogen to be injected into both reactors (one at a time) with the existing plant, and that this will provide enough time for the stocks to be replenished by tanker delivery before there is a need to recommence injection. With regard to automatic initiation (following failure of the PSD system), it is clearly not possible to enhance the system to meet this requirement as it takes at least 30 minutes to prepare the plant before it can inject nitrogen of the required quality. Although the seismic withstand capability is not currently claimed to be to the 10^{-4} per annum event, work is currently in progress to address this.

Hartlepool/Heysham 1

Although the nitrogen can be injected immediately on demand (whether automatically or manually initiated) to provide rapid shutdown, the current plant cannot do so in both reactors simultaneously. As for HPB/HNB, this could be achieved in one of three ways – build a replica plant, add additional plant and segregate, or demonstrate that there is sufficient time post trip to allow nitrogen to be injected into both reactors and replenished by tanker delivery before there is a need to recommence injection. Unlike the HPB/HNB plant, this system could be enhanced to operate automatically if the PSD system fails to insert sufficient negative reactivity. Although the seismic withstand capability is not currently claimed to be to the 10^{-4} per annum event, work is currently in progress to address this. It is also thought that the system may not be fully recoverable. Although there is a means available for ball extraction it may not be possible to remove all the balls.

Possible New Systems

As an alternative to enhancing the current DSD systems, the addition of new DSD systems has been considered. Since the core stiffening study concluded that the best option was to use sacrificed fuel channels, it was clear that these could also be used to provide a means of introducing a DSD system. Two such systems have been proposed, one using boron balls and the other using control rods. Each is described below.

Boron Ball Device

The idea for using sacrificed channels has arisen as a result of the feasibility study carried out on core stiffening options. The two basic components of the system are the in-core stiffening device itself and the plug unit. It would be fairly straightforward to design the in-core stiffening device in a way that would also allow it to act as a receptacle into which boron balls could be dropped and retained. The obvious storage location for the balls would be the plug unit. However, the existing plug units would have to be substantially modified for this, or completely new plug units would have to be designed and built.

The balls need to be stored in the plug unit in a suitable location above the core so that they do not have any effect on reactivity during normal operation. The quantity (mass) of balls that are required will depend on the net worth they must contribute and the overall volume of the channel(s) into which they will be dropped. It may be possible to modify the existing plug unit by replacing some parts with the ball hopper and any other devices. The new plug units will have to incorporate some of the essential features of the existing plug units used in all AGRs, namely closure unit, radiological shield plug, heat shield and a drive mechanism. These can all be seen in Figure 3, which is a section through the Heysham 2/Torness fuel plug unit. The drive unit moves the gag plug up and down to control coolant flow rates. In the proposed boron ball device, a gag unit will not be required, and the drive unit will form part of the ball release mechanism. The lower part of the plug unit, namely the neutron scatter plug, could be replaced by the boron ball storage section, as the boron balls will also act as a very effective neutron shield. The region above the neutron scatter plug (the piston seal housing) can also be utilised for ball storage if required. A schematic of the overall arrangement is shown in Figure 4. Alternatively, a completely new design can be adopted.

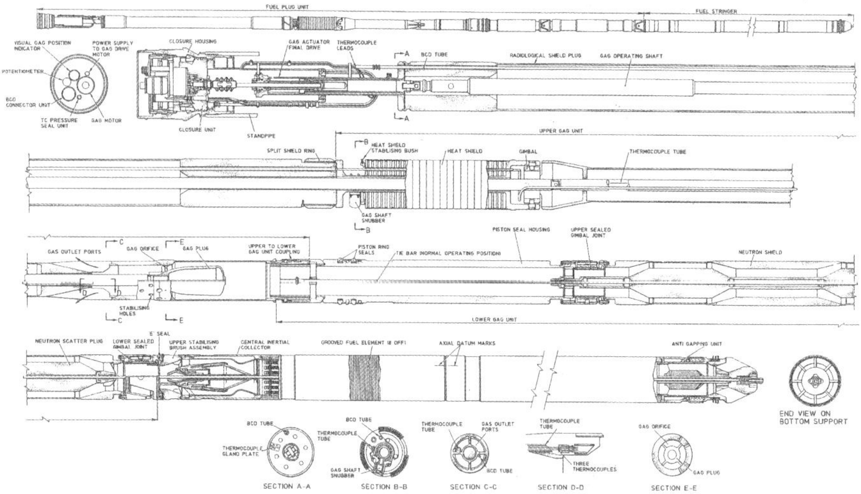

FIGURE 3: Fuel plug unit at Heysham 2/Torness.

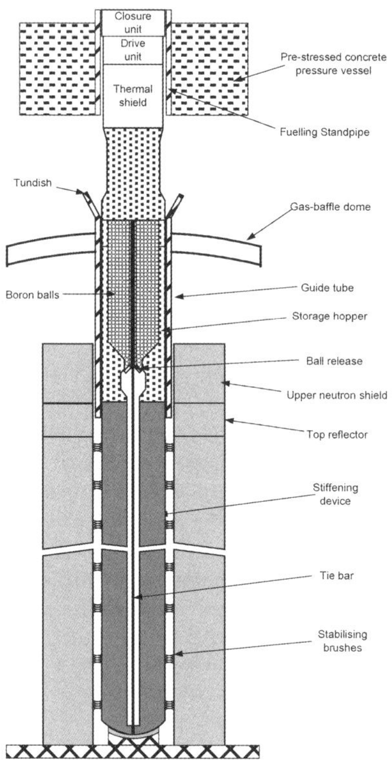

FIGURE 4: Schematic of boron ball device.

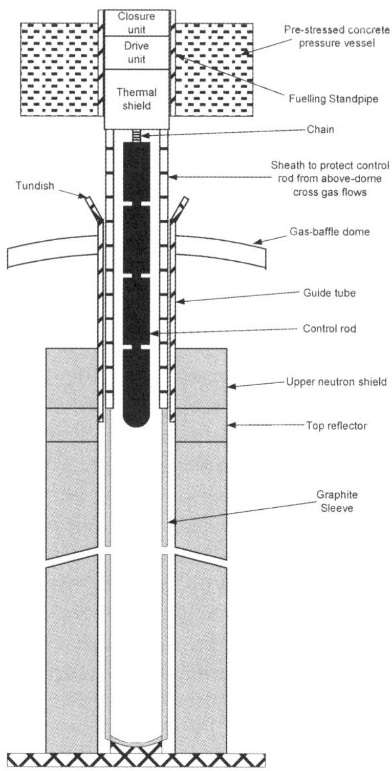

FIGURE 5: Schematic of control rod device.

Although the initial idea was to combine this option with a stiffening device, if it is subsequently decided that core stiffening is not to be undertaken, the boron ball device could still be deployed as a stand-alone modification. This could make the option easier to design and manufacture, and would cost less if fewer sacrificed channels were required for shutdown in comparison to providing the appropriate level of stiffening. The main drawback with this option is that recovery from its operation (whether via a legitimate demand or a spurious trip) will require the affected unit(s) to be removed, the balls extracted and refilled, and the unit(s) reinstalled.

Control Rod Device

This proposal was developed based on the fact that if sacrificing fuel channels is considered to be a viable option, they could be used instead for the installation of black control rods. The much larger channel diameter of a fuel channel will allow a much larger diameter control rod to be introduced, and therefore significantly increase its worth. The control rods would be equivalent to the on-lattice control rods at Dungeness B (DNB). The plug units could be a scaled up version of the existing ones or based on the DNB design.

A major issue to deal with is empty channel flow, similar to that which arises when a fuel assembly is discharged on load. It may be possible to design an extended plug unit that can provide the necessary reduction in flow. If not the channel will have to be sleeved in some way, replicating the effect of the fuel sleeves. This could be in the form of a column of sleeves supported by the fuel support strut. Alternatively a column of sleeves could be suspended from the plug unit in a similar way to the fuel elements. A schematic of such an assembly is shown in Figure 5.

The main benefit of this option over the boron ball option is that recovery from its operation will simply require the rod(s) to be raised back up to their fully withdrawn position. A possible disbenefit is that the sleeves, if installed, will not provide the same level of stiffening as a stiffening beam would, although as mentioned above for the boron ball devices, there may not be a requirement to provide any stiffening.

CONCLUSIONS

Two feasibility studies have been carried out with the aim of making the AGR at HPB/HNB and HRA/HYA cores more tolerant to widespread brick cracking. The first of these looked at the possibility of introducing stiffening devices into specific core channels to limit the potential for brick displacements/column bowing. The second of these looked at enhancing the existing diverse shutdown systems, and introducing new systems. These could be used to safely shut down both reactors simultaneously and to provide long term hold-down, in the event that widespread brick cracking prevented the black control rods (*i.e.* the primary shutdown system) entering the core in sufficient numbers. This paper describes ways by which core stiffening and diverse shutdown and hold-down can be achieved, and shows that the requirement for both can be met using a single (dual-purpose) plant modification. In either case a full assessment of the safety benefits and disbenefits must be fully explored before further investment in the plant infrastructure is considered.

Acknowledgements

The author would like to thank British Energy Generation Ltd (BEGL) for permission to produce this paper. The views expressed in this paper are those of the author and do not necessarily represent those of BEGL.

An Integrated Architecture for Graphite Core Data Analysis

Gordon J Jahn, Stephen D J McArthur and Jim Reed[#]
Dept of Electrical and Electronic Engineering, University of Strathclyde, Glasgow, UK.
[#]British Energy (Generation) Ltd, Barnett Way, Barnwood, Glos., UK.
Email: gordon.jahn@eee.strath.ac.uk

Abstract
This paper presents a method for capturing event data from various different monitoring techniques in an extensible manner. The paper describes some of the data that is gathered and how this data can be used to infer the state of the reactor core before introducing a multi-agent system that can take data and events to produce event data that can be readily examined and interrogated to cross-reference the results of multiple analyses. The paper briefly details the operation of the system at present and shows how the system can be expected to work as development continues. Conclusions are drawn as to how this system will fit into the overall ageing management process.

Keywords
Condition monitoring, multi-agent system, data integration

INTRODUCTION

In order to allow the graphite cores to continue to operate throughout and potentially beyond their design life there is a constant requirement for core condition monitoring to ensure degradation is occurring as predicted and in an operationally safe manner. During commissioning relatively few sensors were built into reactor cores and as a result much of the data available for analysis is gleaned from secondary systems. The analysis and verification of increasing amounts of data and findings from the secondary systems becomes in itself a growing burden that must be actively managed.

This paper initially describes the various continuous and discrete-event data that are currently available from primary and secondary systems relating to the reactor cores before giving an overview of how each of the physical properties that can be measured can affect the physical make-up of the reactor core. Once an outline of the system in development is provided, an analysis of the expected interactions between the measured quantities that include fuel grab loads, neutronic and thermal powers, and control rod heights is given. Through this, it can be seen how these physical interactions manifest themselves within the system.

Throughout this paper, the emerging paradigm of using intelligent agents for power applications (Liu, 2000 & Buse, 2003), and more specifically condition monitoring (McArthur, 2004) is applied to the challenge of analysing the myriad of data available from the nuclear power station. The paper describes and justifies an architecture that has been adopted and explains how this will help achieve a number of key goals for the system development. These goals include the provision of a robust on-line monitoring system that can gather and analyse the various types of data to provide summary information along with an advanced data-mining system that can identify interesting anomalies within the various data streams being analysed.

MOTIVATION

As part of the ageing management of their graphite cores, British Energy has recently instituted Monitoring Assessment Panels (MAPs) at their stations. These panels are designed to formalise the process of corroborating data and findings between the various monitoring functions that already take place.

The current process is based on a reporting system where members of key teams within each station report back to the MAP which is then responsible for reviewing the monitoring data as a whole body of evidence in order to reach a number of conclusions and recommendations. There are, however, around 15 unique items being monitored as part of the MAPs, including the data covered in the 'Available Data' section below. Given that the MAPs will meet every three months and will cover data from control rods positions measured every few seconds up to refuelling traces which are available for each channel every few years, the challenge of analysing this data effectively is a great one. It can therefore be seen that a system that can effectively manage the key events and data with the MAP process will clearly be of great value to British Energy.

AVAILABLE DATA

There is relatively little primary data relating to the reactor core – the only direct online measurements that are normally made inside the core are of temperatures in fuel channels. The thermocouples used for this were built into the reactor at design time. In addition to these, however, there are other direct measurements and data obtained during regular inspections – these include geometry measurements from the Channel Bore Monitoring Unit (CBMU) and television pictures of the channels.

In addition to the raw data from the reactor there are other derived and secondary values relating to core condition that can be obtained. As there may be more of these values as new techniques become available, any system must be able to handle new data and event sources, but the initial version of the integrated condition monitoring system will be capable of capturing the significant data and events from the various processes that feed into the Monitoring Assessment Panels. The data includes fuel handling, control rod and core-follow route power comparison data and events. Each of these is taken in turn to be described below. It should be noted that the main archive will tend to store event data rather than direct data and that this data can then be accessed by contacting the appropriate agent.

Fuel Handling Data

The fuel handling teams on station are responsible for refuelling the reactors on a regular basis and are, as such, perfectly positioned to gather and analyse a significant volume of information that might relate back to the core condition. The main data item gathered by the fuel handling team is fuel grab load traces (FGLTs), but the people involved with fuel operations also have the opportunity to collect and gather information on other events that may be of interest within the overall monitoring programme. These include load change events which are of interest particularly in relation to the core-follow route power comparisons and any deformations to the stringer whilst it has been in the core. In the latter case, although something like a digital photograph of a stringer may be of little use on its own, when used as supporting evidence for another analysis it may allow the possible scenarios to be eliminated.

Control Rod Data

The reactor control rods are constantly being adjusted to trim the reactor power output and periodically the emergency shutdown rods are dropped into the reactor over several seconds to ensure their continuing functionality. Throughout these movements, data is logged on the positions of all the rods within the reactor and as a result a number of key events can be drawn – these include non-standard rod group characteristics where a group of control rods are not operating exactly as expected and non-standard individual characteristics where a single rod is not operating as expected.

The other significant event associated with the control rod system that requires capturing in order to gain a useful picture of rod-related problems is the information on rod replacement. In this case, the system would be looking to pick up on recurrent problems after the rod has been exchanged and this might relate to reused components within the new rod mechanism or the location of the rod channel within the reactor.

Core-Follow Route Power Comparisons

The core-follow route power comparisons are a number of calculations that are made on a regular basis to assess the amount of neutronic power that can be expected to have been released into the reactor based on a number of inputs. The power released by the nuclear fission can then be compared to the amount of energy seen within the coolant cycle (thermal power). These numbers should match within a few percent otherwise an explanation is sought for the difference in power which might result in a hot-spot within the reactor which could then cause other complications.

MULTI-AGENT SYSTEMS

Multi-agent systems have been in development for some time (Wooldridge, 1995) and have seen growing use in condition monitoring applications recently. Three major reasons for implementing large condition monitoring applications in this manner are the simplification of a complex problem, compatibility with existing systems and flexibility for extension.

Ultimately an agent is a software program, but one which is designed to be able to act autonomously. In this way each agent should be capable of setting its own goals and planning and executing tasks in such a way as to meet the goals it sets. Each agent is thus written to be able to perform a number of key tasks which are called behaviours.

The ability to communicate within multi-agent systems is underpinned by two key technologies – the Agent Communication Language (ACL) and the ontology. The ACL is the actual format of messages on the wire and may be a binary encoded version of the messages, XML or any other format. This can be seen as analogous to choosing whether to use e-mail or pen and paper for sending a message. The actual content of the message is defined by the ontology – this would be the words you either type or write within your e-mail or letter. In the case of the system presented here, the ACL is defined by the platform although the ontology must still be designed and implemented – the ontology is described in further detail in the 'System at Present' section below.

Simplification of Implementation
It has already been highlighted that the integrated monitoring system will initially span three sets of experts on station – the fuel handling, control rod and thermal-neutron power teams. In addition to this, there are several groups across British Energy that must be involved to some degree within the eventual operation of such a system. These include the Graphite Core Project Team based at Barnwood in addition to the fuel handling, shift and nuclear safety group engineers at each station. Finally, there is the complication that there are subtle differences between almost all of the AGR stations within the UK that may or may not be significant in terms of monitoring. All of this leads to a complex requirement that can only realistically be achieved by breaking the problem down into smaller, manageable chunks.

Compatibility with Existing Systems
The integrated monitoring programme at British Energy is a relatively young project and as such the data sources that are expected to feed into the system were not designed with ongoing condition monitoring in mind. It is therefore crucially important that the integrated monitoring system is designed to be integrated with the existing systems in as transparent a manner as possible. A multi-agent system is an excellent way of managing the introduction of such a system. A single agent within the multi-agent system can be implemented for each of the legacy systems that require to be interfaced with. In this configuration, these agents are acting as a bridge between the parameters and protocols of the existing system and the ontology that has been defined for the multi-agent system.

Flexibility for Extension
Multi-agent systems offer a useful feature over monolithic systems in that each component of the system can interact with any other agents that can communicate with it in the same language. This is extremely useful in the case of ageing management due to the volume of data that could be gathered. The multi-agent architecture effectively and efficiently allows the system to be built and developed in an ongoing fashion. This will allow additional components to be targeted and prioritised in terms of their expected value – value which may be likely to change with time. In addition to helping with getting data into the condition monitoring system, the multi agent approach also helps once data is in the system. In the event of new analysis techniques or user requirements being developed, this system will be ideally positioned for rapid implementation and rollout.

SYSTEM AT PRESENT

The integrated monitoring system is currently in development. As noted in the 'Motivation' section above, the first stage will be to put together a system for use at British Energy. The initial release will therefore contain just six basic agents:

- A fuel grab load trace event agent.
- A load agent.
- A power comparison event agent.
- An archiving agent.
- An analysis agent.
- A user agent.

In addition to these agents, a Directory Facilitator agent is provided by the platform – this acts as a yellow pages style service within the community to allow agents to find other agents who can perform certain tasks. As noted above, a key requirement for a multi-agent system is an ontology that allows all the participating agents to communicate and interact. The ontology for this system is shown in Figure 1. This diagram shows the various concepts that have been defined for this system. The Agent Action and AID concepts are pre-defined for the Java Agent Development Environment (JADE) system and are used to hold detail behaviours and the addresses of agents respectively. The other concepts seen in the diagram have been defined specifically for this system. It can be seen that the power station has been modelled further than strictly necessary – it is intended that this will help in the future as the system is extended to take in further data points from further plant items.

It should be noted that the JADE system has been chosen for its compliance with the Foundation for Intelligent Physical Agent (FIPA) standards which are endorsed by the IEEE Computer Society. As such, the system being developed is built firmly upon international standards. It can be seen from the diagram that the ontology is built from concepts within which there are a number of slots which might contain values or other objects within the ontology and that the ontology also defines the cardinality of slots. As an example, a fuel grab load trace event ('FGLTEvent') is also a fuel handling event ('FuelHandlingEvent') which is itself also a reactor event ('ReactorEvent') and so on. An additional top-level concept of 'EventGroup' has been defined which can contain many condition monitoring events ('CMEvents'). These concepts are then used by the multi-agent system.

FGLT Event Agent

The FGLT event agent will ultimately be strongly bound to the FGLT analysis system also under development at Strathclyde University. It is intended that this agent will provide information on when certain channels were refuelled and information about the analyses that have taken place (West, 2006). This agent directly interacts with the archiving agent to get an event ID which can then be used to track and trace this event throughout the system. It is also expected that events generated by the FGLT agent will be surrounded by load change events – one bringing reactor power down and one ramping back up. Additionally, it would be expected that following a reactor refuelling, there will be control rod events triggered by the additional fuel being present within the reactor, thus requiring additional control. Although control rod data is not planned for the initial version of the multi-agent system it is expected that this data will soon become available.

Load Agent

The load agent is an agent that will monitor the operating state of the reactor and generators. This information is very useful as a number of other monitoring events might be expected to be associated with load change events. As an example, the refuelling events will take place after the load has been reduced and you would expect to see load increased again after such an event. On either side of these load events, control rod positions would also be expected to move. The status will thus be continuously monitored and logged so the system is capable of using this information in its analyses. Each load change event will be archived with the archiving agent in the same manner as for the FGLT agent, but agents requiring load status information for a particular time in the past contact the load agent.

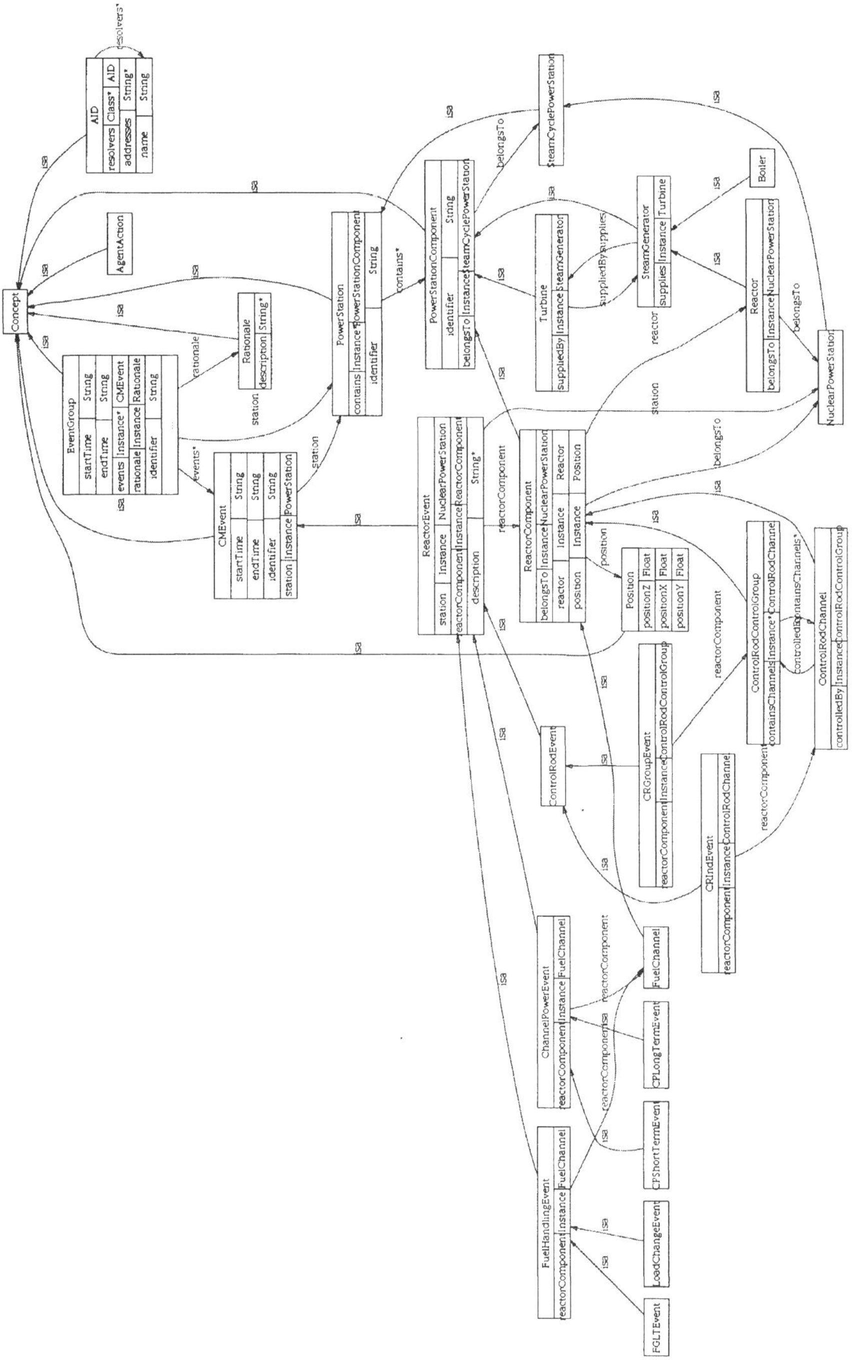

FIGURE 1: Diagram showing the Integrated Monitoring System ontology.

Power Comparison Event Agent

Power comparisons between expected neutronic power and thermal power are regularly undertaken as part of the operational verification of the stations. This process is undertaken by engineers using PANTHER software. It is intended that the Power Comparison Event agent will be able to interface with the PANTHER system and obtain the results from that system which can be incorporated with the other monitoring system results. In the case of power comparisons, it is important to note which channels have been recently refuelled and when there have been load changes as these two events can result in anomalous results within the power comparison calculations. These data interactions will lead to interactions within the multi-agent system.

Archiving Agent

The archiving agent is key component within the multi-agent system. This agent will be expected to play a key role in data and event tracing. The analysis parts of the system have been designed in such a way that they will only accept data that has been "tagged" with a valid ID. The IDs are created and distributed by the archiving agent. This provides two important benefits to the overall system – events can be compared for overlap and duplication at the entry point to the system ensuring that the event list maintained by the archive is correct and a full audit trail can be generated allowing engineers to trace the events back through the system to source in the case of investigations being required. The archiving agent is also responsible for event versioning. This may be important in the case of data being re-analysed for some reason.

Analysis Agent

The analysis agent within the system will form the key component for handling user-requested analyses and automated ongoing analysis work. The agent will be capable of handling analyses in two different manners – it will handle ongoing, routine analysis on an event-driven basis as well as ad-hoc analyses demanded by engineers through the user agent.

User Agent

The user agent within the system is the link between the human engineer and the multi-agent analysis system. This system will initially have a web-based front end that will facilitate easy deployment – in this case an agent runs within the web server. In the long term, the multi-agent system approach allows for this to be replaced with more complex agents that can learn the interests of the engineer.

FURTHER WORK

Over the coming months, the basic multi-agent system for pulling together and managing event data for British Energy's AGRs will be completed. At this stage, it is anticipated that the system may be rolled out to a single station for testing. There are clearly then other data types that can be incorporated into the system – the next being control rod data. Ultimately, it is believed that the system will be of more benefit if it is capable of pulling together evidence across stations too and as a result the project will continue to look at decentralising the system across the various stations in the British Energy fleet. At this time it is hoped that the multi-agent systems on each station will be capable of interacting with agents at other stations and with specialised user-agents at Barnwood that will aid with overall ageing management. Following this, the aim is to start looking at the available information and data in more novel ways to gain as much benefit as possible from the data. It is further hoped that the experience of this project may be of some use in helping define the measurement and monitoring technologies that might be specified in future nuclear builds.

CONCLUSIONS

This paper has presented a method for combining event data from multiple sources within a multi-agent system. The resulting integrated monitoring system will provide key benefits for ageing management in helping to identify common patterns and rule out events that might otherwise require significant time to investigate.

Acknowledgements
The authors would like to thank British Energy Generation Ltd for financial support. The views expressed in this paper are those of the authors and do not necessarily represent those of the British Energy Generation Ltd.

References

Buse, D.P. et al (2003). Agent Based Substation Automation. *IEEE Power and Energy Magazine*, Vol 1, No 2, pp50-55.

Liu, C.C., Jung, J., Heydt, G. T., Vittal, V., Phadke, A. G. (August 2000). The Strategic Power Infrastructure Defense (SPID) System: A Conceptual Design. *IEEE Control Systems Magazine*, pp40-52.

McArthur, S. D. J., Strachan, S. M., Jahn, G. J. (2004). The Design of a Multi-Agent Transformer Condition Monitoring System. *IEEE Transactions on Power Systems*, Vol 19, No 4, pp1845-1852.

West, G. M., Jahn G. J., McArthur, S. D. J. (2006). Graphite Core Condition Monitoring Through Intelligent Analysis of Fuel Grab Load Trace Data. *Proceedings of the Ageing Management of Graphite Cores Conference*

Wooldridge, M., Jennings, N. R. (1995). Intelligent Agents Theory and Practice. *The Knowledge Engineering Review*, 10 (2), pp115-152

Graphite Core Condition Monitoring through Intelligent Analysis of Fuel Grab Load Trace Data

Graeme M West, Gordon J Jahn, Stephen D J McArthur and Jim Reed[#]
Advanced Electrical Systems Research Group, University of Strathclyde Glasgow, UK.
[#]British Energy (Generation) Ltd, Barnett Way, Barnwood, Gloucester, UK.
Email: graeme.west@strath.ac.uk

Abstract
As a graphite core ages, there is an increased requirement to monitor thc distortions within the core to permit safe continued operation of the station. In addition to existing monitoring and inspection, new methods of providing information relating to the core are being investigated. Two systems are described in this paper to support the use of Fuel Grab Load Trace (FGLT) data, which has been shown to provide information relating to core distortions and brick cracking. The first system was designed and implemented to provide support to analyzing large volumes of fuel grab load traces for patterns and thereby increase the understanding of how FGLT relates to physical core distortions. The second system utilizes this developed knowledge to provide a method of automatically assessing new FGLT data for the presence of abnormal behaviour that is symptomatic of brick cracking.

Keywords
Condition monitoring, nuclear reactor refuelling, intelligent systems

INTRODUCTION

Information relating to the current condition of the core is very valuable as the nuclear stations within the UK enter the latter stages of their operational lives. There is a desire by the station operators to extend the station operational lifetimes beyond the initial predictions in order to drive more value out of them. This is balanced by the obligation to ensure that the station is being run safely. This paper describes the utilisation of fuel grab load trace (FGLT) data to provide additional information relating to the current condition of the core. FGLT data is captured and retained as part of station records for protection purposes, particularly for detecting faults such as a failure of the tie-bar load mechanism, ledging and false-bottoming of the fuel assembly within the channel during refuelling. An expert system was designed to analyse the last 50 cm of FGLT data to assess the success of fuel touchdown and is reported by Steele (2003). It has been shown, however, that this data can also provide information relating to distortions within the core. There are two main objectives of this research, firstly to develop a method for exploring the underlying relationship between the data and physical core condition and secondly to develop a system for automatically assessing new FGLT data in a structured and repeatable manner so that any anomalous behaviour can be quickly identified, based on this initial understanding of the data. This paper is divided into four main sections. The first section provides a brief overview of the refuelling process and how FGLT data can provide information relating to core condition. The second section describes a process of data-mining the FGLT data in order to develop an understanding of how the data relates to the core condition, and also to develop an understanding of what a "normal" refuelling event should look like. This data-mining approach is supported by the development and implementation of a software system which will also be described. The third section describes taking this newly acquired understanding and implementing it within a system that can automatically interpret new fuel grab load trace data. Finally, the last section discusses the future direction of this work, with reference to potentially supporting future core safety cases.

BACKGROUND

It is assumed that the reader is familiar with the design of AGR graphite cores. This section provides a very brief overview of the refuelling process and provides an explanation as to why the FGLT data can provide information relating to the reactor core brick condition.

Refuelling Background

AGR refuelling takes place on a regular basis with about 60-70 channels per reactor refuelled each year. A single refuelling event involves the removal of the old, spent fuel and the replacement with new fuel. During the entire refuelling process, a load cell measures the apparent weight of the fuel assembly as it traverses the core. Traditionally this has been recorded using a paper plotter although there is a move towards recording and storing this information digitally.

Relating FGLT Data to Physical Core Condition

It was recognised that the physical features, such as the piston seal bore, could be related to features within the fuel grab load trace. Further investigations revealed that changes in the channel diameter are also reflected in the trace. The fuel assembly usually has two sets of brushes which guide it through the reactor during refuelling. These brushes form an interface with the channel wall and changes in channel diameters are reflected in the FLGT data. It should be noted that there are slight differences in design of fuel assembly and is dependent upon the station. This work focuses on Hunterston B and Hinkley Point B stations, although the same principles are equally applicable to most of the other AGR stations. As the reactor ages, the graphite bricks will deform which may ultimately lead to brick cracking. It is currently thought that brick cracking will be reflected in one of two ways within the FGLT data, either as a distinct peak in the data, or as a step change in load across an entire brick. As FGLT data was not specifically collected for condition monitoring purposes there is not the underlying years of experience with interpreting it for these purposes. Figure 1 shows a comparison between the FGLT data and a set of data measurements of channel diameter taken during an outage using a channel bore monitoring unit (CBMU). The similarities in the data can be clearly seen and in particular the peaks in the FGLT data correspond to the reductions in diameter at the brick layer interfaces. In addition, a suspected crack in the CBMU data can also be seen reflected in the FGLT data. This diagram also indicates the difference in data quality with the CBMU data producing a cleaner profile of the channel, whereas the FGLT clearly has an element of noise introduced. However, the CBMU data is only gathered from a limited number of channels over an infrequent time period whereas the FGLT is gathered on a much more regular basis.

Types of Distortion

The objective of this research is to develop a system for detecting abnormal core distortions from the FGLT data. This process is difficult for two main reasons. Firstly, there are very few instances of brick cracking and secondly the exact nature of how these defects form and manifest themselves within the FGLT data is not fully understood. It is currently thought that there are two main modes of brick cracking, namely circumferential cracking and primary cracking. Circumferential cracking refers to cracks that propagate round the inner surface of the channel wall as shown in Figure 2. The stresses built up within the brick cause it to distort at the crack interface causing a restriction at the crack over time. This manifests itself as a peak within the FGLT data and has been corroborated with known cracks from CBMU data.

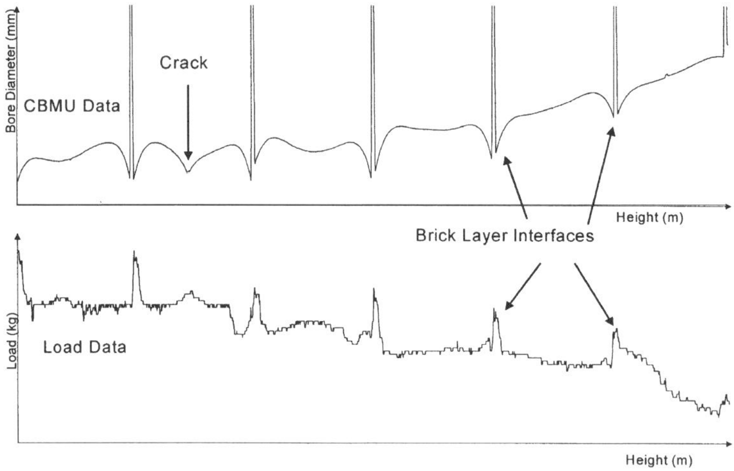

FIGURE 1: Diagram comparing FGLT data with channel diameters.

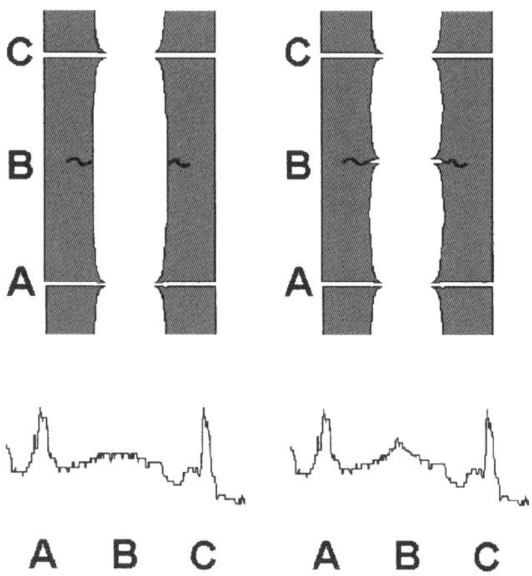

FIGURE 2: The development of a circumferential crack. The upper two diagrams show a cut-away of a graphite brick with points A and C illustrating the interfaces between two bricks. The lower two diagrams show the equivalent FGLT response.

Primary brick cracking occurs when the brick splits down its entire length. In addition, it is hypothesised that a second full-length crack can develop breaking the brick into two parts. The results of primary brick cracking are not fully understood, though it is thought that both of these cracks would cause a step change in expected load across the entire brick layer. It is hypothesised that a single primary crack will cause the brick to open out slightly producing a slightly wider channel diameter. As a result there will be less friction on the fuel assembly and overall a drop in load. In the situation of a double cracked brick, there may be no visible change if the brick is held in place by the surrounding bricks. However, if there were a shearing force between the two halves of the brick, then there would be a resulting reduction in the diameter seen by the fuel assembly, resulting in a step change in load across the brick layer. It should be stressed that this understanding at present is still being developed and while there are very limited instances of brick cracking the behaviour will be restricted to theoretical and experimental results.

Availability of Data

FGLT data has been recorded for every refuelling event undertaken at every station and is kept as part of station records. Though there is a move to make the capture of this data electronic, the majority still is stored as paper traces. In an effort to provide a benchmark of FGLT data for future analysis, and to allow a greater number of these traces to be compared, a substantial number (approaching 1000 events) have been converted to electronic files through a process of scanning and digitisation. This has provided a large volume of data, but as a result, automation of the analysis is required in order to prevent the analyst becoming overloaded with data. The following section describes the data-mining approach adopted for analysing this data.

A FGLT PROFILING SYSTEM USING A DATA-MINING APPROACH

A profiling system was developed to explore the FGLT data, which allows engineers to identify outlying refuelling events from a user-defined set of events. It also allows the user to define a set of good behaviour based on a predefined set of known, acceptable refuelling events. In addition, relationships between the channel location, load status, date of refuelling and status of fuel stringer can be explored. The resulting profiling system is easy to use, provides a consistent method of comparing FGLT data, allows several hundred traces to be examined simultaneously and allows models of expected behaviour to be developed.

Data-Mining Approach

The approach adopted for the development of the system, and therefore the analysis of the FGLT data, follows the general Knowledge Discovery in Databases (KDD) process proposed by Fayyad (1996). The key stages of this are summarized in Figure 3, and can broadly be separated into a pre-processing stage, where the data is selected and cleaned and a data-mining stage where the data is examined for underlying patterns.

Data Pre-Processing

The step of understanding the domain and selecting the appropriate data was straightforward, as there was only FGLT data available. It was, however, important to ensure that as much of the event specific information was preserved with the actual data. This included the station, reactor and channel relating to the refuelling event, as well as the date of refuelling. In addition information as to whether the reactor was on-line or off-line was recorded as well as whether the fuel stringer had previously been shuffled from another channel, as it is suspected that this will have an effect on the load trace. Cleaning the data involved removing spurious data points from the digitised data during the digitisation process and also removing stoppages, where a fuel assembly is paused during refuelling to inspect the seals before being shuffled to another channel. Also, where relevant, the data was rescaled to a common sample rate to allow two or more events to be compared on a point-by-point basis. Finally, the data was segmented into individual brick layers to allow a more detailed brick by brick analysis to be undertaken. The resulting cleaned, pre-processed data was then stored in a database to permit fast access and searching, using the event information as the indexing terms.

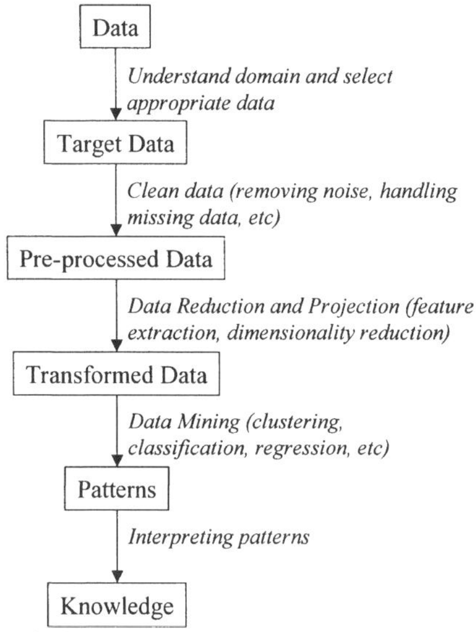

FIGURE 3: The key stages of the knowledge discovery in databases process.

Data-Mining

Data-mining is the process of discovering new patterns and knowledge within datasets. For detecting abnormalities relating to circumferential cracking, brick layers were analysed on an individual basis. In addition, the shape of the brick profile, rather than the absolute values of load, was found to be the important factor. Also, in order to identify a trace as abnormal, some measure of normality was required. As there was no definite knowledge of what a good and bad brick layer trace should look like, it was felt that selecting one event as a benchmark would not necessarily be the most optimal solution. Instead, it was decided that an average trace developed from all the traces in the data set would provide a good reference. The assumption here is that most of the events represent good behaviour and that there is only one single mode or shape relating to good behaviour. For the purposes of exploring the FGLT data, the method shown in Figure 4 was developed and implemented within the profiling system.

The steps from generating the average trace through to identifying the least similar traces were automated within the profiling software. Human intervention is still required in order to make the judgement of which traces should be included in the initial collection of traces and also to assess those traces that are identified as least similar to the majority. The measure of similarity chosen was the root mean squared error between a trace and the reference trace. However, the manual tasks of plotting the graphs and calculating the similarities have been automated resulting in a very quick way of comparing many traces. The repeatable nature of the software also makes it useful for providing a consistent, auditable measure of an anomaly as opposed to a more subjective, manual analysis.

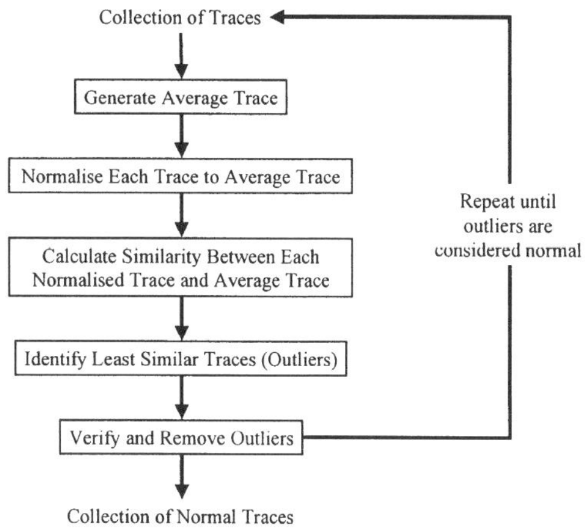

FIGURE 4: Method of analyzing FGLT data.

Implementation and Use of the Profiling System

The system was implemented using a client-server based architecture. The system is accessed through a web-based front end where the user selects the appropriate files for analysis using a number of search criteria such as station, reactor and refuelling date. This dataset is then queued on the server for analysis which is undertaken by the processing engine at the next opportunity. The user can monitor the progress of the analysis from a web-page and is informed when the analysis is complete. When the software is used to perform the analysis, as shown in Figure 4, there are two outcomes. The first is identification of those traces that differ significantly from the majority of the other traces which can then be examined in greater detail, perhaps in conjunction with other sources of data such as inspection records from outages or with further historical data from the station. The second is to identify a set of traces that the user considers to be normal. This data set can then be used to derive a set of envelopes of expected behaviour against which new refuelling events can be compared. In this way, a set of envelopes for each reactor in each station can be defined.

Case Study

This case study demonstrates the use of the profiling system to define a set of envelopes for brick layer 7 in Reactor 3 for Hunterston B, and four screenshots from this process are shown in Figure 5. Firstly, the user enters the search criteria and is presented with a list of matching files. This is shown in the top left screenshot. The user then selects those files to include in the analysis. The analysis is then queued on the server and undertaken automatically when it reaches the top of the queue. The top right screenshot shows a plot of the raw data, while the bottom left screenshot shows the data normalised to the average trace, indicated by the thick black trace. The bottom right screenshot shows a plot of the RMS errors from this average trace for each trace. An ordered list of traces from highest to lowest RMS error is also provided. Finally, the maximum and minimum envelope along with the average trace can be saved to an external file for incorporation into the intelligent analysis system.

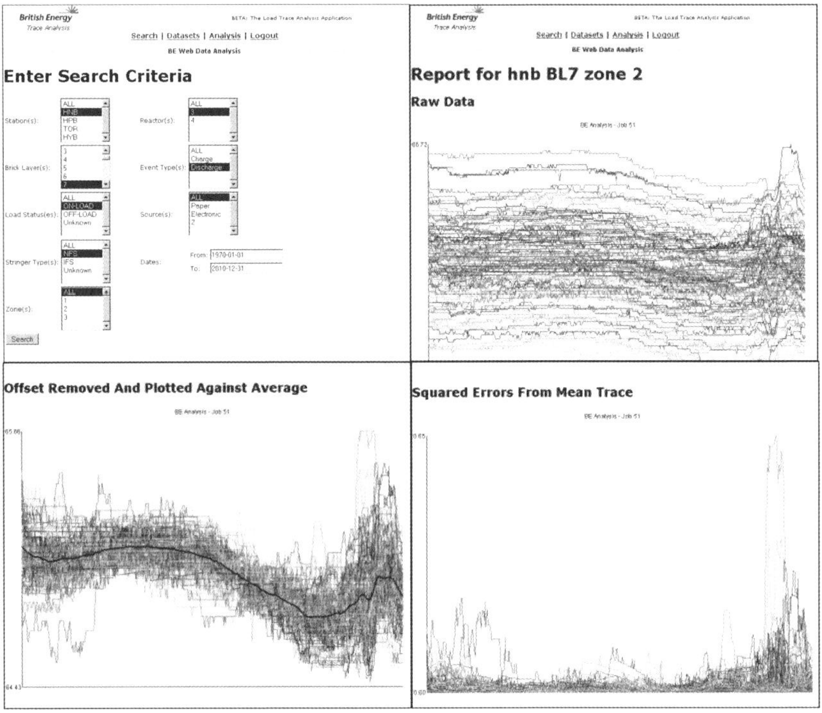

FIGURE 5: Four stages of using the profiling system.

AUTOMATED INTELLIGENT ANALYSIS SYSTEM

In addition to the profiling system for exploring the FGLT data, a second system was developed to automatically analyse FGLT data taken from new refuelling events. This system uses sets of acceptable envelopes derived using the profiling system to make an assessment of any new trace. In addition to assessing the shape of each brick layer, thus identifying possible circumferential cracking, it also assesses the average load of each brick layer and compares it with a set of expected brick averages. A step change in load across a brick layer, therefore a similar step change in average load value for the brick layer may indicate the presence of primary brick cracking. The automated intelligent analysis system has a simple front end. The user selects a refuelling event file or files to analyse. The system obtains some key configuration information from the filename, such as the station and reactor, and loads the appropriate rules and envelopes of expected behaviour. The raw data is then segmented into individual brick layers by identifying the brick interfaces characterized by peaks in the data. This identification is undertaken using a rule-based engine that classifies each peak within the data according to its shape, relative location within the load data stream and its associated height value, if available. These underlying rules are generic to all the stations although the individual limits are specific to the station to which the event being examined belongs. Once the brick layers are identified, they are compared to the pre-defined envelopes in order to identify any that breach the bounds of the envelope. A report is automatically generated which provides the results of each brick layer analysis, both in terms of any breaches of the expected envelope and also a comparison to the expected average load. Summary results of the analysis are also stored within a database to be retrieved by future analyses of that channel.

FUTURE WORK AND CONCLUSIONS

Both systems are installed and are undergoing an evaluation period at the Engineering Offices at British Energy. Currently, the analysis undertaken by the automated intelligent analysis system is restricted to making statements about whether the expected envelopes have been breached or not. A further area of research is to characterise these deviations and this will hopefully become possible as deeper understanding of FGLT data applied to core condition monitoring is developed. There is also scope for including historical analysis of tie-bar load trace data with the ultimate goal of deriving a model of how graphite core bricks distort over time based on previously observed behaviour. This raises interesting questions about how and where the data should be stored.

This paper has described the use of FGLT data to provide information relating to core condition from a source not originally intended to provide condition monitoring information. This has resulted in the development and implementation of two supporting systems. The first system importantly allows a deeper understanding of the behaviour of FGLT data to be developed through a data-mining approach. It provides the means to visualize and explore large volumes of FGLT data in a quick and repeatable manner and as a result define models of expected behaviour. The second system uses these models of expected behaviour to automatically assess FGLT data gathered from new refuelling events quickly identifying potential instances of brick cracking. The user is provided with a detailed summary of the analysis on a brick by brick basis. Both these systems allow existing data sources to be leveraged to provide information relating to the condition of a channel following a refuelling event, which occur with a far greater frequency than the existing detailed inspections undertaken during outages. In addition, gathering existing and historical data in a single location and through the use of the profiling software, patterns and relationships between different refuelling events can be rapidly and consistently be explored to determine whether time, location, stringer condition and load conditions have an effect on the trace, ultimately increasing the understanding of the existing, and possibly future, condition of the core.

Acknowledgements
The authors would like to thank British Energy Generation Ltd for financial support. The views expressed in this paper are those of the authors and do not necessarily represent those of the British Energy Generation Ltd.

References

Fayyad, U., Piatetsky-Shapiro, G. and Smyth, P. (1996) “From Data Mining to Knowledge Discovery in Databases” AI Magazine 17(3): Fall 1996, 37-54.

Steele, J. A., Martin, L. A., McArthur, S. D. J, Moyes, A., McDonald, J. R., Howie, D., Elrick, R. and Yule, I. (2003). "An Intelligent System for Interpreting the Nuclear Refuelling Process within an AGR". Proceedings of the Institution of Mechanical Engineers Part A, Journal of Power and Energy, Vol. 217, 2003.

Application of Whole Core Modelling Methodology to Life Extension of AGR Reactor Graphite Cores

Neil McLachlan and Dave J Shaw[#], Hassan Salih[#]
British Energy (Generation) Ltd, Barnett Way, Barnwood, Gloucester, UK.
[#]AMEC NNC Limited, Booths Park, Chelford Road, Knutsford, Cheshire, UK.

Abstract
The ageing effects in an AGR graphite core manifest themselves in a variety of ways including physical changes corresponding to dimensional changes and ultimately distortion of the individual components and initial geometrical configuration of the core. In addition, the ageing process of the graphite core may lead to the inducement of through thickness full length axial cracking of the fuel bricks and perhaps the possibility of double cracking of the fuel bricks. For life extension of the AGR reactors, it is necessary to have a whole core modelling methodology that can be used to predict the consequences of the ageing effects on the geometrical layout of the core. Such a methodology has been developed and evolved over many years and the application of this methodology to test rig arrangements and reactor core configuration in recent times is shown in this paper.

Keywords
Whole core monitoring, life extension, modelling

INTRODUCTION

At the start of reactor life, the design of an AGR graphite core is arranged so as to provide unimpeded access of the fuel assemblies and control rods and adequate cooling of the fuel and core. For safe reactor operations, such unimpeded access and cooling are required for the duration of reactor life. However, reactor operations involving prolonged irradiation of the core graphite components result in ageing effects. The ageing effects manifest themselves in a variety of ways such as loss of weight, changes in material properties and physical changes corresponding to dimensional changes and ultimately the distortion of the individual components and initial geometrical configuration of the core. In addition, the ageing process of the graphite core may lead to the inducement of through thickness full length axial cracking of the fuel bricks and perhaps the possibility of double cracking of the fuel bricks.

For safe reactor operation there is also a need to evaluate and determine the potential distortion of the reactor core and in particular the displacements of the bricks that make up the fuel and control rod channels. To this end, a whole core modelling and analysis methodology has been developed to simulate the complex kinematics behaviour of the core. Such complexity stems from the fact that the core is made up of a larger number of discrete bricks keyed together to form a structure consisting of multi-layers of graphite components.

This paper presents the details of the whole core modelling methodology and its application to an intact, but distorted, AGR graphite core. The application of the methodology to a core containing graphite bricks with single cracks and doubly cracked bricks is also covered. The aim is to show how the peak displacements of the bricks are evaluated when account is taken of the postulated numbers and distribution patterns of cracked bricks in the various layers of the core. As prelude to the above, the whole core modelling methodology is also applied to test rig arrangements that have been used in the validation of the modelling methodology.

Details of the developments of the methodology and the extent of the validation carried out are given by Shaw *et al.* (2006).

WHOLE CORE FE MODELLING METHODOLOGY

The whole core finite element modelling and analysis methodology has been implemented via AGRIGID, which is a FORTRAN code that takes as input a description of the geometry, keying arrangement and the restraint of an AGR graphite core (or test-rig) including ageing effects due to irradiation, and generates an ABAQUS model input file suitable for static non-linear solution using ABAQUS/Standard. The finite element (FE) based model corresponds to a rigid body and spring model that is used to obtain static displacement results for a given loading for an array of graphite bricks, a section or a full AGR core. The rigid bodies correspond to fuel and interstitial bricks, while the springs represent key/keyway joints and steelwork restraints. In addition to the FE model generation, AGRIGID may also be used to impose constraints, boundary conditions and loadings.

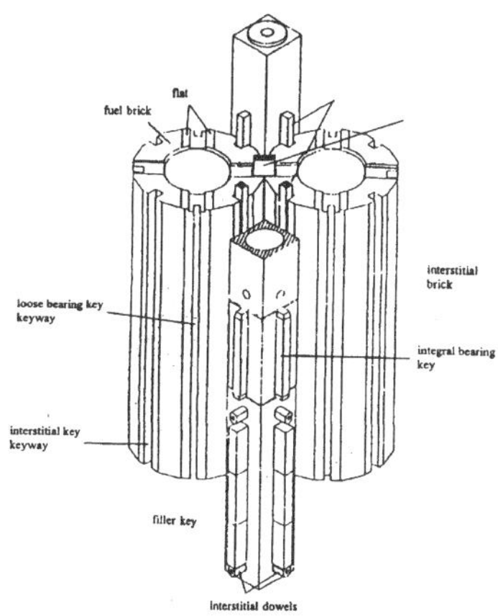

FIGURE 1: Typical core arrangement to be represented using AGRIGID.

Intact Arrangements

The movements of the graphite bricks in an AGR core are governed by the keying arrangement of the bricks. Thus in any mathematical modelling of the core, the simulation of keying system is the dominant required feature. A typical AGR graphite core consists of layers of fuel bricks and interstitial bricks. Adjacent bricks are keyed together via key/keyway arrangements involving loose and integral keys, as shown in the typical arrangement in Figure 1. Gaps of various sizes exist in the loose and integral keying systems. The fuel bricks are also keyed to each other vertically in each column via the end-face keying system. Inter-layer keying system has also been used in some of the AGR graphite cores.

The AGRIGID code simplifies the 3-D geometric modelling of the core bricks so that they are represented by beam and spring elements, and at the same time retains the essential features for modelling the various keying arrangements. With such a representation, the potential movements of the core bricks and their constraint by interaction via the keys/keyways are fully simulated.

The perimeters of the fuel bricks at their bottom, top and intermediate heights are represented by 8 rigid beams to form an octagon, whilst the perimeters of the interstitial bricks are represented at their mid-height by 4 rigid beams to form a square. The three perimeter modelling levels of the fuel bricks are connected by axial rigid elements located along the central axes of the bricks, see Figure 2 for the FE model of a unit cell of fuel and interstitial bricks.

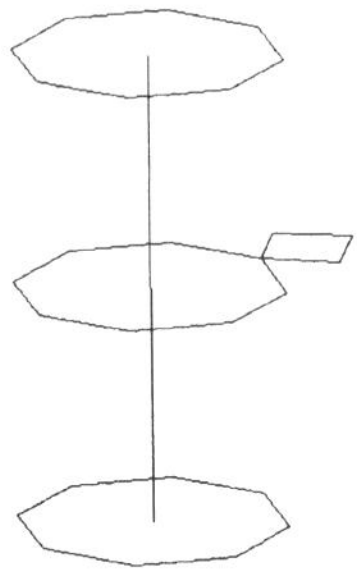

FIGURE 2: Unit cell generated using AGRIGID.

For the HPB/HNB reactors, the loose and integral bearing keys/keyways are represented with non-linear JOINTC springs at fuel brick mid-height level. JOINTC is an ABAQUS/Standard multi-degree-of-freedom spring element. These springs have a low stiffness ('ether') for the key/keyway gap and a high stiffness ('graphite') for key/keyway contact. Therefore, for an intact bearing key/keyway there will be an 'ether' stiffness defined for a gap in both tangential directions and a 'graphite' stiffness beyond the gap, *i.e.* the spring properties are tri-linear. Adjacent to the top and bottom of the fuel bricks there are spacer and filler keys, which are represented in a similar manner to the bearing keys. In order to simplify the code, the spacer and filler keys are only included if the vertically adjacent fuel brick is cracked; for intact bricks these keys do not become active. For the axial keys/keyways, *i.e.* end-face keys/keyways, the rotational component about the vertical axis 'z' is included with an appropriate JOINTC spring. Note that AGRIGID uses springs only of type JOINTC, which is a six-degree of freedom element.

Irradiation will produce dovetailing in the intact keyways due to shrinkage of the graphite material. Dovetailing is the reduction in the width at the tip of the keyway in comparison with the value at the base of the keyway. The limiting translation constraint of the key/keyways is calculated based on a dovetailed keyway and includes the pure translation component, *i.e.* the key/keyway tip clearance, and for loose keyways, the additional translation component due to rotation of the keys.

Primary Cracked Arrangements

Primary cracked fuel bricks, where the crack is located in a keyway, are introduced into the FE model by modifying the spring properties that connect the cracked fuel brick to its surrounding fuel/interstitial bricks. The key/keyway gap for the cracked keyway is increased to simulate the primary crack opening gap. When a fuel brick cracks, the position of the intact keyways of the cracked brick will alter compared to its neighbours, thus producing an incongruity in the brick shape, *i.e.* a C-shape is produced from the initial circular shape. Both the circumferential and radial positions of the keyways are altered. Therefore, the spring properties of the intact keys/keyways of a primary cracked brick are modified, by introducing a displacement offset, in the circumferential direction away from the cracked keyway, in the force/displacement curve of the springs representing the key/keyways. This offset is calculated automatically by AGRIGID by assuming that the offset varies linearly with the circumferential angle from the cracked keyway to the keyway diametrically opposite.

The FE modelling of the incongruity results in a model that is initially not in equilibrium, *i.e.* some key/keyway gaps will be over-closed and therefore subjected to forces due to the initial definition of the model; this causes numerical problems if analysed. Therefore to overcome these problems, the first step in the analysis of an arrangement of graphite bricks with bricks having primary cracks is to fix the displacement of all the bricks in order to initialise the model and generate reaction forces that are in equilibrium with the restoring forces generated due to over-closed gaps. In the second step, the constraints of the bricks are removed and the effect of incongruity of the cracked bricks takes place, *i.e.* the restoring forces will reduce to zero as the bricks rearrange themselves to account for the effect of crack opening of the primary cracked bricks into a C-shape.

Doubly Cracked Arrangements

An axially doubly cracked fuel brick is modelled by splitting the octagon perimeters of the fuel brick at the three levels into two halves, as shown in Figure 3. The tri-linear characteristics of the key/keyway spring at a primary crack are replaced with two springs with bi-linear characteristics, *i.e.* the 'graphite' stiffness on one side of the tri-linear spring property is replaced with the 'ether' stiffness resulting in a loss of constraint on that side of the keyway. The two halves of the doubly cracked fuel brick are also connected with bi-linear springs to represent brick to brick contact. Note that there is no shear constraint between the brick halves other than offered by the surrounding keying arrangement.

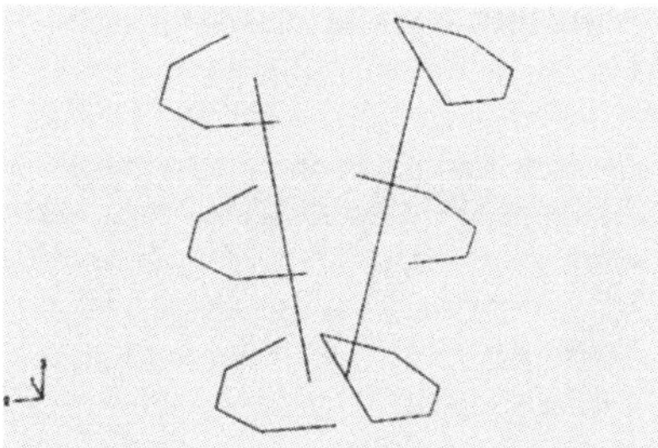

FIGURE 3: Doubly cracked brick represented using AGRIGID.

APPLICATION OF AGRIGID TO TEST-RIG ARRANGEMENTS

The whole core modelling methodology using AGRIGID has been validated against the following test-rigs:

- 5x5x1/2 slice rig.
- 4x4x1/2 slice rig.
- 3x3x1/2 slice rig.
- Box-kite development rig, Phase2, 3x3x[1/2+1/2].
- Box-kite development rig, Phase 3, 3x3x[1/2+1+1/2].
- Box-kite development rig, Phase 4, 3x3x[1/2+2+1/2].
- Box-kite rig 4x4x[1/2+3+1/2].

- 1/8th scale rig.

A selection of this validation work is presented henceforth.

Slice rig

This test-rig is an example of a rig containing intact and primary cracked slices and which is loaded externally using an applied displacement to a single intact brick. Figure 4 shows a plan view of the deformed shape of the array at the end of the analysis with the undeformed array superimposed. Figure 4 also shows the reaction force history and clearly indicates where the reaction force suddenly increases that the array has locked-up, *i.e.* key/keyway clearances have been taken up. A comparison of various intact and primary cracked configurations for various key/keyway clearances and loading directions are shown in Figure 5. The FE models generated using AGRIGID show good agreement with the test-rigs.

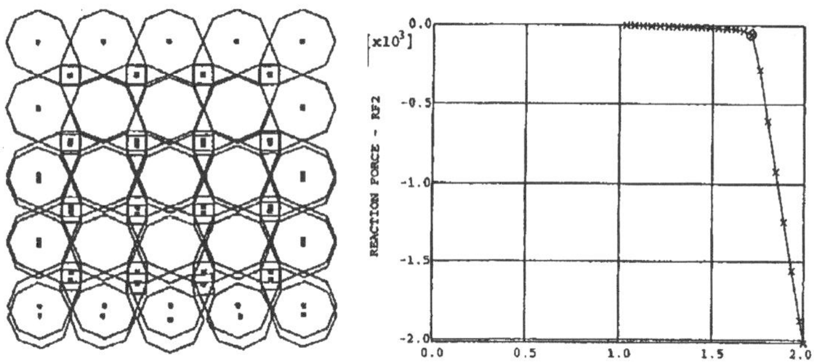

FIGURE 4: 5x5x1/2 slice rig FE results.

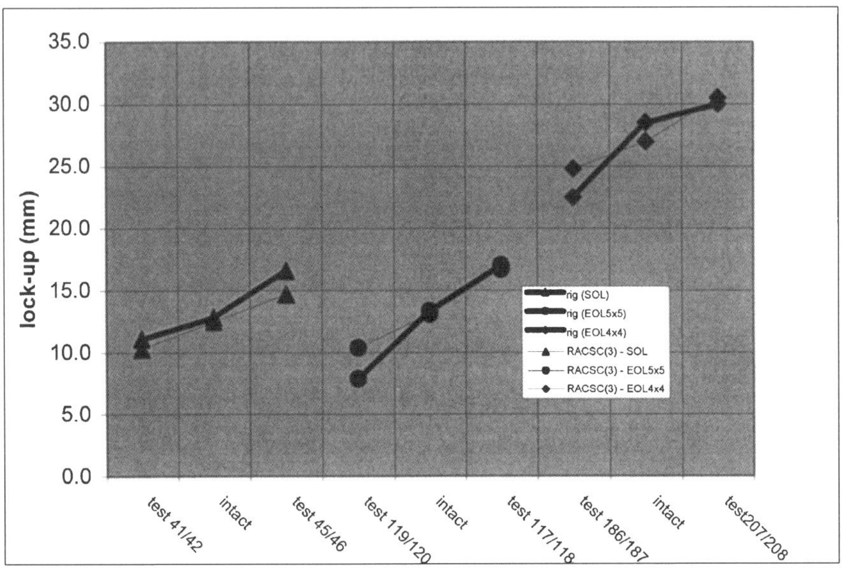

FIGURE 5: Slice rig and FE lock-up comparison.

Box-Kite Rig

This test-rig is an example of a rig containing intact and doubly cracked bricks and which is loaded externally using an applied shear or separation force to a specified doubly cracked brick. Figure 6 shows an elevated view of the deformed shape of the array at the end of the analysis. Figure 6 also shows the resultant displacement histories of the loaded doubly cracked brick halves due to, in this case, a separation loading. These displacements are plotted at the top and bottom of the brick and also at the bearing key height of the brick (shown in bold).

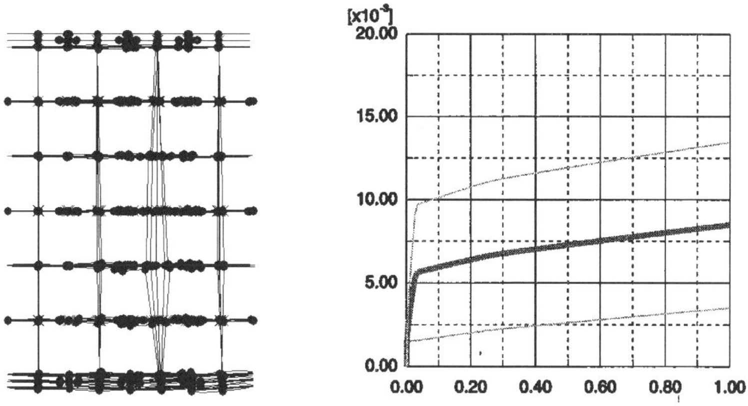

FIGURE 6: Boxkite rig FE results.

Comparisons of the lock up displacements of the doubly cracked brick configurations used in the various Box-kite development rigs are shown in Figure 7. The FE models generated using AGRIGID show good agreement with the test-rigs.

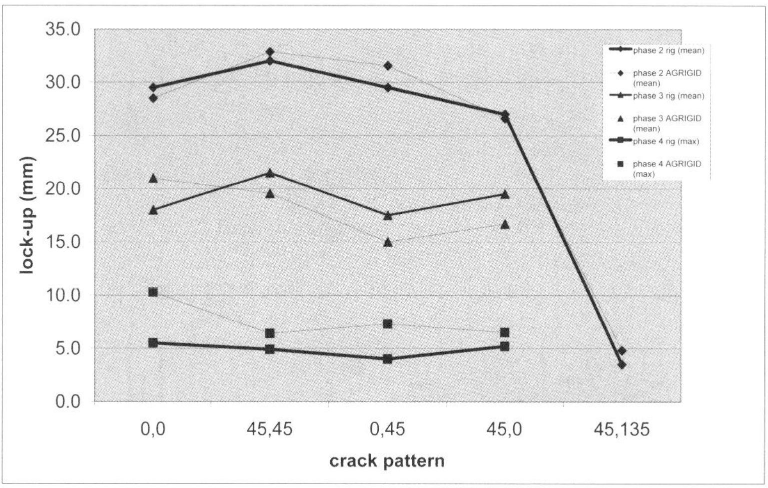

FIGURE 7: Box-kite rig and FE lock-up comparison.

APPLICATION OF AGRIGID TO AGR REACTOR CORES

Following on from the successful validation work, a number of large size FE core models have been generated using AGRIGID with postulated primary and axially doubly cracked fuel brick distributions, see Figure 8.

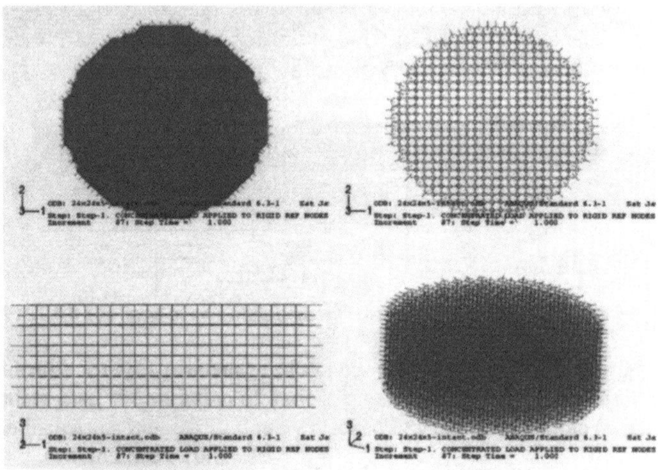

FIGURE 8: FE model of core generated using AGRIGID.

The shear/separation movements of the doubly cracked bricks, global movement of the fuel bricks and various other parameters due to a lateral gravity type loading have been determined from the ABAQUS/Standard analysis of these arrays and a sample of the results are presented. This type of loading was chosen in order to distort the core to its bounding position, *i.e.* minimum potential energy. The doubly cracked fuel bricks shear histories and their distributions are presented in Figure 9. The history plot shows that the core stops distorting at approximately 10% of the loading step. At this point, the key/keyway clearances have been taken up and the array is constrained from moving further. The distribution of doubly cracked brick shear displacements shows distinct peaks, which correspond to discrete key/keyway constraints.

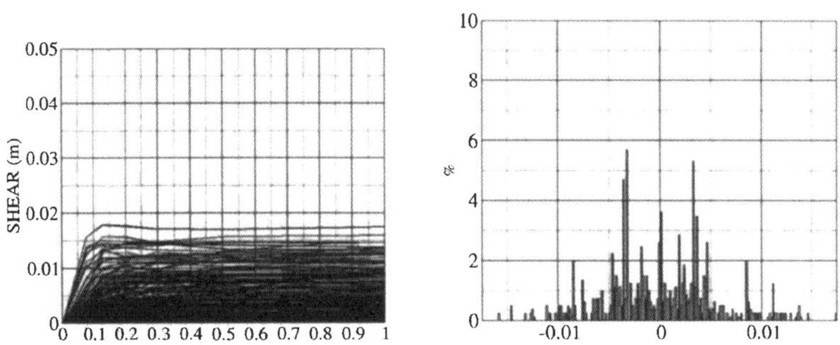

FIGURE 9: Doubly cracked brick shear FE results.

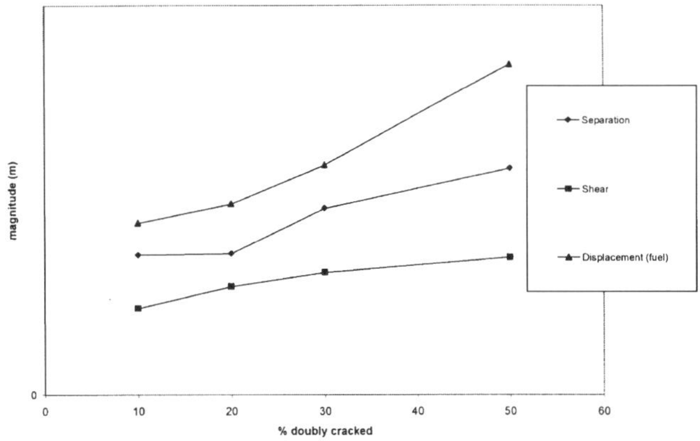

FIGURE 10: Trend of core distortions with increasing % of doubly cracked bricks.

In order to identify a target percentage of doubly cracked bricks to be used for a particular AGR core safety case, the variation of various parameters are predicted from the FE models containing an increasing number of doubly cracked bricks. An example of the trend of such predictions is shown in Figure 10.

CONCLUSIONS

Details of the whole core modelling methodology that has been developed and evolved and validated over a period of many years are presented. The application of the methodology via the AGRIGID code to test rig arrangements and reactor core configurations is presented. The predicted results for the simulation of the test rig arrangements are shown to be in good agreement with the measured values.

Acknowledgements
The authors would like to thank British Energy Generation Ltd for permission to present this paper.

References

Shaw, D J, Salih, M H S and McLachlan, N (2006). Development and validation of whole core modelling methodology for AGR graphite cores. Presented at the Ageing Management of Graphite Reactor Cores Conference, 28 to 30 November 2005, Cardiff, UK (also this book).

A Strategy for Monitoring Bore Cracking in AGR Graphite Cores

Philip R Maul and Peter C Robinson
Quintessa Limited, Dalton House, Newtown Road, Henley-on-Thames, UK.
Email: philipmaul@quintessa.org

Abstract
The monitoring of the physical integrity of graphite cores in AGRs is a critical part of the safety case for their continued operation. Too many graphite bricks with two full-length bore-distorting axial cracks could compromise the integrity of the core. There are now sufficient data available to develop models for the evolution of the number of such cracks as a function of time in order to calculate the period of safe reactor operation. To date a number of bricks with one such crack and three bricks with two such cracks have been observed. Quintessa has developed novel statistically-based models for cracks that are initiated at the bore, and these have been used as part of an overall monitoring strategy. Details of the strategy are presented along with an example of its application.

Keywords
AGR, graphite cores, statistical models

INTRODUCTION

The monitoring of the physical integrity of graphite cores in Advanced Gas-cooled Reactors (AGRs) is a critical part of the safety case for their continued operation. This paper gives details of a strategy developed by Quintessa for the monitoring of bore cracks using statistical techniques. The approach provides a transparent and auditable justification for the safe operation of the reactors until the next inspection. The paper is structured as follows:

- Background information is given on core monitoring and safety cases for continued reactor operation.
- An overview of the monitoring strategy is then given.
- A summary is then presented of the models used for representing brick cracking.
- This is followed by an example application of the strategy for the inspection at Reactor 1 at Heysham 1 in September/October 2005.
- Finally, the status of the strategy and plans for its future development are discussed.

CORE MONITORING AND SAFETY CASES

The core monitoring strategy has to date been employed for the eight oldest reactors at Hinkley Point B (HPB), Hunterston B (HNB), Hartlepool (HRA) and Heysham 1 (HYA). The first two (HPB and HNB) are of the same design and are considered together; HYA and HRA are considered as a second group. Analysis of the monitoring data confirms that it is valid to 'pool' the data in the two reactor groups, but not to treat both groups together.

Monitoring of the graphite core involves TV inspections, Channel Bore Measurement Unit (CBMU) measurements, and trepanning. No cracks initiated at the bore were observed prior to 1994, but since then several have been found. Most of these cracks have been found just above the middle of the channel. Axial, circumferential and lasso cracks have been observed.

A discussion of the significance of the different types of cracks is outside the scope of this paper, but in the calculations presented, 'significant' cracks have been taken to be those that are already full-length axial and bore-distorting or which could progress to such a state. It is necessary to define the number of bricks with two significant cracks that can be present in the reactor whilst maintaining core integrity. Experimental and theoretical studies undertaken for British Energy indicate that this figure is around 200. In this paper bricks with one 'significant' crack are referred to as 'singly-cracked' and bricks with two 'significant' cracks are referred to as 'doubly-cracked'.

OVERVIEW OF THE STRATEGY

The strategy is based on a statistical analysis of the available data. Bricks in the layers considered to be vulnerable to bore cracking are currently assumed to be statistically identical, although extensions to the current models could be considered where cracking probabilities are different for different types of brick. An important principle is that the approach must be capable of extrapolation; our aim is not to find a single 'best' model, but to delimit an envelope of models that are consistent with the available data. These models will behave differently when extrapolated, enabling us to quantify the level of confidence that we have in predictions and to select conservative assumptions to feed into a safety case. The main stages in the methodology are:

1. Define models and parameter values that are consistent with the available data.
2. Predict the future behaviour of the system using the feasible models/parameter values to provide conservative estimates of the safe operating period.
3. Design a future monitoring programme that will provide the most useful information for refining the safety case.
4. Predict the various possible outcomes at the next inspection.
5. Calculate the impact on the safety case of these possible outcomes.

The methodology is iterative as each new inspection provides additional data. A key feature is the ability to compare predictions made before the inspection with actual observations; this gives confidence in the use of the models, and enables models that cease to be consistent with the data to be rejected. Some of the steps in the methodology can be undertaken analytically in a spreadsheet, while others need a more sophisticated computer code. Wherever possible, both approaches are applied to cross verify the results.

MATHEMATICAL METHODS

Hazard Functions and Transition Probabilities

A given brick is assumed to be in one of three states: 0 (uncracked), 1 (singly-cracked) and 2 (doubly-cracked). In defining these states, only significant cracks are considered. The probability that a brick in state 0 at time t changes to state 1 (single cracking) in the small time interval $(t, t+dt)$ is given by $h_0(t)dt$, where $h_0(t)$ is the single cracking hazard function. The probability that a brick in state 1 at time t changes to state 2 (double cracking) in the time interval $(t, t+dt)$ is given by $h_1(t)dt$, where $h_1(t)$ is the double cracking hazard function. If the probability of a brick being in state i at time t is denoted by $P_i(t)$ then the following equations, often referred to as the forward equations (*e.g.*, see Cox and Miller, 1965), can be derived:

$$\frac{dP_0}{dt} = -h_0(t)P_0, \quad \frac{dP_1}{dt} = -h_1(t)P_1 + h_0(t)P_0, \quad \frac{dP_2}{dt} = h_1(t)P_1. \qquad [1]$$

If a brick is known to be in state i at time t_A these equations can be solved for the transition probability P_{ij} that it is in state j at time t_B. Further progress cannot be made without making some assumptions about the nature of the hazard functions. We will consider the situation where we have a constant ratio between the single and double cracking hazard functions, with these being expressed in the following simple forms:

$$h_0(t) = u\,f(t); \quad h_1(t) = v\,f(t). \qquad [2]$$

The transition probabilities can then be written analytically provided that $f(t)$ can be integrated to give the quantity T:

$$T = \int_{t_A}^{t_B} f(\tau)\,d\tau. \qquad [3]$$

Restricting consideration to a simple power law functional form for $f(t)$ is appropriate with the currently available data; little would be gained by introducing more complex forms. In any case, very different patterns of bore cracking can be obtained by different choices of the function $f(t)$ within this class of functions. It is therefore assumed that:

$$f(t) = (t - t_0)^p\,H(t - t_0), \qquad [4]$$

where H is the Heaviside step function and t_0 is the time when the hazard commences.

The two simplest two-parameter models considered are: a 'uniform' model, where the hazard is constant ($p=0$) after a specified time, and a 'linear' model, where the hazard increases linearly ($p=1$) after a specified time. The hazard initiation time has been taken to be the time when the fuel management regime changed at the reactor in question. The two parameters of these models are u and v of Equation [2]. If the time when the hazard commences is not fixed, then we have a three-parameter model. For example, if we take $p=0$ but consider t_0 to be variable (taking any value after the fuel management regime change), we have a 'delayed uniform' model. Finally, if p is also allowed to vary, we have a 'general' four-parameter model.

The Log-Likelihood Function

The log-likelihood function is central to our approach to determining appropriate parameters for a model and to determining whether a model is adequately consistent with the data. It is generally not know precisely when any observed cracks actually occurred so we must use the specific model hazard function to provide probabilities for each possible outcome on each inspected brick. Assuming all the bricks are independent, the likelihood is then the product of the probabilities for the outcomes that were observed. The log-likelihood is then simply the sum of the logarithms of the probabilities.

If θ is a k-dimensional parameter, $L(\theta)$ the likelihood of the model under consideration, $l(\theta)$ the corresponding log-likelihood and $\hat{\theta}$ the maximum likelihood estimate, then if the model is correct, $2[l(\hat{\theta})-l(\theta)]$ has an approximate chi-squared distribution with k degrees of freedom, denoted χ_k^2 (*e.g.*, see Davison, 2003 or Meeker and Escobar, 1998), and this can be used to define confidence intervals/regions for the model parameters. It is straightforward to write down the log-likelihood function for the three state system. Likelihood functions include a multiplicative combinatorial term that cancels out when ratios of likelihoods are considered; for simplicity this combinatorial terms has been omitted. If there are n_{ij} bricks initially in state i and subsequently observed to be in state j, then the authors obtain:

$$\ell = -\sum_{i=1}^{n_{00}} u\,T_i^{00} - \sum_{i=1}^{n_{11}} v\,T_i^{11} + \sum_{i=1}^{n_{01}} \log\left[\frac{u(\exp(-u\,T_i^{01}) - \exp(-v\,T_i^{01}))}{v-u}\right]$$
$$+\sum_{i=1}^{n_{02}} \log\left[1 - \frac{v\exp(-u\,T_i^{02}) - u\exp(-v\,T_i^{02})}{v-u}\right] + \sum_{i=1}^{n_{12}} \log\left[1-\exp(-v\,T_i^{12})\right]. \qquad [5]$$

In general the peak of the log-likelihood function has to be found numerically to give maximum likelihood estimators for u, v, t_0 and p, although good analytic approximations can be derived for two-parameter models where t_0 and p are fixed.

Goodness of Fit

The maximum likelihood estimator provides the 'best' choice of parameter values to match the data in the current model. However, it provides no guarantee that the model is actually a good fit for the data. The deviance of a model M is defined by (*e.g.*, Davison, 2003):

$$D_M = 2(l_S - l_M), \qquad [6]$$

where l_S is the maximum log-likelihood under the saturated model S and l_M is the maximum log-likelihood under model M. A saturated model has as many degrees of freedom as the data allows. For the data that we have the saturated model would have a parameter for each different type of observation, where a different type means either a different initial state, different time of previous inspection (if any) or different inspection time. If we have two models M_1 and M_2, where M_1 is a simpler (nested) version of M_2, the difference between their deviances is given by

$$\Delta D = 2(l_2 - l_1). \qquad [7]$$

If the simpler model M_1 is true, this difference in deviances has an approximately χ^2 distribution with n_2-n_1 degrees of freedom, where n_i is the number of degrees of freedom for model M_i. If the deviance is large then the model M_1 is inadequate compared to M_2.

If observations with similar times in the saturated model are grouped together to give an 'aggregated' model, this reduces the number of degrees of freedom of model M_2. Comparison of maximum likelihood values with threshold values obtained from the 'aggregated' model gives one way of comparing the performance of models of different complexity, with different numbers of free parameters. Various information criteria can also be employed to undertake such comparisons; *e.g.* the Bayes' Information Criterion (Davison, 2003). With the currently available data it is generally found that two-parameter models perform at least as well as the three- and four-parameter models. As more data become available, it is likely that the use of more sophisticated models with more parameters will be justified.

Information from Monitoring

When an observation is made, the information it provides on the single and double cracking rates depends on the outcome. In a model with a single parameter θ and log-likelihood $\ell(\theta)$, the observed information is defined to be (*e.g.*, Davison, 2003)

$$J(\theta) = -\frac{d^2\ell(\theta)}{d\theta^2}. \qquad [8]$$

In the region of the maximum likelihood estimator, the log-likelihood function can be approximated by a quadratic as a function of θ, and the curvature given in Equation [8] is then a linear function of the number of observations available. The higher the value of the curvature, the less uncertainty is associated with the maximum likelihood estimator and the greater the available information. Before an observation is made, it is generally not known what information will be obtained, but the expected (*i.e.* average) amount of information can be defined to give the quantity I known as the Fisher information. When a model has more than one parameter, the Fisher information becomes a matrix. When making decisions about monitoring strategies it is helpful to consider how the expected information will change if different strategies are pursued.

Projections for a Planned Inspection

To calculate the distribution of bricks in each state at a planned inspection additional simulations are performed for the bricks to be inspected. These are conditioned on any already observed states for these bricks. The numbers of occurrences of each combination of counts in each state are combined from all samples using the likelihood as a weighting. If **y** are the available data, $\pi(\theta)$ is the uniform prior over a volume V of parameter space and $\pi(\theta \mid \mathbf{y})$ is the posterior distribution we have:

$$\pi(\theta \mid \mathbf{y}) = \frac{\pi(\theta)L(\mathbf{y} \mid \theta)}{\int \pi(\theta)L(\mathbf{y} \mid \theta)d\theta} = \frac{L(\mathbf{y} \mid \theta)}{\int_V L(\mathbf{y} \mid \theta)d\theta} \qquad [9]$$

If $p(\mathbf{z} \mid \mathbf{y}, \theta)$ is the projected distribution of data at some time in the future using the model with specified parameter values we have:

$$p(\mathbf{z} \mid \mathbf{y}) = \int p(\mathbf{z} \mid \mathbf{y}, \theta)\pi(\theta \mid \mathbf{y})d\theta = \frac{\int_V p(\mathbf{z} \mid \mathbf{y}, \theta)L(\mathbf{y} \mid \theta)d\theta}{\int_V L(\mathbf{y} \mid \theta)d\theta} \qquad [10]$$

Equation [10] is evaluated numerically, although for some situations analytical approximations can be found that provide useful checks on the numerical calculations. The choice of prior distributions is a controversial topic, but calculations are only sensitive to this choice if the shape of the prior distribution varies significantly over the region of the parameter space where the likelihood values are concentrated. To date, the authors have used uniform priors throughout.

EXAMPLE CALCULATIONS

Examples are given here of the calculations involved in applying the inspection strategy to the inspection at Reactor 1 at Heysham 1 in September/October 2005. Calculations are given both before and after the inspection were undertaken. This was the second full application of the strategy for the HRA/HYA group; the first being for HYA Reactor 2 in May/June 2005.

Pre-Inspection Calculations

For this inspection, the log likelihood for the Aggregated Model was calculated to be -78.62 with 16 degrees of freedom. From this, a threshold log-likelihood can be calculated at the 5% level for models with various numbers of degree of freedom. Comparison of the maximum log-likelihood found for each model against this threshold gives an indication of how good the model is against the current data; these comparisons are shown in Table 1.

TABLE 1: Maximum log-likelihoods for the pooled data compared with the Aggregated Model.

Model	Degrees of freedom	Maximum log likelihood	Distance above threshold as logarithm	Relative likelihood at threshold
Uniform	2	-89.21	1.26	28.4%
Linear	2	-86.23	4.24	1.44%
Delayed Uniform	3	-85.20	4.61	2.65%
Delayed Linear	3	-84.99	5.26	1.45%
General	4	-84.99	4.14	8.18%

The Uniform Model is the worst on the basis of the distance above threshold, and is close to being rejected (and since it is also the most optimistic model it was dropped from further analysis). The General Model is next worst in terms of the relative likelihood at threshold, but is still an acceptable model. The Delayed Linear Model is best in terms of distance above threshold with the Linear Model just best in terms of the relative likelihood at the threshold. Using the safety case criterion of 200 doubly-cracked bricks criterion, failure time calculations can be undertaken. These predictions are probabilistic, with the uncertainty over the parameter values and the stochastic behaviour of the models taken into account. In Table 2 a conservative lifetime estimate is given by the 5th percentile, while the median gives a more realistic estimate. The percentage failing by June 2008 is also shown.

TABLE 2: Estimates of the safe operating period and probability of core failure by June 2008 for HYA 1 R1 prior to the September/October 2005 inspection.

Model	5th Percentile	Median	June 2008 Failure Probability
Linear	Jun 2012	Feb 2016	0%
Delayed Uniform	Jun 2016	Apr 2025	0%
Delayed Linear	Apr 2009	Mar 2012	0.62%
General	Sep 2009	Nov 2015	0.59%

Thus, the Delayed Linear and General models give relatively early estimates, with the other models being much later. For both the Delayed Linear and General models a non-trivial fraction of the cases give failure before June 2008.

The expected outcomes at the inspection are presented in terms of the number of additional singly- and doubly-cracked bricks that will be seen. Each model calculates the probability of seeing any combination of numbers of new singly- and doubly-cracked bricks. These results are ordered, with the most probable first. Each is then given a score according to the total probability of all outcomes that are at least as unlikely, so the most likely scores 100%. Table 3 gives example calculations for the linear model, with outcomes shaded (above 50% mid-grey, above 5% pale-grey and above 1% dark-grey). Although there are differences in detail between the models, the general patterns for the other models are rather similar. Up to 8 additional singly-cracked bricks and 2 additional doubly-cracked bricks would be totally in line with the current models. As many as 12 or 13 additional singly cracked bricks would not be totally surprising provided at most about 2 additional doubly-cracked bricks were observed. Similarly, 4 or 5 doubly cracked bricks would not be totally surprising provided at most about 10 additional singly-cracked bricks were observed.

TABLE 3: Linear Model complementary cumulative probability of outcomes.

	D0	D1	D2	D3	D4	D5
S0	7.0	4.7	3.3	1.6	0.5	0.1
S1	27.9	21.1	12.0	6.5	2.6	0.7
S2	53.6	46.2	26.1	10.3	3.6	1.5
S3	86.6	66.0	37.3	17.0	5.4	2.4
S4	93.1	81.0	42.9	15.9	7.8	1.9
S5	100.0	70.7	34.6	18.1	7.4	2.0
S6	75.8	61.8	32.0	12.9	4.4	1.2
S7	57.6	49.8	22.7	9.5	2.9	1.7
S8	40.1	29.9	13.9	6.1	2.3	0.9
S9	24.3	19.5	8.9	4.1	1.2	0.5
S10	14.9	11.1	5.1	2.1	1.0	0.4
S11	8.3	5.8	3.8	1.1	0.6	0.2
S12	3.1	2.7	1.4	0.2	0.2	0.0
S13	1.3	1.0	0.5	0.3	0.1	0.2

Post-Inspection Calculations

The inspection identified four bricks with significant axial cracks with one of these having two such cracks. This outcome (S3, D1) is completely in line with the predictions made before the inspection. Pre- and post-inspection results can be compared in terms of the log-likelihood versus parameter value. The log-likelihood is plotted relative to the threshold determined by the aggregated model. Calculations for the linear model are shown in Figure 1; the post-inspection results are the shaded contours, with the pre-inspection results as black lines for comparison.

It is clear that there has been little change overall. The double cracking rate peak is slightly higher and the single cracking rate peak slightly lower. The value at the peak is higher, meaning that the new results are in reality less uncertain even though the area inside the zero contour is about the same. This is confirmed by the information as defined in Equation [8] which shows a 38% increase for the single cracking parameter and little change for the double cracking parameter. The revised failure time predictions are shown in Table 4. For both the Delayed Linear and General models a non-trivial fraction of the cases give failure before June 2008, but this is now less than 0.1% giving a confidence of 99.9% that the safe operating criterion will be met at June 2008.

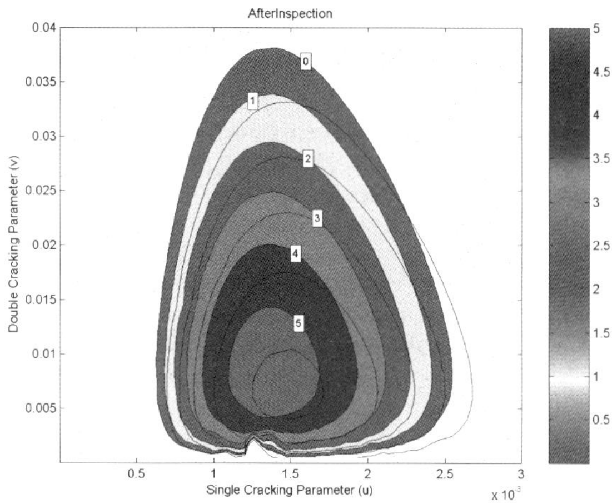

FIGURE 1: Log-likelihood above threshold for the Linear Model.

TABLE 4: Estimates of the safe operating period and probability of core failure by June 2008 for HYA R1 after the September/October inspection.

Model	5th Percentile	Median	June 2008 Failure Probability
Linear	Dec 2012	Feb 2016	0%
Delayed Uniform	Feb 2017	Sep 2024	0%
Delayed Linear	Sep 2009	Aug 2012	0.07%
General	Mar 2010	Mar 2018	0.09%

DISCUSSION AND CONCLUSIONS

The bore monitoring strategy provides a transparent and auditable justification for the safe operation of the reactors until the next inspection. Confidence in the process is enhanced by comparing model calculations before each inspection with the actual data obtained at the inspection. To date it has not been possible to use calculations of expected information to help specify the optimum monitoring strategy for forthcoming inspections, but it is hoped to be able to do this in 2006. Because of the limited data that are currently available, little advantage is obtained by using models more complex than the 'uniform' and 'linear' two-parameter models. However, each new inspection provides significant new information, and it may soon be possible to justify the use of more complex models, possibly differentiating between 'high' and 'low' risk bricks dependent on their location in the reactor and history of fuel management.

Acknowledgements
The work described in this paper was undertaken for British Energy Generation Ltd. The authors would like to thank Dr Jim Reed, Dr Mark Bradford and Mr John Stephenson of BEGL for their input to the development and application of the graphite monitoring strategy. The views expressed in this paper are those of the authors and do not necessarily represent those of the British Energy Generation Ltd.

References

Cox D R and Miller H D (1965). The Theory of Stochastic Processes. Chapman Hall, London.
Davison A C (2003). Statistical Models. Cambridge University Press.
Meeker Q M and Escobar L A (1998). Statistical Methods for Reliability Data. John Wiley and Sons.

Development and Validation of Whole Core Modelling Methodology for AGR Graphite Cores

Dave J Shaw, Hassan Salih and Neil McLachlan[#]
AMEC NNC Limited, Booths Park, Chelford Road, Knutsford, Cheshire, UK.
[#]British Energy (Generation) Ltd, Barnett Way, Barnwood, Gloucester, UK.

Abstract
The AGR reactors have graphite cores and some of these reactors have been operating for a significant part of their design life. As a consequence ageing effects on the graphite of the core are manifesting themselves in distortion of the core and the likelihood of cracking of the fuel bricks. Accordingly, a whole core modelling methodology has been developed in order to predict the movements of the core components and the geometry of the core late in reactor life. The development of this methodology, its validation against test data from different test rigs and its verification against alternative modelling methods is described in this paper.

Keywords
Whole core modelling, kinematic analysis, 3-D rigs

INTRODUCTION

At the design phase of the graphite core of the AGR reactors, some 30 years ago, there were no detailed modelling and analysis methods capable of whole or part core modelling. At this time there was no demand for such methods because designers did not anticipate significant distortion of the core geometry from the initial design configuration during the design life of the reactor. This perception was based on the knowledge and understanding at the time of the ageing effects on the graphite of the reactor core. But even if it were known that significant distortion of the core would arise late in reactor life, it would not have been possible to undertake detailed whole core modelling and analysis owing to the lack of suitable computational codes. This is in addition to the lack of appropriate computers to perform the analysis in a meaningful time scale and costs.

This situation has persisted until recent years, when better understanding of the effects of graphite ageing arising from long term operations of the AGR reactors was gained. This was achieved from improvements in inspection and monitoring of the condition of the reactor cores, and the testing of trepanned graphite specimens taken from core components during regular statutory outages. The testing of such specimens has, by definition, provided physical and mechanical properties of the core graphite consistent with the irradiation and oxidation levels that have been experienced by the core components.

With the current state of knowledge of the effects of graphite ageing, it is now predicted that distortion of the geometry of the AGR reactor cores from their initial design configurations will occur late in life of the reactor. In addition, the core distortion may also be accompanied by axial cracking of the graphite fuel bricks, and such cracking may even develop into the double cracking of the fuel bricks into halves. For safe reactor operations, it is essential that there is unimpeded access of the reactor control rods and the fuel assemblies, and adequate cooling of the fuel and core. Accordingly and as a consequence of the graphite ageing effects, revised safety cases for continued operations of the AGR reactors have been prepared by

British Energy (BE) over the last few years. Such safety cases are known as the 'Revised AGR Core Safety Case', or RACSC, with RACSC3 having been completed.

One of the requirements of the RACSC is for whole / part core modelling to be carried out to determine the potential movements of the core components arising from the graphite ageing effects. Hence, a few years ago a major effort was initiated by BE to develop, verify and validate a methodology for whole core modelling and analysis. This paper outlines the development of such methodology and in particular the extent of validation against measured data from test rigs specifically designed and commissioned to produce data suitable for validation of the whole core modelling and analysis methodology.

DESCRIPTION OF A TYPICAL AGR CORE

Typically, an AGR graphite core is a multi-layered arrangement of discrete graphite bricks that are loosely keyed to each other using a system of keys located in keyways. There are two types of bricks, of different sizes and geometry, namely fuel bricks and interstitial bricks. The hollow fuel bricks are stacked in layers (typically 12) to form the fuel channels and similarly the hollow interstitial bricks are also stacked to form the control rod and other utility channels. The stacked bricks of a channel are keyed to the bricks of adjacent channels using a combination of loose and integral keys located in keyways. Similar key / keyway arrangement is also used at the end faces of the stacked fuel bricks. An inter-layer keying system has also been used in some of the AGR graphite cores. A typical arrangement of the core components is shown in Figure 1.

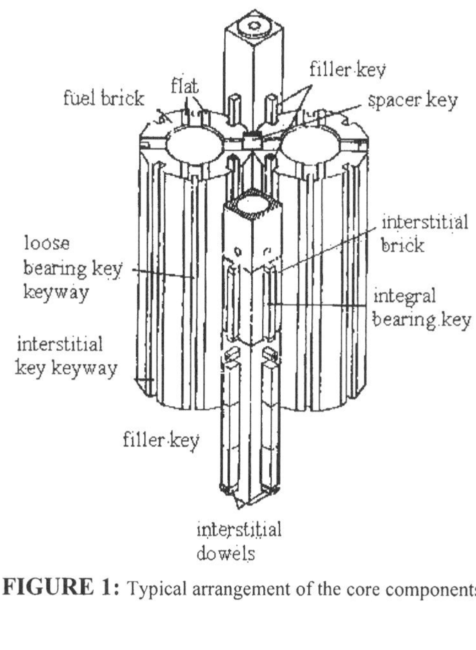

FIGURE 1: Typical arrangement of the core components.

DEVELOPMENT OF WHOLE CORE MODELLING

From the description of a typical AGR core, it can be seen that the reactor core is made up of a large number of graphite components loosely keyed to each other. The kinematics behaviour of the whole core is rather complex as individual components have the freedom to move relative to each other until the clearances in the keys / keyways have closed. Thereafter, the local components move together as a sub-assembly and such behaviour is rather progressive and hence an incremental solution is required to obtain the movements of the individual graphite components of the core.

To meet the need of whole core kinematics analysis, a modelling methodology has been developed and evolved over a period of many years during which the modelling progressed from 2-D, as per a single layer of the core, to 3-D whole core analysis that accounts for and include ageing effects associated with prolonged reactor operations. Ageing effects are included such as core distortion due to irradiation and the potential for cracking of the fuel bricks.

FE Solid Modelling

The movements of the graphite bricks in an AGR core are governed by the keying arrangement of the bricks. Thus in any mathematical modelling of the core, the simulation of the keying system is the dominant required feature. To demonstrate such behaviour and to validate the modelling method, a 2-D test rig was devised and in which 100 mm thick slices of full size graphite bricks were used; this rig is referred to as the slice rig. The test rig arrangement is shown in Figure 2 and comprises a 5x5 array of fuel bricks, the corresponding interstitial bricks and the associated loose and integral keys. A finite element model of this arrangement was created using 8-node solid elements, and the interactions of the keys in the keyways were modelled using surface-to-surface contacts; the FE model is shown in Figure 3. Analysis of the FE model for a pull displacement applied to the middle brick of the first row was carried out and the displacement at which the array locks up was determined. This value was found to correlate well with the corresponding measured value in the slice rig.

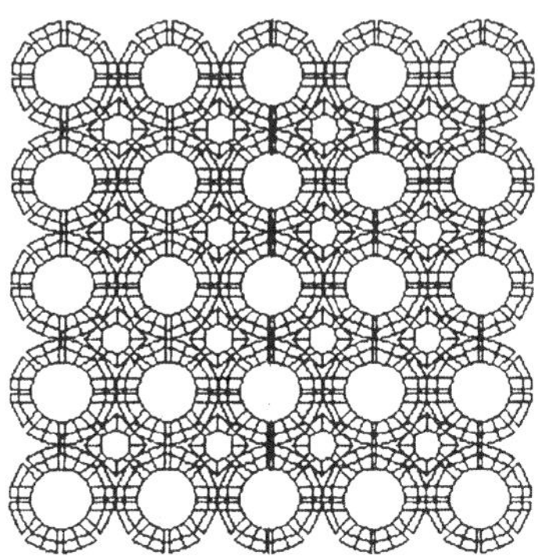

FIGURES 2 and 3: Schematic and FE model of 5x5 array of fuel bricks.

Examination of the incremental FE solution showed that the predicted lock-up displacement was governed by the magnitudes of the key / keyway clearances and the contribution to the lock up displacement of the deformation due to the elasticity of the brick slices and the keys was negligible. This suggested that the FE solution is independent of the FE mesh used, and for all practical purposes the brick slice could have been modelled as rigid objects. The dominant feature of the FE solution is the surface-to-surface contacts as the keys interact in the keyways.

Alternative FE Core Modelling

Although the solid FE model produced the desired results, the generation of the FE mesh and the analysis were expensive in time and computational demanding. Hence considerations had already been given to an alternative finite element modelling computationally less demanding yet retained the dominant feature of the solution, namely the interaction of the keys in the keyways. In this alternative FE modelling, the brick slices were modelled as rigid objects using line elements that maps to the perimeter of the brick slices, with each line starting and ending in a key / keyway location. As there are eight keys associated with the fuel brick, the shape mapped by the perimeter lines is octagonal. Similarly, the interstitial bricks with four key / keyways were mapped as square objects. The interactions of the keys in the keyways were represented by non-linear springs having bi-linear load deflection curves. The first linear line of the characteristic represents the behaviour whist the key is moving within the clearance of the keyway, whilst the second linear line represents the behaviour when the key is in contact with the keyway.

The alternative FE model of the slice rig arrangement is shown in Figure 4, and far simpler than the solid FE model shown in Figure 3. The alternative FE model was analysed incrementally for a pull displacement applied to the middle brick of the first row and the predicted lock-up displacement was in agreement with the measured value in the slice rig. The solution of this model was obtained in a fraction of the CPU time of the solid model. Also the simple pattern of the FE model lent itself to being generated automatically using a script file as an input to a FORTRAN routine.

3-D Fuel Brick Model

The slice rig model in Figure 4 is essentially 2-D, and the 3-D representation of a fuel brick was achieved by retaining the 2-D representation of the keys / keyways at a horizontal plane located at the height of the keys / keyways, with the length of the fuel brick represented by a rigid beam; such a representation is referred to as the stick-spring model. A stack of fuel bricks and the associated interstitial bricks can therefore be represented by stick-spring model for each brick with the ends of the sticks coupled together, using spring elements, to maintain continuity and compatibility of the stacked fuel bricks.

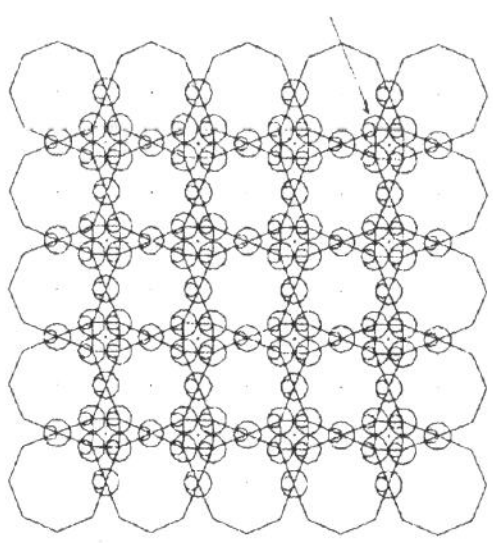

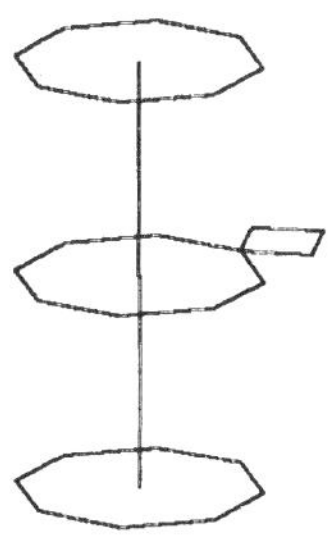

FIGURES 4 and 5: Alternative FE model of the slice rig arrangement and 3-D representation of a single fuel and interstitial brick.

To allow for the representation of the inter-layers keying system, a modelling arrangement similar to that of the radial loose and integral keying system is used at the upper and lower ends of the fuel bricks. The resulting representation of a fuel brick and the associated interstitial bricks is shown in Figure 5. This unit cell representation is the basic modelling block, such that an array of the graphite core is generated from the assembly of such unit cells.

For the modelling of fuel bricks having a single full length axial crack and doubly cracked fuel bricks, as may arise from the consequences of the ageing of the graphite core, the unit cell model was developed further to allow for the representation of such cracks. This is done by the inclusion of additional nodes and connecting springs at the cracked locations within a brick. A full description of the modelling of cracked bricks together with a schematic of the FE model is by McLachlan *et al.* (2006). The generation of the core model and the corresponding input file suitable for analysis using the FE code ABAQUS/Standard is achieved by a specially written FORTRAN code called AGRIGID. Details of this code are given by McLachlan *et al.* (2006).

Validation

Extensive validation of the stick and spring whole core modelling methodology for an intact core and a core with cracked fuel bricks has been carried out. As the computational model is intended to assess core geometry and includes certain simplifications, the aim is to show that it is either in good agreement or bounds real behaviour. Specially designed and commissioned test rigs have been constructed and measured data from a large number of 2-D and 3-D tests have been used to validate the predictions from the whole core modelling methodology. The four sources of validating the whole core modelling methodology are:

- Measured data in the 2-D slice rig.
- Measured data in the 3-D IMC rig.
- Measured data in the 2-D and 3-D 1/8 scale rig.
- Measured data from the 2-D and 3-D box-kite rig.

The 2-D slice rig is the first of the test rigs in which many tests were carried out; a typical arrangement of this rig is shown in Figure 2. A 5x5 array and 4x4 array of fuel brick slice with the associated interstitial bricks and the loose and integral keying system were constructed. Tests involving different key / keyway clearances, different brick-to-brick spacing, and different loading directions were carried out. In some of the tests, bricks with an open crack were used to simulate a single crack in a fuel brick. Different combinations of such cracked bricks involving different crack orientations and locations with the array have been used. The rig arrangements, constraint and loading were simulated using the stick and spring whole core modelling methodology and the predicted results were compared with the corresponding measured displacements in the rig. The correlation between the results was good.

The 3-D IMC rig comprised different arrays of one, two and three layers of full size graphite bricks. Tests involving pulling and tilting of one or more of the bricks were carried out and the movements of few of bricks were monitored. A typical arrangement of the IMC rig is shown in Figure 6. The test arrangements were simulated using the whole core modelling and good agreement was found between the predicted and the measured displacements of the bricks.

FIGURES 6 and 7: IMC rig and 1/8th scale rig.

The 3-D 1/8th scale rig comprises 1/8th scale bricks made from acetyl, and is equivalent to an AGR core, being up to 24 columns in diameter and 10 layers in height. The key / keyway clearances are similar to those in an AGR core in order to obtain measurable displacement results. Loading of the rig is by tilting its base plate to an angle so that gravity causes displacements of the bricks. The general arrangement of the rig is shown in Figure 7. Many tests with all bricks intact and with some of bricks doubly cracked were carried out and the test arrangements were simulated using the whole core modelling methodology. The predicted displacements from the FE models were either in good agreement or bounding to the measured displacements in the tests.

The 3-D box-kite rig was designed for the purpose of investigating and measuring the movements of doubly cracked bricks and in particular the shear and separation of the two halves of the cracked bricks. The 2-D version of the box-kite rig is an assembly of full size bricks slices made from acrylic material, as shown in Figure 8; this arrangement is similar to that of the slice rig in Figure 2. In this rig, many tests were carried out and in which doubly cracked slices in different locations were considered. Measured displacements of the shear and separation of the brick slices in a 2-D configuration, as per a layer of an AGR core have been obtained and used in the validation of the stick and spring core modelling methodology.

The 3-D arrangement of the box-kite comprised full height bricks constructed from three acrylic slices connected together using aluminium channelling inserted into the keyways for form a representation of a whole or doubly cracked fuel bricks. Combined with keys cut from MDF, a comparatively lightweight rig can be readily assembled in a variety of configurations suitable for examining shear and separation movements of doubly cracked fuel bricks; a typical arrangement of the box-kite rig is shown in Figure 9. Many tests involving doubly cracked bricks in different locations were carried out, and typical of such arrangements is shown in Figure 10, in which clusters of doublet, triplet and quadruplet doubly cracked bricks are shown.

The various arrangements tested in the box-kite rig were simulated by the stick and spring whole core modelling methodology and the predicted shear and separation movements of the cracked bricks were found either in good agreement or bounding to the measured values in the rig. The outcome from this comparison is an important validation of the whole core modelling methodology for the case of doubly cracked fuel bricks, as might arise in an AGR core from ageing effects on the graphite of the core bricks.

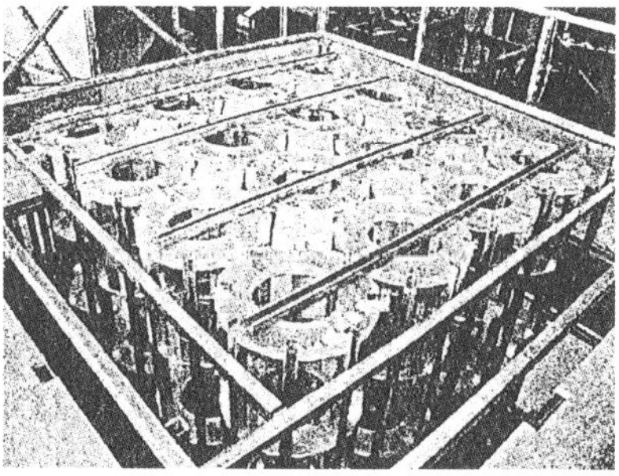

FIGURES 8 and 9: 2-D and 3-D version of the box-kite rig.

The need for validation of the whole core modelling methodology was recognised as an important part of the development of the methodology and the commitment to this aim is ongoing. Hence, currently a full core rig comprising $1/4^{th}$ size bricks is being designed and commissioned to produce test data for further validation of the whole core modelling methodology.

Verification

Verification of the stick and spring whole core modelling methodology of intact core and with singly cracked bricks has been carried. Three independent sources involving different solution methods were used, namely:

- The use of alternative FE models such as solid modelling
- The use of rigid body dynamic code (Working Model) in quasi-static simulations
- The use of the a model based on linear optimisation (CPLEX)

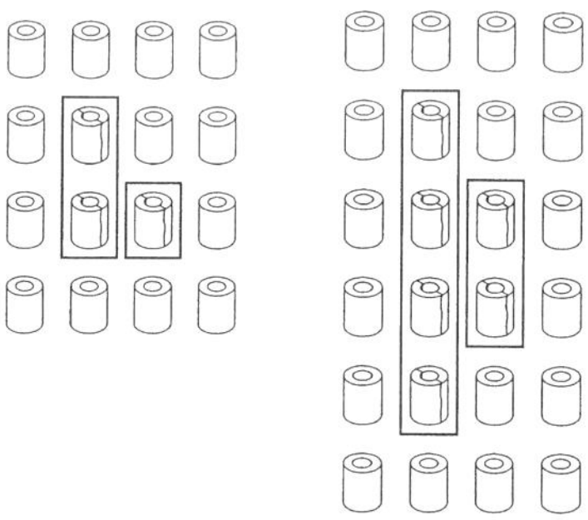

FIGURE 10: Typical arrangements of doubly cracked bricks in box-kite rig.

The verification has been carried out by performing cross comparisons of the same arrangement solved in one or more of the above alternative methods. Consistent results were obtained to those predicted by the stick and spring whole core modelling methodology.

Applications

The applications of the whole core modelling methodology to simulate the arrangements of a number test rig are given by McLachlan *et al.* (2006). Also given in this reference is the application of the methodology to an actual AGR reactor configuration.

CONCLUSIONS

A whole core modelling methodology has been developed and evolved over a period of many years for the purpose of predicting the movements of ageing AGR graphite cores. The methodology has the capability of addressing the consequence of cracked fuel bricks on the geometry and configuration of the core. Extensive validation of the methodology has been conducted against measured data from different 2-D and 3-D test rigs. Also verification of the methodology against alternative modelling methods has been carried out. This paper has presented a description of the development, validation and verification of the stick and spring whole core modelling methodology.

Acknowledgements

The authors would like to thank British Energy Generation Ltd for permission to present this paper.

References

McLachlan, N., Shaw, D. J. & Salih, M. H. S. (2006). Application of whole core modelling methodology to life extension of AGR reactor graphite cores. Presented at "Ageing Management of Graphite Reactor Cores" Conference, 28 to 30 November 2005, Cardiff, UK (also this book).

Method for Determining if Control Rods and Fuel Stringers can be Inserted / Removed from an AGR Core

Alistair F Thomson, Martin J Ripley and Neil McLachlan[#]
AMEC NNC Limited, Booths Park, Chelford Road, Knutsford, Cheshire, UK.
[#]British Energy (Generation) Ltd, Barnett Way, Barnwood, Gloucester, UK.
Email: alistair.thomson@amecnnc.com

Abstract
The cases for the safety and functionality of the AGR cores are based on the requirements for unimpeded movement of both fuel and control rods and cooling of fuel and core. These requirements are principally related to the geometry of the fuel and control rod channels. A computer code (LIMBO) has been developed to assess conservatively whether control rods and fuel stringers could be inserted into a core with specified distortion and hence the maximum distortion that can be tolerated can be calculated. The method uses three-point contact for fuel and control rods, and also assesses articulation angle criteria for control rods.

Keywords
Fuel stringer, control rod, distortion

INTRODUCTION

British Energy is revising the core safety cases for the AGR stations to enable them to be operated with cracked fuel bricks (the Revised AGR Core Safety Case, or RACSC). This will be done in stages, accounting for the range of cracked brick scenarios which might arise, where the cracks are essentially axial and run through the brick aligned with the radial keyways. In the first stage of the revised safety cases, it will be shown that the AGRs can be safely operated with predominantly one crack per fuel brick (RACSC2). In the second stage case it will be shown that the AGRs can be safely operated with at most a limited number of doubly cracked bricks (RACSC3). RACSCs are being developed first for Hinkley Point B / Hunterston B (HPB/HNB) as these are the lead stations.

The cases for the safety and functionality of the AGR cores are based on the requirements for unimpeded movement of both fuel and control rods and the provision of adequate cooling of the fuel and moderator under normal operation and under fault conditions. These requirements are principally related to the geometry of the fuel and control rod channels. As part of the assessment of the unimpeded movement of both fuel and control rods and the provision of adequate cooling of the fuel, the following route has been established:

1. Assess envelopes on fuel and control rod channel distortions using computational core models.
2. Determine the amplitudes of a "geometric sequence" of 23 channel shapes which would just fit into these envelopes, and add allowances as necessary.
3. Assess the ability of control rods and fuel to move within these maximum amplitude channel shapes.

Stage 1 has been performed separately, see Shaw *et al.* (2006), but Stages 2 and 3 of this route have been substantially automated in terms of a computer code, LIMBO.

ASSESSMENT OF THE PROBLEM

The requirement to be satisfied is that control rods and fuel stringers must be able to pass unimpeded through the core. The typical control rod consists of eight sections of approximately equal length. Each section is connected to the adjacent section by a mechanism which allows the two sections to move relative to each other provided the angle at the joint is less than a specified value. This angle is called the articulation angle. Note that in some AGRs there are sensor rods which are identical to the normal rods except that three sections are connected together to form one long section.

The fuel stringer is typically made up of eight sections (elements), each of which has an exterior graphite sleeve. There is no equivalent of the articulation angle, but the relative movement of the sections may cause gaps to appear between the sleeves. There are limits on the gap size: too large a gap will allow too great a coolant gas flow to bypass the lower fuel elements.

The HPB/HNB cores consist of twelve layers of graphite bricks which are held in place by various constraints. The graphite bricks also have keys which fit in slots (keyways) to ensure that the layers remain interlocked. However, due to construction tolerances and changes to the brick dimensions due to irradiation of the bricks and resulting cracks, the bricks can move relative to each other. Detailed assessment of the core can calculate the maximum possible movement of each layer of bricks and this gives the core envelope, see Figure 1.

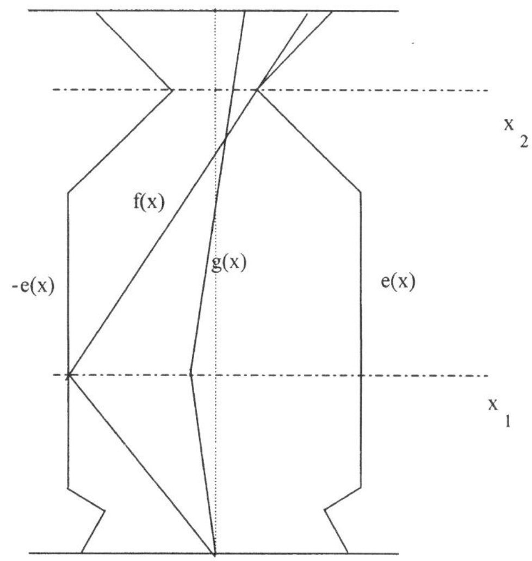

FIGURE 1: Core envelope and channel shape.

This does not given any information on the position of bricks in a single vertical column and thus on the degree of distortion that may exist in individual control rod or fuel stringer channels. In order to assess if the core still permits unimpeded movement of both fuel and control rods all possible forms of distorted channels have to be considered. It is not practicable to consider every possible option, so to achieve a high degree of coverage it was proposed to specify a large number (23) of different shapes of channel distortion. These shapes are specified such that the maximum deviation from the centre point is 1 mm to give a normalised channel distortion.

Method for Assessing Channel Shapes

Maximum Achievable Shape Amplitude
Having defined a shape of the normalised channel distortion, g(x), a new channel distortion shape, f(x), is defined such that

$$f(x) \equiv Ag(x) + Bx \qquad [1]$$

In which B is chosen such that A is the largest possible value given that f(x) must lie within the core envelope, e(x), Figure 1. The maximum value of A is known as the "Maximum Achievable Shape Amplitude" for a channel based on the normalised shape g(x). The 23 normalised channel shapes have been chosen to represent a range of possibilities, from a simple distortion like that given in Figure 1 to much more complex shapes which have as many as 10 corners. See Figure 2 for a selection of channel shapes used.

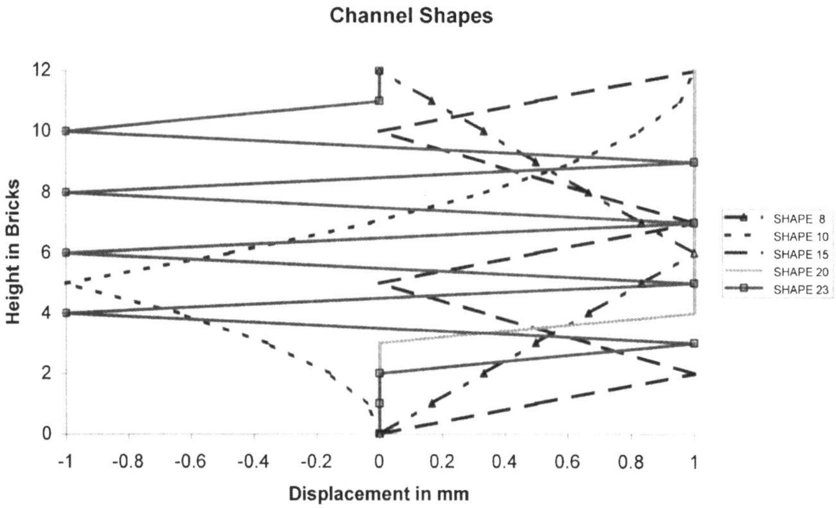

FIGURE 2: Selection of normalised distorted channel shapes.

The bricks in some cores are also interlocked using filler / spacer keys which are placed in slots in the bricks. These keys, as well as interlocking horizontally adjacent blocks, also tie the brick layers together in a manner which restricts the angle at the brick joints. This is equivalent to a limit on $\frac{df(x)}{dx}$. For simple shapes, *e.g.* Shape 8 (see Figure 2), the envelope restriction is the more restrictive, but for more complicated shape, *e.g.* Shape 23 (Figure 2), the restriction due to filler / spacer keys is more significant.

Maximum Tolerable Shape Amplitude

Having calculated the "Maximum Achievable Shape Amplitude" for each normalised shape, a method was required to determine if the control rod and fuel stringer could be inserted into a channel with this degree of distortion. It is desirable to perform the calculation in a manner which not only specifies whether the rod can move unimpeded through the core, but also gives an estimate of the margin that is still available. To do this, the "Maximum Tolerable Shape Amplitude" for each normalised shape is calculated. The "Maximum Tolerable Shape Amplitude" is the amplitude of the shape at which the rod would stick in the channel.

The first check is to consider if a single rod section or fuel element could move unimpeded through the core. This requires it to be demonstrated that three-point contact will not occur. For a channel with a single corner, the "Maximum Tolerable Shape Amplitude" can be calculated using the standard three-point contact equation.

$$A = \left| \frac{\frac{4G}{L}}{(S_2 - S_1)} \right| \qquad [2]$$

where		
	A	is maximum tolerable shape amplitude.
	S_1	is the slope of the channel below the corner with unity amplitude.
	S_2	is the slope of the channel above the corner with unity amplitude.
	G	is channel clearance, channel diameter minus rod diameter.
	L	is the length of the rod section.

For the three-point contact criterion, corners can be considered independent if the distance between two corners is greater than L/2. If the separation is less than this then equations can be developed for two corners, but it depends on whether the two corners form a U shape or a Z shape and results in a further two equations. If a third corner is introduced, a further four combinations are introduced thus requiring seven different equations to cover all the possible combinations. However, for the majority of cases for three-point contact, Equation [2] is used as the rod section size being typically less than two brick lengths and the minimum distance between corners is one brick length. The remaining six equations are only used in the case of the sensor control rods which do have one long section.

Having calculated the "Maximum Tolerable Shape Amplitude" for a single rod section or fuel element, this now being the upper limit, it must now be assessed whether the articulation angle restriction would reduce this value further. Although the rod consists of many sections, it was decided to base the method for calculating the "Maximum Tolerable Shape Amplitude" on two rod sections. However, it is recognised that in the case of a channel with multiple corners there will exist a degree of interaction between corners. A number of different methods have been derived for calculating the shape amplitude for the articulation angle criterion, see below. The different methods are obtained by applying different constraints on the rod end points and on whether a section is in contact with the corner. In order to decide which method should be used each method needs to be checked to ensure that it is valid. A method is valid if the constraints it applies to the rod at a corner are compatible with the constraints required at adjacent corners. Using these conditions it is possible to calculate the minimum distance that two corners can be apart for a given method to be valid. The method that yields the largest amplitude is used provided the method is valid. Note that if the constraints are not compatible then it cannot be guaranteed that the shape amplitude will not be limited by four point contact for a two section rod or five point contact for a three section rod. These methods ensure that the Maximum Tolerable Shape Amplitudes calculated by LIMBO are either equal to or less than the actual Maximum Tolerable Shape Amplitude.

It is not possible to give details of all methods used as there are five basic types of equation, and each type can have several variations which give in total 17 different methods to cover the whole range from corners being independent to the case where the distance between the corners is less than the rod section length. The method used depends on the distant that corners are apart and whether corners form a U or a Z shape. For a multi-section rod with a maximum articulation angle,θ, the equation for a single corner is:

$$A = \frac{\left(\frac{(L_1 + L_2)}{D_1 L_2} G + \theta\right)}{(S_1 - S_2)\frac{(L_1 + L_2 - D_1)}{L_2}} \qquad [3]$$

$$D_1 = \sqrt{\frac{(L_1 + L_2)G}{A(S_1 - S_2)}} \qquad [4]$$

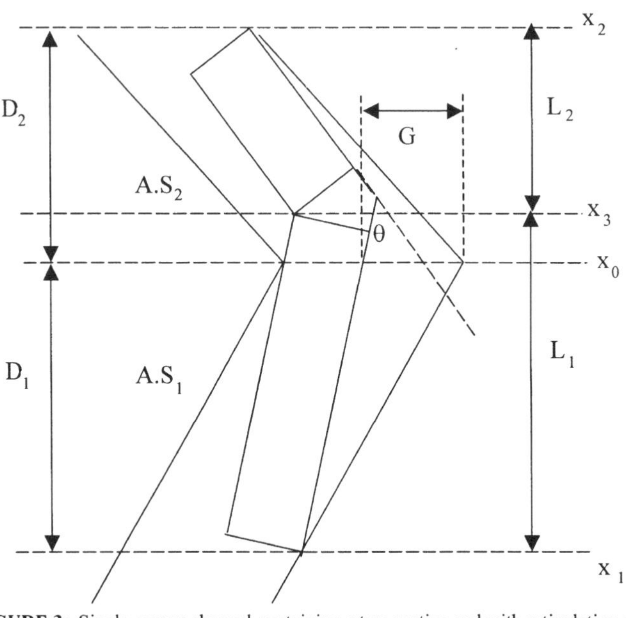

FIGURE 3: Single corner channel containing a two section rod with articulation angle θ.

Where
- A is maximum tolerable shape amplitude.
- S_1 is the slope of the channel below the corner with unity amplitude.
- S_2 is the slope of the channel above the corner with unity amplitude.
- G is channel clearance, channel diameter minus rod diameter.
- L_1 is the length of the shortest rod section.
- L_2 is the length of the longest rod section.

Note that all rods considered consist of either all sections of length L_1 or one section of length L_2 and all remaining sections of length L_1.

In the case of the articulation angle criteria, corners have to be very far apart before they can be considered independent of each other. In the above equations, it is assumed that the start and finish points of the rod's two sections are in contact with the outer edge of the channel. If a second corner exists, and for the sake of simplicity this is an identical corner, and the two corners form a Z shape, then identical conditions are applied to this corner. It can then be shown that the corners require to be between $2L_1 + L_2$ and $3L_1 + L_2$ apart before the two corners can be considered to be independent. For channels where this restriction cannot be satisfied an alternative method of calculating the Maximum Tolerable Shape Amplitude is required.

For Z-shaped pairs of corners, if the boundary conditions are changed such that the bottom of the first section of the rod touches the outer edge of the channel, and the top of the second section touches the inner edge of the channel then the equation used is:

$$A = \frac{\left(\frac{(L_1 + L_2 - D_1)}{D_1 L_2} G + \theta\right)}{(S_1 - S_2)\frac{(L_1 + L_2 - D_1)}{L_2}} \qquad [5]$$

and this equation is valid provided the corners are more than $L_1 + L_2$ apart. The next method considered is for a Z-shaped corner for which the boundary conditions are changed such that the bottom of the first section of the rod and the top of the second section are held in the centre of the channel then the equation used is:

$$A = \frac{\left(\frac{(L_1 + L_2)}{D_1 L_2}\frac{G}{2} + \theta\right)}{(S_1 - S_2)\frac{(L_1 + L_2 - D_1)}{L_2}} \qquad [6]$$

This equation is valid provided the distance between the corners is greater than $L_2 + 0.4142L_1$. Note that for the above three methods, account must be taking of a third corner so even if the above conditions are valid the method may not be, because of the constraints of the presence of additional corners. A similar set of equations can be developed for U-shaped corners or combinations of Z-shaped and U-shaped corners if three or more corners are present.

Further sets of equations have been derived for conditions in which corners are less than L_2 apart so that all channel shapes can be assessed. However, if none of these equations are valid then the Maximum Tolerable Shape Amplitude is set equal to the shape amplitude for which a clear straight path exists through the channel.

Testing

In order to check the results of LIMBO, a program was produced that could check that a five section rod could pass through a distorted channel of specified magnitude. In all cases of a distorted channel with amplitude equal to "Maximum Tolerable Shape Amplitude" calculated by LIMBO, the program proved that the rod could pass through the channel with maximum articulation angle less than that used in the LIMBO calculation.

RESULTS

The HPB/HNB cores have been assessed and the result for three-point contact and articulation angles of 3 and 6 degrees assessed. For all channel shapes there is a significant margin available when the "Maximum Achievable Shape Amplitude" is calculated taking into account both the lateral envelope constraint and the effects of the spacer / filler keys. The lateral envelope constraint is all that is required for most shapes (1 to 17). For shapes, 18 to 23, which all have the common feature that the distortion changes from zero to maximum in a single layer, the lateral envelope constraint on its own is not sufficient. The effect of the spacer / filler keys is required to demonstrate a margin between tolerable and achievable amplitudes.

It should be noted that the calculations which produce the core envelope do not take account of every effect. In particular the following two effects are ignored:

1. The elasticity of the core layer.
2. The freezing in of earlier core movements.

These effects will tend to increase the Maximum Achievable Shape Amplitude, but when an allowance is included for these terms, the conclusions given are not affected.

CONCLUSIONS

The LIMBO code provides a quick and reliable method of assessing whether the control rods and fuel stringers have a clear path through a core in which the degree of distortion is known. The code can also be used in reverse to calculate the degree of core distortion that would be required to occur before it was possible that a clear path through a channel may not exist. Thus it is a useful tool in assessing the effects of core distortion.

Acknowledgements

The authors would like to thank British Energy Generation Ltd for financial support. The views expressed in this paper are those of the authors and do not necessarily represent those of British Energy Generation Ltd or AMEC NNC Limited.

References

Shaw, D. J., Salih, M. H. S. and McLachlan N. (2006). Development and validation of whole core modelling methodology for AGR graphite cores. Presented at "Ageing Management of Graphite Reactor Cores Conference", Cardiff, 28-30th November 2005 (also this book).

PART E

Lessons Learned from Management of Ageing Plant

Experience of Ageing Processes in Nuclear Power Reactors

Brian Eyre
University of Oxford, Oxford, UK.

Abstract
A major factor in the ageing of nuclear power plants is the time dependent deterioration of materials performance with the associated potential for premature failure. The paper discusses the major factors involved in ageing from this materials perspective. It will consider the lessons to be learnt by referring to some specific examples.

Keywords:
Experience, ageing processes, nuclear power reactors

INTRODUCTION

Nuclear power generation has had an impressively successful record since its introduction in the 1950's. Of the 550 reactors constructed over this period, 80% (440) are still operating and a further 24 power units are under construction. However, construction has been at a much lower level in recent years and about 75% of operating reactors now have more than 20 years service. The original design lifetimes for commercial power reactors ranged from 20 years for the early Magnox reactors to 40 years for the more recent light water reactors (LWRs). However, lifetimes of reactors have already been extended significantly and there is a drive to extend the lives of current reactors by as much as 20 years to 60 years in the case of the more modern pressurised water reactors (PWRs).

There is therefore a major incentive to manage and where possible reduce the effects of ageing-related degradation in reducing component lifetimes. A key factor in such degradation is the deterioration in material properties. A number of factors are involved and these will be considered in the paper. We also consider specific examples of component ageing processes potentially resulting in premature failure or restrictions on life and the lessons to be learnt. An important generic lesson is the need to understand the underlying mechanisms of ageing in order to develop improved models for lifetime prediction. This is required to provide a sound basis for ageing management with minimum impact on plant safety and availability and for developing improved materials and designs.

FACTORS INFLUENCING MATERIALS AGEING

For reliable and safe operation of nuclear power plants there is a need know the degree of ageing-related deterioration of components. In particular, there is a requirement to have a basis for assessing residual lifetimes and fitness for continuing service. In considering options for managing the consequences of ageing, components can be classified as being life limiting, such as reactor pressure vessels and many core structural components, or as replaceable such as circuit components and cabling. In practice, establishing in service the degree of deterioration and therefore being able to define a forward management strategy has proved to be extremely difficult. Continuous monitoring of changes is a desirable, but as yet a largely unrealised objective. Thus periodic inspection is required as an essential part of the safety justification and considerable progress has been made in improving the reliability and precision of inspection methods, particularly ultrasonic NDT.

Central to component ageing is the time dependent changes in materials properties which are often due to subtle changes at the micro-structural level. This is determined by a number of intrinsic factors including composition and thermal and mechanical treatments. These control parameters such as grain-size, precipitate and dislocation distributions, all of which impact on the initial mechanical, physical and chemical properties. Materials ageing in plant components are then dependent upon extrinsic factors linked to the construction and operating characteristics of the plant including such as:

- Temperature and particularly thermal stresses.
- Internal stresses, for example in the vicinity of welds.
- Coolant chemistry.
- System stresses.
- Irradiation, resulting in micro-structural damage and changes in diffusion and chemical properties.

Most of these factors apply to any process plant, but irradiation effects are unique to nuclear plants and they can work synergistically with the other parameters to exacerbate ageing effects. An important factor in successfully managing plant ageing is to have validated models for predicting component lifetimes. As will be discussed later, considerable effort has gone into developing such models with varying degrees of success.

Practical experience has shown that there are a number of top level factors involved in the management of component degradation due to ageing including:

1. **Design,** for example through the choice of materials and their condition, configuration, fabrication, accessibility, operating conditions, environmental conditions, *etc.*
2. **Quality Control,** during construction to ensure compliance with the design intent, for example surface finishes, heat-treatments, welding and other fabrication processes.
3. **Inspection and Monitoring,** including validation to establish confidence in the detection limits.
4. **Operating Conditions,** particularly flexibility of control, *e.g.* coolant chemistry and operating temperatures.
5. **Replacement Strategies** and other recovery strategies such as thermal treatments.
6. **Vigilance and Clear Communication,** such as key requirements that apply to all the above criteria.

EXPERIENCE OF AGEING PROCESSES AND THEIR MANAGEMENT IN NUCLEAR POWER PLANTS

In this section, three examples are considered of materials ageing processes in reactors that have the potential to seriously affect availability and in one case at least be life limiting. The examples include two where a long term strategy for management of the ageing process has been effective. The third example relates to a long term issue in LWR's where management of the ageing effects has required, in extreme cases, component replacement. These three examples are by no means exhaustive, but they serve to illustrate the essential elements of a successful plant ageing management strategy.

Embrittlement of Reactor Pressure Vessels

At a very early stage in the development of nuclear power, it was recognised that irradiation embrittlement of reactor pressure vessels (RPV) could pose a threat of unstable fracture causing an extremely severe accident and obviously be life limiting. Such a failure is most serious for the pressurised water reactors but it was also of concern to the early Magnox reactors which have ferritic steel pressure vessels. Thus, from the mid 1950's major programmes were mounted to investigate irradiation embrittlement in ferritic steels. Initially, the focus was on interpreting the mechanical property changes in terms of irradiation induced point defect damage, although valuable work was carried out on the interactions of interstitial carbon and nitrogen in the steel due to trapping of intrinsic defects (vacancies and self interstitials).

An important development occurred in the late 1960's when results from surveillance testing of samples of PWR RPV steels showed greater embrittlement in weld metals with high copper content. This stimulated fundamental research into understanding the effects of Cu and other elements on micro-structural changes in RPV steels due to irradiation and their role in causing hardening.

The application of high resolution analytical transmission electron microscopy and 3D atom probe showed the presence of copper precipitates whose distribution is influenced by irradiation. This led to the development of a mechanism based model in which the hardening is due to the combination of two effects: irradiation induced copper precipitates and intrinsic irradiation damage in the form of point defect clusters. The essential elements of that model still stand, but over the last 20 years or so detailed studies have revealed much more detail about the copper precipitates themselves, their interactions with other alloy elements in the RPV steels such as manganese and their distribution as a function of irradiation conditions. However, the intrinsic irradiation damage has proved to be more intractable in terms of obtaining direct observational evidence. The best evidence is on relatively pure iron irradiated in accelerators with heavy ions and more recently with reactor neutrons. This shows that interstitial loops can form during irradiation and that they can have a distinctive geometry compared to other BCC metals in that both $a\langle 100\rangle$ and the lower energy $a/2\langle 111\rangle$ (Burgers vectors) types can form. In addition, less direct measurements using positron annihilation and small angle neutron scattering have given information on the distribution of intrinsic damage clusters and analysis of the data has indicated the form of the vacancy component of the damage. The structural studies have been complimented by systematic mechanical property studies elucidating the dependence of hardening on the copper rich precipitates and intrinsic irradiation damage. Research has also been carried out on the potential role played by impurity segregation to grain boundaries causing an irradiation induced temper embrittlement effect. An important aspect here has been to determine the boundary conditions for this to occur during long term irradiation and to establish the potential for it to occur during service in commercial RPV steels.

Thus, the high quality fundamental research into irradiation hardening and embrittlement of RPV steels has resulted in a well founded understanding of the mechanisms involved and has enabled the development of quantitative predictive models that incorporate the mechanistic processes. From a practical viewpoint, the early realisation of the effects of Cu resulted in much tighter specifications for both the vessel steels and the weld metals used. In addition, the trend curves for the embrittlement of RPV steels on operating reactors taking into account the mechanistic understanding both reduce the degree of conservatism in the predictions and can be used to extrapolate forward with much greater confidence.

Void Swelling in Fast Reactor Components

The realisation that void swelling posed a problem to the lifetime of fuel element components in fast reactors came from observations on samples of stainless steel fuel following irradiation in a fast reactor core, namely the Dounreay Fast Reactor. In the mid 1960s, transmission electron microscopy showed the presence of voids in type 316 steel samples that had been irradiated to about 2×10^{22} $n \cdot cm^{-2}$, equivalent to about 25 displacements per atom (dpa) and a fuel burn-up of 4%. The distribution and sizes of the voids depended on dose and temperature and thus fuel element components, particularly the fuel wrappers that are subject to variations in these parameters would distort due to differential swelling and require to be removed from the core before they became jammed. While the problem is not life limiting to the reactor itself, replacement of fuel element before the fuel had achieved its target burn up threatened load factors for fast reactors.

The initial observations of void swelling and a recognition of its practical implications stimulated major programmes in all of the countries with fast reactor programmes: UK, France, Germany, USA, Japan and (the then) USSR. The work was aimed at both understanding the mechanisms and developing materials with greater resistance to void swelling. It was quickly realised that voids form due to partitioning of the irradiation induced vacancy and interstitial point defects resulting from the competition between the different processes determining their eventual fate. Specifically, most of the point defects created by irradiation are lost due to recombination. Due to a first order size effect, referred to as the bias term, surviving interstitials are preferentially attracted to dislocations while the excess surviving vacancies associate with insoluble He gas atoms to form small bubbles that grow to voids due to an excess flow of vacancies. The swelling associated with the presence of voids is in fact due to the plating out of the excess interstitials on the dislocations. Thus, the basic mechanism for void swelling was established at an early stage and rate theory was used to develop quantitative predictive models. But in applying such models to alloys in which solute element interactions with the intrinsic point defects play a key role in their distribution and survival, assumptions were required on the nature of such interactions and these become very complicated on moving to more complex materials characteristic of commercial alloys used in practice.

Thus, experimental studies played a key role in both the calibration of adjustable parameters in the models and in the development of alloys with compositions optimised for resistant to void swelling. The key to this is to minimise the survival of vacancies and interstitials by stimulating recombination, for example, by trapping at solutes or precipitates, or loss of both defect types at sinks - dislocations and boundaries. A particularly interesting and practically important result from this work was the apparently intrinsic void swelling resistance of ferritic alloys. The reasons for this are not fully understood, but as referred to previously, ferritic steels form a<100> loops as well as the energetically more favourable a/2<111> loops. Thus, having two dislocation sinks for use with different strengths of the bias term could result in all of the interstitials and vacancies being lost at the two loop types (with interstitials attracted preferentially to the a<100> loops and the excess vacancies diffusing to the a/2 <111> loops) resulting in zero net growth.

On a more practical note, as a result of the very extensive research programmes, a range of both austenitic and ferritic alloys were developed that had good void swelling resistance over the required dose range to obtain optimum performance from fuel elements in fast reactor cores. Thus, nickel based alloys such as Nimonic PE16 and ferritic stainless steels were shown to have very good swelling resistance to doses in excess of 20 dpa. For the mixed oxide (typically UO_2-20%PuO_2) fuel used in sodium cooled fast reactors, the optimum burn-up is in excess of 200 GWd/t_e. In 1974, the target burn-up for fuel in the Prototype Fast Reactor (PFR) at Dounreay was 7.5%, equivalent to 75000 GWd/t_e, mainly limited by the swelling of the wrappers, but by the late 1980's, when the reactor was approaching its end of life, fuel burn-up in excess of 20% were being achieved from fuel elements clad in a Nimonic alloy PE16 with ferritic steel wrappers, both having very good swelling resistance. This was one of the major successes of the fast reactor programme.

Corrosion Environmentally Assisted Cracking (EAC) of Reactor Materials

Light waters reactors have been beset by corrosion related problems associated with interactions between reactor components and the coolant water. For core internal components there is the added effect of neutron irradiation that can impact on both materials structure and properties and on water chemistry. In extreme cases, degradation due to EAC has only been managed by replacement of major components such as steam generator units (SGUs) and RPV heads on PWRs and core shrouds on boiling water reactors (BWRs). Three examples are summarised below with particularly serious consequences for the reliable operation of PWRs.

Steam Generators

Historically SGUs in commercial PWR units have been particularly problematical. Experience has shown that degradation can result from a range of corrosion and wear processes including wastage, pitting, denting, stress corrosion cracking (SCC), fretting and fatigue. Considerable progress has been made in improving specifications and control over water chemistry and this has been effective in reducing degradation due to wastage, pitting and denting. However, inter-granular attack and SCC, particularly in regions where local chemistries may differ considerably from the bulk water chemistry such as in crevices at tube sheets and tube support plates, remains a major problem area. A fundamental cause of many of these corrosion problems is that Alloy 600 used for the tubing is less corrosion resistant than originally thought and this has been exacerbated by crevice corrosion effects.

Management of SGU degradation due to SCC and inter-granular attack remains a difficult area. SGU tube integrity is a safety critical issue as well as having a major impact on plant availability and economics. For example, SGU tubes account for 90% of the primary coolant pressure boundary and failure can result in significant releases of radiation to the atmosphere. Operators have had to resort to tube plugging and repair or in extreme cases to replacement of SGUs exhibiting tube failures. On average about 7000 tubes were plugged or repaired on PWRs world wide between 1997 and 1999, at considerable cost including loss of efficiency. Uncertainties in the NDE methods for detecting of cracks have resulted in considerable conservatism and there is evidence that a large proportion of tubes were plugged unnecessarily. Up to June 2000, some twenty five plants in the US had already replaced their original SGUs at a basic hardware cost of $150M plus a roughly equivalent amount due to loss of availability per PWR unit. Replacement steam generators have Alloy 690 tubes containing a higher chromium content to confer improved corrosion resistance. There is insufficient service experience with this material yet to be certain that it offers a long term solution to the SCC problem. It is interesting to note that PWRs built by KWU of Germany used alloy 800 (which also has a higher Cr content) for SGU tubing and they have encountered much less tube cracking due to SCC.

Thus, regarding the more effective management of SGU degradation in existing plants, there is a need to develop improved NDE methods to provide a sufficiently accurate characterisation of the cracking to allow continued operation of SGUs containing defected tubes. In addition there is a focus on obtaining a better understanding of coolant chemistry, particularly the relationship between bulk and crevice chemistry with the aim of enabling improved chemistry management.

More generally, there is a need for a more complete understanding of SCC in SGU materials. This is required to advance more effective ageing management in existing SGUs through the development of mechanistically based residual life prediction models and also to lay the basis for improved design and materials selection in new SGUs for replacement components and new PWR plants. There is a particular need to understand the processes occurring at the crack tip and to elucidate the roles played by the environment and material microstructure in the susceptibility to SCC.

Reactor Pressure Vessel Head Penetrations (VHPs)

One of the most potentially serious examples of a failure to effectively manage a materials ageing problem on PWRs was the reactor vessel head degradation found in 2002 on the Davis Besse Nuclear Power Station (DBNPS) plant in the US. There was a considerable body of experience and information prior to this event that had not been adequately taken into account. There were key lessons for both the regulator and the plant operator.

In most PWRs, Alloy 600 is used for VHPs and instances of leaks have been found at other Alloy 600 penetrations since the mid 1980's. The VHP penetrations include control rod drive mechanism (CRDM) tubes that are shrunk fit and then welded to the RPV steel. Thus there are high residual stresses and there is a transition weld between the two materials which adds to the complexity. Primary water SCC (PWSCC) has been experienced at CRDM's after about 15 to 20 years operation, the first case of a through wall failure being found during a pressure test on a French PWR at Bugey in 1991. Cracking initiates in the region of the J welds on the primary water side and generally propagates in an axial direction. Reliance had been placed on external visual inspection of the vessel heads for detecting failures at VHPs, but following the Bugey finding and examples of PWSCC on other VHPs, France and Japan embarked on a major programme of RPV head replacement at considerable cost. In contrast in the mid 1990's, the United States Nuclear Regulatory Commission (US NRC) while recognising cracking of VHPs as being of long term safety significance, were prepared to rely on vigilance in the form of visual inspection programme as adequate defence. Plant operators started to introduce remote inspection based on eddy current and ultrasonic measurements and interacted with vendors in developing susceptibility models based on experience at their plants. However, in 2000, boric acid residues were found on the vessel head of Oconee Unit 1 in the vicinity of CRDM tubes containing axial cracks due to PWSCC. Then in 2002, significant wastage of the reactor vessel head base metal was found on the DBNPS and this was due to boric acid leakage from CRDM tubes containing axial cracks. The wastage reduced the RPV head thickness to about 6.5 mm and the area of wastage was about 160 cm^2, with the primary pressure being effectively contained by the inner wall cladding over this area. Clearly the cladding is not designed for this. Had the cavity not been found by chance while repairing cracks in the nozzle, continued operation of the reactor could have resulted in a loss-of-coolant-accident. Thus the DBNPS can be regarded as a near miss in terms of a threat to plant safety.

In a lessons learnt exercise instigated by the NRC, it was concluded that the degradation event was preventable and particularly:

- NRC, DBNPS and the nuclear industry failed to adequately review, assess and follow-up on relevant operating experience.
- DBNPS failed to ensure that plant safety issues would receive appropriate attention.
- NRC failed to integrate known or available information into its assessments of DBNPS's safety performance.

The generic lesson from this example is similar to that for the SGU issue in the importance of having validated residual life prediction models based on a mechanistic understanding of the failure processes, in this case both PWSCC of the tubes and boric acid corrosion of the vessel head base material.

Flow Assisted Corrosion

A third example of a corrosion related problem in LWRs is wall thinning due to flowing water in secondary pipe work. This came to the fore in the US in the mid 1980's when two instances of pipe ruptures occurred in secondary carbon steel pipes operating under single phase flow conditions. The first was on the Trojan PWR plant after 9 years operation and resulted in one plant worker being injured. The second was on the Surry PWR unit 2 after 14 years. In this case, rupture was a catastrophic 360° circumferential break. A large piece of the pipe was ejected some distance from the break and there were four fatalities. In both cases, failure was associated with extensive wall thinning due to mechanical and chemical interactions with the flowing water, *i.e.* corrosion / erosion. In both events, the fluid medium was single phase pressurised sub-cooled water at nominally 175 °C and the flow in the ruptured locations was turbulent due to the local piping and fittings configurations. Historically, in both nuclear and fossil plants wall thinning has been experienced under two phase conditions. Thinning under the single phase conditions was not expected.

Following the Trojan and Surry incidents assessments of LWRs in the US showed extensive wall thinning in feed-water / condensate lines had occurred and that it was more prevalent in PWR's compared to BWR's. The timing of the failures in the Trojan and Surry plants was used to estimate upper and lower bounds for pipe rupture in single phase conditions.

As result of this experience, inspection and monitoring programmes and replacement criteria were tightened as an effective means for managing the wall thinning issue in secondary pipe-work in PWRs and BWRs in the US and elsewhere. Nevertheless, a further major pipe rupture due to wall thinning occurred on the Mihama PWR unit 3 in 2004, again resulting in the fatality of four plant workers. Subsequent examination showed the failure process to be corrosion / erosion in turbulent flowing water. Post accident review showed that inspection programmes had not been carried out on the ruptured pipe section and examination of other

PWR plants operated in Japan showed similar wall thinning although the extent was not at the critical level for rupture.

All of this experience highlights another example of an unexpected degradation process on a plant component for which rupture is a design base event. The consequences were exacerbated by a lack of vigilance in terms of inspection and monitoring. It also highlights the importance of effective communication of experience and lessons learned and the need for rigour in applying mitigating measures.

As with the other examples, there is also a need to obtain a mechanistic understanding of the degradation process. Pipe wall thinning in flowing water involves a complex interaction between a number of factors relating to material, water chemistry, temperature and flow conditions – velocities and turbulence. There is a need to develop a physically based model to predict thinning rate as a function of water flow conditions in order to both optimise inspection and replacement schedules and to develop preventative measures such as improved materials or protective coatings to enhance component lifetimes and enable the design base to be satisfied.

LESSONS LEARNT AND CONCLUSIONS

Plant ageing related degradation processes in nuclear power plants impact on both safety and availability and can be very costly to plant operators. There is a considerable incentive to have an effective strategy for their management, aimed at reducing the incidence of such processes to a tolerable level. Regarding the top level factors listed in 'Factors Influencing Materials Ageing' it could be added the need to anticipate the unexpected. This emphasises the need for validated models to predict damage accumulation and failure processes thus enabling anticipation as opposed to disruption that comes from unexpected failures. Part of the validation process is fitting the model predictions with existing data in order to calibrate adjustable parameters. But it is essential that the models are based on a mechanistic understanding if they are to reliably predict future behaviour. This is exemplified by the examples considered in this paper, particularly the contrast between the positions on RPV embrittlement and void swelling with that for corrosion related processes.

With regard to irradiation embrittlement of RPV steels, considerable progress has been made in identifying and understanding the basic processes and this has been used to reduce the degree of conservatism in the assumptions used in the development of trend curves to predict changes in fracture transition behaviour. But there remains a need to improve our understanding of intrinsic irradiation damage component and particularly its effect on mechanical properties. There is also a need to understand the redistribution of solute atoms under irradiation conditions. Nevertheless, it is expected that for vessels manufactured using clean steels and weld consumables, RPVs should have a lifetime of 60 years subject to continuing vigilance in inspection and surveillance on the plant and in the timely interpretation of research data to identify any unexpected behaviour that could impact on the lifetime assessments.

Although fast reactors are now on the back burner of nuclear technology development in much of the world, a significant increase in nuclear power could renew the drive for closing the fuel cycle to optimise the energy extracted from uranium (and thorium). Recycling fuel in fast reactors enables some hundreds times more energy to be extracted from the natural resources of uranium and thorium compared to recycling in thermal reactors. The work carried out on void swelling in the structural materials used for fast reactor fuel elements illustrates particularly well the importance of basic understanding in successful technology development. In particular, void formation and growth requires point defect survival and an imbalance between the vacancy and interstitial flows to voids and other sinks. Minimising void swelling has been achieved through control of point defect flows to sinks through the manipulation of alloy compositions and microstructures. This example is particularly pertinent to the irradiation induced changes, particularly dimensional changes, in graphite. There is again a need to understand the interactions between the irradiation induced point defects and the microstructure including its dependence on composition and fabrication history. Such an approach will provide a basis for understanding the batch to batch variations in behaviour that is an important characteristic of graphite and for developing more reliable models for residual life prediction.

Corrosion related issues have proved to be the most troublesome of ageing related phenomena in LWRs. They have been extremely costly in terms of repair and component replacement and in lost availability. Safety has also been threatened, as highlighted by the Davis-Besse event. Progress has been made in managing corrosion related degradation through control of water chemistry and in the case of SGUs and VHPs, through changes in alloy (from 600 to 690) for replacement components, although in this latter case, operating experience is still limited. Experience has also demonstrated a lack of vigilance in monitoring and detecting degradation processes before they reach a critical level and as a result inspection and monitoring procedures have been tightened. However, at a more fundamental level there is an inadequate understanding of the mechanisms of EAC and thus we do not yet have well founded models for residual life predictions and for optimising alloy microstructures to be EAC resistant.

The general lessons can be represented by two quotations by the 17th century Spanish philosopher, Baltasar Gracian (1601-1658):

> *'When in doubt there is safety in sticking with the intelligent and prudent, for sooner or later they catch up with luck'*
>
> *'Technology without understanding is compounded madness.'*[13]

Acknowledgements:
I thank Dr's Colin English and Christine Brown for helpful comments on the paper.

[13] Gracian actually referred to science rather than technology.

The Importance of Understanding of Basic Principles in the Management of Ageing Plant

Michael Burdekin
Formerly School of Mechanical Aerospace and Civil Engineering, University of Manchester.
Email: mburdekin@aol.com

Abstract
The paper compares a number of problems that have been tackled by the Technical Advisory Group on the Structural Integrity on Nuclear Plant (TAGSI) for pressure vessel steels with problems tackled by the Graphite Technical Advisory Committee (GTAC) in relation to graphite. The potential application of the TAGSI framework for Incredibility of Failure (IoF) safety cases to other high integrity situations is discussed. The role of monitoring and inspection in the management of ageing plant is also discussed. It is concluded that prediction of future continued deterioration of plant requires a sufficient understanding of the basic mechanisms involved, supported by monitoring and inspection, to confirm that the predictions are being realised and that no unexpected phenomena are occurring. Areas where a better understanding of deterioration mechanisms would be beneficial are highlighted.

Keywords
Empirical models, materials degradation, understanding principles

EXPERIENCE OF AGEING IN STEEL COMPONENTS

TAGSI

The Technical Advisory Group on Structural Integrity of Nuclear Plant (TAGSI) was established by AEA Technology Ltd in 1991 as the successor to the Light Water Reactor Study Group which had been influential in demonstrating the benefits of high quality design, materials and construction on the safety of pressurised water reactor systems. The remit of TAGSI was broadened to cover a wide range of structural integrity issues raised by its sponsors, who over the years have included the licensees, AEA Technology, MoD, NII and UKAEA. Membership of the TAGSI Main Committee has consisted of an independent Chairman, up to ten independent members and one representative from each of the sponsor organisations. A formal report is produced for the sponsors and in many cases this leads to the publication of a paper in the open literature. Three topics considered by TAGSI relevant to ageing in steel components and safety cases are described below.

Effects of Radiation on Steels

A number of issues concerned with the effects of radiation on steels have been considered by TAGSI. The major concern with ferritic steels is strengthening and embrittlement leading to an increase in the transition temperature for brittle fracture to occur. It was concluded by TAGSI (Knott *et al.*, 1999) that the overall form of a dose-damage relationship appropriate for ferritic steels up to about 350 ^{0}C is as follows:

$$\Delta T_{40J} = A_m F_T \sqrt{D} + B_{Cu} + C(D^n - D_0^n) \quad [1]$$

where ΔT_{40J} is the shift in 40 Joule transition temperature in the Charpy test, A_m is the matrix damage sensitivity, F_T is the irradiation temperature dependence term, B_{Cu} is the saturated copper precipitation term, C is a measure of susceptibility to intergranular fracture, D and D_0 are the dose and threshold dose in units of (dpa x 10^5). These three terms indicate different mechanisms of embrittlement operating on the matrix, precipitates and grain boundary effects and include a delay term in one case.

Re-heat Cracking

Re-heat cracking is a form of cracking that can develop at welded joints in steels exposed to high temperatures either as a post weld heat treatment (PWHT) or in service. Problems have occurred in some non post weld heat treated thick section stainless steel components in service and TAGSI was asked to review the understanding of this phenomenon. The problem arises predominantly from the accumulation of creep strain under the action of applied and residual stresses, in a triaxial stress state, until the multi-axial creep ductility is exhausted.

TAGSI considered that for prediction of the potential occurrence of re-heat cracking, the accumulation of creep strain and stress relaxation for each increment of time, detailed finite element analyses are carried out (Knott *et al.*, 2005). The first stage is to relate the instantaneous equivalent (von Mises) creep strain rate to the instantaneous value of the equivalent (von Mises) stress. The second stage is the estimation of the creep ductility under a multi-axial stress state, using the Spindler equation to relate the (equivalent) fracture strain in a multi-axial stress state to that in uniaxial tension. The key to reducing the likelihood of the phenomenon occurring in service is to use materials with increased available creep ductility. In practice this was achieved by replacing components in type 316H steel with ones in type 316L. The uniaxial creep ductility in type 316H stainless steel had been found to give a lower bound of 0.9% at very low strain rates, whilst that of type 316L steel was found to be significantly higher than this. The reason for the difference was attributed to the availability of the extra carbon in 316H to precipitate carbides during the high temperature exposure.

Incredibility of Failure

One of the major principles of safety in nuclear plant is the requirement for 'Defence in Depth'. This requires there to be a number of layers of protection such that if failure occurs in one layer the remaining layers would ensure that there would be no harm to individuals or to the general public at large. Where possible, 'Defence in Depth' is sought through diversity, redundancy and segregation to give a robust, fault tolerant design. There are some components where it may not be possible to provide such requirements by provision of physical means of containment. Where there are judged to be unacceptable consequences for public safety from the potential failure of a single component, 'Incredibility of Failure' (IoF) needs to be demonstrated. In the mid 1990s, the NII asked TAGSI to provide advice on what constitutes an IoF safety case and how can the 'strength' of an IoF safety case be judged?

The results of the TAGSI discussions led to the publication of an authoritative report (Bullough *et al.*, 2001) that has been used as a basis for structuring a number of IoF cases. In its response, TAGSI concluded that where 'Defence in Depth' did not exist by multiple physical barriers, an IoF case could be sought in a 'Conceptual Sense'. A Conceptual Defence in Depth required the application of multiple arguments (legs of the case) and/or diverse sources of information. Robustness is achieved through the application of multiple legs, with due regard paid to redundancy, diversity and the extent to which there is a degree of independence between separate legs. Legs are constructed from elements, where an element is a separate and discrete item. An element may be present in more than one leg, but may have different importance in different legs. TAGSI considered that it was unlikely that the legs of an IoF case would be completely independent, but the important requirement was that together they should confirm an adequate level of confidence.

TAGSI concluded that the strength of an IoF case depended on the confidence that the probability of failure is negligible (taken as at most 10^{-7} failures per year). TAGSI also concluded that the use of probabilistic analyses, scoring systems and expert elicitation were potentially suitable methods for examining qualitatively or quantitatively the strength of a safety case and its legs. TAGSI also recognised that some of the legs of a safety case may change with time, when updating of a safety case is carried out. This is particularly the case for legs where information from monitoring or inspection in service may be provided on the presence or absence of defects and on changes in material properties with ageing.

EXPERIENCE OF AGEING IN GRAPHITE COMPONENTS

GTAC

The Graphite Technical Advisory Committee (GTAC) was established by the HSE Nuclear Installations Inspectorate (NII) at the beginning of 2004 to provide a source of independent advice on a structured list of some twenty specific questions on graphite. The Committee consists of nine independent members chosen to have a mixture of experience of graphite in nuclear applications or of structural integrity assessments from other fields. For each question on which NII seeks advice, the terms of reference and concerns are set by NII and then each of the licensees gives their perspective view on the matter. A sub-group of members of the main committee then considers the issues in detail, studying relevant references of work carried out by the industry and holding discussions with technical experts from the licensees before producing a draft report. This report is then discussed at full meetings of the whole committee. As a final stage, the draft report is sent to the licensees to enable them to comment on any issues of fact or accuracy. When any such comments have been taken into account a final report on each question is submitted formally to NII. Some examples of topics considered by GTAC are described below.

General Points on Behaviour of Graphite in Nuclear Reactors

For Gilsocarbon type graphite as used in AGRs, the properties of graphite for use in prediction of behaviour are given in a series of data and method sheets known as GCDMC Paper 28 (British Energy Generation Ltd., 2003). The data are essentially derived from the results of Materials Reactor Tests (MTR). These data sheets are reviewed and updated on a regular basis as further data become available. Recent reviews have been carried out by Bradford on strength and dimensional change (Bradford 2002, 2003). For PGA graphite, as used in MAGNOX reactors, the main reference for material properties is by Marsden and Wilkes (2001). A review of the dimensional change behaviour of PGA graphite was carried out by Payne (2004).

As far as structural integrity is concerned, the key properties of graphite are dimensional change (rate) with irradiation, thermal conductivity, co-efficient of thermal expansion (CTE), creep behaviour with irradiation, modulus of elasticity, Poisson's ratio and strength. Because of the gradients of dose and temperature across graphite bricks, internal stresses are set up that, in the later stages of life for AGRs, would be expected to be tension at the keyway regions with consequent risk of cracking. Considerable effort has been made by the licensees to try to predict if and when 'primary' cracking of bricks from keyways might occur.

However, some cracking of bricks has occurred in AGRs at an earlier stage than predicted for keyways and this is now termed early life bore cracking (ELBC). As a result, the thrust of safety cases has changed to a fitness for purpose basis, seeking to demonstrate the extent of brick cracking that can be tolerated in a reactor without affecting safe operation and ability to re-fuel and shut down the reactor safely.

Confidence in Empirical Models of Materials Behaviour

GTAC was asked to respond to the following questions concerning current models of materials behaviour in graphite.

> *GTAC Question 3: "What confidence can be ascribed to empirical models of materials behaviour? How important is mechanistic understanding of irradiated materials behaviour in particular where extrapolation and interpolation is necessary?"*

> *GTAC Question 4: "What confidence can be ascribed to the material properties data obtained from Materials Testing Reactor samples and from samples removed from irradiated cores? How may these data be used?"*

The committee reviewed the individual properties of elastic modulus, strength, dimensional change (rate), thermal expansion, thermal conductivity, creep, elastic Poisson's ratio and density in terms of available data, and the mechanistic understanding of these. GTAC's confidence ranges from moderate to high for the different properties, except for creep and Poisson's ratio, in which confidence is low or undecided.

The empirical equations for modulus and strength given in British Energy Generation Ltd. (2003) are based on the work of Bradford (2002, 2003) as follows:

$$E_d = {}_0E_d.P.S.W \qquad [2]$$

$$\sigma_{ox,i} = \sigma_o.f.\sqrt{P.S}.W \qquad [3]$$

where σ_o is the unirradiated and unoxidised virgin strength, P is the Young's modulus saturated pinning term, S is the Young's modulus structure term, W is the weight loss term given by W = A exp ($-\lambda$ x), where x is fractional weight loss, with A and λ constants that may be different for modulus and strength and for different sources of graphite. The effect of weight loss on the structure term is taken into account by the concept of an 'effective dose' γ_{eff}, which allows for a delay in the effect, defining:

$$\gamma_{eff} = \gamma - \gamma(x) \qquad [4]$$

The key properties of dimensional change rate, modulus and strength are controlled by the 'pinning term' P, the 'structure term' S and the weight loss term' W (including the concept of the threshold level for dose). Although GTAC found the basic explanations of these effects plausible, the Committee was uncertain as to the extent to which they had been rigorously established.

A particular concern of GTAC was the justification for use of the product relationship to allow for the effects of irradiation and weight loss on strength and modulus. It was accepted that within the range of available data, the product rule could be supported. However, most of the MTR data on strength was obtained on irradiated material (0 - 160 x 10^{20} n.cm^{-2}) without weight loss or on material up to 43% weight loss (but with low accompanying dose estimated at 0 - 40 x 10^{20} n.cm^{-2}). Continued validation of the empirical product rule is quite strongly reliant on its agreement with monitoring data.

An important conclusion reached by GTAC was that the various empirical models should be used only for interpolation within the range of data underlying the models. Their use for extrapolation is not considered to be appropriate because there is uncertainty in understanding of the mechanisms which may take place for combinations of high irradiation dose and weight loss.

The Determination of Stresses in Graphite

GTAC was asked to respond to the following questions concerning current models of determining stresses in graphite.

> *GTAC Question 8: "What confidence may be ascribed to predictions of brick internal stresses? How may brick internal stress predictions be validated? Under what conditions may such predictions break down?*

In the early stages of irradiation, the bore regions try to shrink more than the surrounding regions and the bore experiences tension stresses whilst the outside of the brick and the keyways experience compressive stresses. As irradiation proceeds, the bore regions reach turnaround in dimensional change rate first, and the bore then starts to expand again relative to the surrounding parts of the brick. The bore will see reducing tension stresses progressing towards compression with further increases in dose, whilst the outside of the bricks and keyway regions will see reducing compression moving towards tension.

At any position within a brick, the total strain ε^{tot} will be the sum of the strains due to dimensional change ($\varepsilon^{dimchange}$ due to shrinkage and growth caused by irradiation), thermal strains ($\varepsilon^{thermal}$ due to non uniform CTE and temperature gradients), the relaxation due to creep (ε^{creep} caused by irradiation and consisting of primary and secondary terms) and the strains due to stresses (ε^{stress}).

$$\varepsilon^{tot} = \varepsilon^{dim\,change} + \varepsilon^{thermal} + \varepsilon^{creep} + \varepsilon^{stress} \qquad [5]$$

It should be noted that thermal strains arise from two sources, namely the presence of temperature gradients across a brick and from gradients of CTE caused by dose and temperature. Temperature gradients are present in steady state operation, but also occur during transients, *e.g.* temperature excursions or shut down.

British Energy and British Nuclear Group use different interpretations of Equation [5] to determine internal stresses. In both approaches, the actual calculations are carried out using standard finite element analysis programs with specific materials properties sub-routines to allow for the changes in properties with dose, temperature and weight loss. Analyses are carried out incrementally with increasing dose (usually expressed in terms of full power years (fpy)). They are also carried out with the reactor at full power and after shutdown. Clearly, these analyses must involve some iterative routines to converge on consistent distributions of dose, conductivity, CTE, temperature, dimensional change (rate) and modulus. Furthermore, the increment of creep strain for each increment of dose is dependent on the stress level initially determined and its relaxation. The analyses are extremely complex and the industry is to be complemented on the progress that has been made in carrying out such work.

However, these analyses did not predict the occurrence of early life bore cracking in AGRs. To some extent, the occurrence of ELBC can be explained by invoking additional effects not initially included in the analyses, such as the effects of methane holes, eccentric fuel stringers and fuel end dose effects. Furthermore analyses on Magnox reactors suggest that stresses above the strength of the graphite are predicted in some cases, but no cracking has been reported from bore inspections carried out at any of the reactors concerned.

GTAC concluded that there was high confidence in the basic principles and methods used for calculating stress distributions in graphite bricks, but in view of variable levels of confidence in the prediction of key material properties and the additional uncertainties in field variables (temperature, weight loss and dose), sensitivity studies to explore the effect of reasonable variations are necessary. However, GTAC has low confidence in the ability to predict stresses accurately at specific time / dose levels and locations in reactors and considers that validation of stress predictions can only be achieved in general terms by confirmation of trends in dimensional changes and statistical comparisons of occurrence of cracking.

Criteria of Failure for Graphite

GTAC was asked the following question on criteria of failure:

> *GTAC Question 12: "What confidence can be ascribed to current brick failure criteria for AGR and Magnox graphite bricks? How may alternative brick failure criteria be developed?"*

GTAC concluded that use of critical tensile strength as a fracture criterion is only valid for bulk specimens of similar volume and geometry to the component under consideration. Graphite is a brittle material that contains a range of different inherent flaws and as expected, demonstrates a dependence of strength on stressed volume. This suggests that an approach such as the Weibull approach to strength should be applicable. However, although the basic principles of this approach appear to be appropriate, there are complications specific to graphite that mean further work is required to treat stress concentration regions satisfactorily. This applies to both keyway regions and trepanned holes. It is recommended that such work should be pursued. However, GTAC was satisfied that the approach of using component tests containing the geometry of concern was a satisfactory method of determining the component strength. To reflect the effects of ageing, GTAC supported the consistent use of ratios of strengths of degraded to undegraded material in simple bend test pieces from small to larger sizes as the best surrogate for brick loadings in-service.

For multi axial stress states, there is an interaction between the principal stresses that can reduce the effective strength in the direction of the maximum principal stress. The most appropriate theory for this appears to suggest an effective strength of approximately 0.8 x the uniaxial strength for biaxial stress conditions. The effects of the dimensional change shrinkage / expansion occur in both axial and circumferential directions, so that away from the ends of the bricks, biaxial stresses are set up. However, further work is necessary to understand these effects properly.

Creep Behaviour of Graphite
It became evident during the unfolding of the GTAC programme of questions that the creep behaviour of graphite was a critical aspect of determining stresses and that creep behaviour was not well understood. It is apparent that the level of stresses that would arise from dimensional change gradients and thermal stresses alone are such that they would exceed the strength of graphite by a significant amount. However, these stresses are relaxed by creep and the final level of stress after any increment of dose depends on the increment of creep strain that occurs. There are significant uncertainties over the interactions between creep strain and the coefficient of thermal expansion, dimensional change, the effect of radiolytic weight loss, and the variation of Poisson's ratio. GTAC is considering an extensive additional question concerned with the confidence that can be ascribed to the methodologies of creep assessment and predictions in AGR and Magnox reactors. There is clearly a need for further work to improve the understanding of the mechanisms of creep and the significance of interaction effects on other properties.

MANAGEMENT OF AGEING PLANT

The examples described above from topics considered by TAGSI and GTAC have some aspects in common that are relevant to management of ageing plant. It can be seen that the effects of neutron irradiation on embrittlement of steel occur by three separate mechanisms that may occur at different rates and depend on the detailed composition and microstructure of the material. It is not unreasonable to expect that in graphite there may well be separate effects of irradiation on the crystallite structure and on the surrounding matrix. Hence the proposal for a 'pinning' term that saturates for the crystallite and a 'structure' term operating on the remaining structure is reasonable. However, further work is necessary to understand the exact mechanisms and effects on other microstructural variables.

The example of reheat cracking in steels show that unpredicted forms of cracking can occur. In steels, where these lead to leakage or potential breach of the pressure containment, they are unacceptable and immediate action must be taken to shut down the plant. The graphite moderator bricks are a highly redundant set of components, and cracking of one or more bricks is not necessarily a threat to the safety or operation of the reactor. Essentially, the effect of cracks in graphite bricks needs to be considered in relation to potential effects on ability to insert control rods to shut the reactor down, ability to remove and replace fuel stringers and potential effects on flow of gas for cooling of the fuel. The major consideration in graphite is likely to be the potential movement of the parts of bricks if more than one crack occurs in a brick.

In order to avoid unplanned outages of plant, it is necessary to understand all potential modes of failure and their consequences. This has to take account of the effects of predicted deterioration in service and any known incidences of problems in related plant. It is also necessary to be aware that unexpected problems may arise and to take account of this in critical plant with in service inspections. The best way of implementing such a strategy is by periodic safety and availability reviews (PSARs). This involves a component by component check of the initial design requirements throughout the plant in the light of current knowledge of potential degradation mechanisms, including any information from monitoring of plant operational parameters and from in service inspections. The same basic multi-legged approach of the TAGSI recommendations for structuring IoF cases can be used for any safety case. Once the plant has entered service, the emphasis has to be placed on legs concerned with failure analysis and monitoring/inspection.

One way of implementing a PSAR on the lines of this approach is to select a target reliability for each component dependent on the consequences of failure. This can be done on a qualitative risk assessment matrix basis, for example a 3 x 3 matrix of low, medium and high assessments of probability of failure versus consequences. If sufficient data are available, it may be possible to carry out a quantitative risk assessment. A typical diagram of probability of failure versus time would show increasing probability as material properties are predicted to degrade. However, in the failure analysis there are bound to be uncertainties in material properties to be used, particularly if forward predictions are being made without a sound understanding and modelling basis. By incorporating the results of monitoring tests, it is possible to calibrate the models of materials behaviour and reduce the uncertainties at least at the time of the assessment. Where the models are almost entirely empirically based and test data are not available to cover the next phase of service life, it is necessary to be very cautious about future predictions. Every effort should be made to develop sufficient understanding that it can be safely assumed that there will not be a sudden sharp change in behaviour.

It may be possible to take advantage of a functional testing leg of a safety case if it can be demonstrated that one particular set of loadings is more severe than others and inspections are carried immediately after the occurrence of this loading. For example, in the case of the graphite bricks, there is a difference in severity of stresses between the reactor at power and after shut down. If the shut down conditions are shown to generate higher stresses than at power, it might be possible to demonstrate that the margin between the severity conditions can be traded against further deterioration to help to justify the next period of operation.

CONCLUSIONS

Safe operation of ageing plant requires a sound understanding of the potential mechanisms of deterioration of all components and assessment of the potential consequences of failure. Future prediction of degradation requires modelling of the materials behaviour and demonstration that no sudden change in properties is about to occur. Empirical models alone, based on curve fitting existing data are not suitable for extrapolation beyond the range for which data are available unless there is a real understanding of the mechanisms involved. This means that continued safe operation of ageing plant will inevitably require monitoring of service parameters and periodic in-service inspection to check that expected behaviour is being achieved and that no unexpected events are occurring.

One of the main areas where improved knowledge of the behaviour of graphite is required is creep behaviour and its interaction with other properties essential to the analyses used for the prediction of stresses. Other areas include the development of models for the combined effects of neutron irradiation and radiolytic weight loss on modulus, strength and dimensional change, and the behaviour of Poisson's ratio and lateral creep coefficients under combined conditions.

Acknowledgements

The contributions by colleagues on TAGSI and GTAC to reports quoted from the work of these organisations are gratefully acknowledged.

References

Bradford M. R. (2002). AGR Generic. Proposed changes to the endorsed data for moderator graphite strength, British Energy Generation Report E/REP/PSGP/0005/AGR/02, 2002.

Bradford M. R (2003). AGR Generic: Proposed changes to the endorsed data for moderator graphite dimensional change. E/REP/PSGP/0004/AGR/03, January 2003.

British Energy Generation Ltd., GCDMC/P28, Compendium of CAGR Core and Sleeve Data and Methods.
(a) – (i) Data Sheets C1 to C9, 2003

British Energy Generation Ltd., GCDMC/P28, Compendium of CAGR Core and Sleeve Data and Methods.
(a) – (m) Method Cover Document C1/1 March 1995 and Method Sheets C1 to C13, 2003.

Bullough R., Burdekin F. M., Chapman O. J. V., Green V. R., Lidbury D. P. G., Swingler J. N. and Wilson R. (2001). The demonstration of Incredibility of Failure in structural integrity safety cases, *Int. J. Press Vessels and Piping*, Vol 78, No.8, 539-552.

Hall G., Marsden B. J., Fok A. and Smart J. (2003). The relationship between irradiation induced dimensional change and the coefficient of thermal expansion: A modified Simmons relationship. *Nuclear Engineering and Design*, 222, 319-330.

Knott J. F. and English C. A. (1999). Views of TAGSI on the principles underlying the assessment of the mechanical properties or irradiated ferritic steel reactor pressure vessels. *Int. J. Press Vessels and Piping*, Vol 76.

Knott J. F. *et al.*, (2005). TAGSI response to British Energy questions on reheat cracking, TAGSI/P(98)156 Issue 4 to be published, 2005.

Marsden B. J. and Wilkes M. A. (2001). Irradiation damage and radiolytic oxidation of pile grade A graphite, AEA/16423596/R/01; CSDMC P133/A/1, 2001.

Payne J. F. B. (2004). Prediction of the dimensional change of PGA graphite. NSTS/GEN/REP/0241/03, Issue 2.

An Overview and Summary of the Technical Papers in Parts B, C and D

Peter E J Flewitt, Brian McEnaney[*], Barry J Marsden[#] and Gareth B Neighbour[$]

University of Bristol, [*]University of Bath, [#]University of Manchester, [$]University of Hull

Email: g.b.neighbour@hull.ac.uk

INTRODUCTION

The first international conference on the management of ageing processes in graphite reactor cores was held during 28th-30th November at the University of Cardiff, United Kingdom, in which there were three substantive technical sessions. For each session, the substantive points, questions, comments and discussions were noted and are summarised below in an overview. The primary aim is to allow the scientific community to have a crystallised view of the current state of the art with respect to the management of ageing graphite reactor cores.

ADVANCES IN THE FUNDAMENTAL KNOWLEDGE OF GRAPHITE BEHAVIOUR UNDER IRRADIATION

The presentations in this section covered aspects of nuclear graphite behaviour investigating the relationship between the microstructure and properties from the nano-scale through to the bulk property scale. The aim of the research in this area is to advance the understanding of the influence irradiation microstructural changes (due to both fast neutron damage and radiolytic oxidation) on the bulk properties of irradiated graphite components. There was an excellent presentation by Professor Malcolm Heggie on first principles atomistic modelling properties of structures formed by irradiation damage in graphite. At the microscale, colleagues from Manchester gave a review of the pioneering work on the use of micro X-ray tomography and Electronic Speckle Pattern Interferometry (ESPI) used to study the influence of microstructural changes particularly in fracture and the analysis of the disc tests using a continuum damage model. Professor Brian McEnaney looked at the influence and possible mechanisms for the influence of atmosphere pressure and temperature on graphite properties, Dr Milorad Novovic presented analysis of the failure strength of graphite with different microstructures. This was followed by a presentation by Dr Gareth Neighbour concerning the development of novel computational methods to better understand the inter-relationship between various graphite properties. Dr Martin Metcalfe discussed the influence of porosity and irradiation on property changes and the use of the "product rule". The presentation by Dr Mark Davies and Dr Mark Bradford illustrated the in-depth thinking that is going into the development of new models for irradiation damage in graphite by the operating companies, including a new model for irradiation creep. Finally Dr Ernest Eason gave a presentation on the use of modern statistical modelling, pattern recognition and curve fitting that can be used to investigate irradiation graphite datasets with many variables to determine the statistically significant variables and factors which should be taken into account and in so doing developing models and indicating areas where further research is required. The presentations under this topic point the way forward to an exciting future.

ADVANCES IN UNDERSTANDING OF GRAPHITE-COMPONENT BEHAVIOUR

In his paper presented during the Introductory Session, Dr Graham Heys stated that there were four inter-related themes that should be considered in relation to the objectives of this conference: understanding; monitoring; mitigation and prediction. In this session, there was one paper that related to the first theme and three papers for each of the second and fourth themes.

The first paper in the session that was presented by Dr Andrew Hodgkins from the University of Manchester reporting some recent advances in our understanding of the fundamentals of graphite cracking using R-curve analysis. Specifically, the paper addressed the question of what micromechanical factors contribute to energy dissipation during the fracture of graphite. A particularly valuable aspect of this work was the deployment of new techniques such as high resolution x-ray tomography and optical strain mapping. As also shown by others, Hodgkins and his colleagues find that the initial stage (Stage I) of crack development is accompanied by a rising R-curve. This is associated with the development of crack bridging in the wake zone. For some ceramics the second stage of crack growth (Stage II) is characterised by a plateau in the R-curve, but, in this work, a continuous rise in the R-curve is found. Microstructural features associated with Stage II are a bowed crack front and further pinning of ligaments up to 40 mm behind the crack tip.

Dr Steve Preston of SERCO Assurance and Mr Bill Ellis of Nexia Solutions presented an overview of some of the techniques used to measure properties of irradiated graphites removed from reactors as installed or trepanned samples in commercial reactors or installed samples in Materials Test Reactors, MTR. The paper began by reviewing the problems of measuring graphite properties on specimens of small size. The size of MTR specimens are constrained by the volume of high flux regions in MTR and the installation space while trepanned specimens are severely limited in size by the scope of the trepanning operation and the need to sub-divide the samples for some measurements. For such small samples standard techniques, *e.g.,* ASTM methods, cannot be used and so special test methods must be devised and validated. This was illustrated by the development and application of test methods for measurement of bend strength, dynamic Young's modulus and coefficient of thermal expansion on small graphite samples.

The second paper on monitoring presented by Mr Matthew Brown of Nexia Solutions gave a more detailed account of the considerable effort made over many years to develop improved measurement techniques for Young's modulus and strength of small graphite samples. A special apparatus was developed for measurement of dynamic Young's modulus using the ultrasonic pulse method. A 1 MHz ultrasonic pulse was used and a transit time of about 2 ms was obtained using 6 mm thick graphite specimens. It was found that the use of soft rubber probes improved the reproducibility and coupling efficiency. Several alternatives to the three-point bend method for measuring graphite strength were also presented including the ring burst method and the extended tensile test. Both the measurement of dynamic Young's modulus and strength were adversely affected when high weight loss graphite samples were used. In the case of the ultrasonic pulse method the measurement are affected by increased wave scatter and absorption while strength measurements are affected by crushing of the graphite at load points. Work is ongoing to address these problems.

The first paper on prediction was presented by Mr Chris Jones of SERCO Assurance which surveyed the structure of Finite Element Analyses, FEA, developed at SERCO to predict the development of stresses in core graphite components, the effects of brick cracking and bowing of bricks. It is the development of computing power in recent years that has made possible such sophisticated analyses. However, there are still issues connected to this work that require further attention. In relation to inputs to the FEA models, there is the quality and extent of input data. Originally, these data were derived from MTR experiments, but now there is an extensive and growing database of properties obtained from post-irradiation examination. These data can have two roles, either as input property values to supplement or replace the MTR data, or as field measurements to compare with output predictions from the models used in the FEA. Another input issue is the suitability of the algorithms that link properties, *e.g.* stresses and strains in the FEA models, especially those that model the influence of irradiation. Regarding outputs from the models, it is necessary to validate them, for example by using measurements on test rigs. Where validation is not possible, or difficult, then other confidence building measures should be considered, see below.

One method of confidence building in the use of FEA is the creation of new and independent FEA whose structures and outputs can be compared to those from established codes. This subject was addressed in the paper presented by Dr Derek Tsang from the University of Manchester. The codes use a constitutive materials model for irradiated nuclear graphite. A series of constitutive equations are used for conventional elastic and thermal strains and the additional strains due to irradiation and radiolytic oxidation. The codes are used to derive stress profiles through AGR core bricks from the keyway to the core as a function of irradiation and oxidation.

The use of FEA to model stresses in Magnox core bricks as a result of thermal transients, as at shutdown, was described by Dr Colin Elcoate of Frazer-Nash Consultancy. The system was modelled using three FEA codes. FEAT was used to predict temperatures and stresses in a single brick, PANTHER was used to predict boundary conditions for a single brick in a single channel and MACE was used to determine the boundary conditions for the PANTHER predictions. Plant operating data, such as bulk gas inlet and outlet temperatures, were used as inputs to the models and the temperatures predicted by FEAT were validated using temperature measurements from thermocouples embedded in the graphite core bricks.

Jaap van der Laan of NRG, The Netherlands, provided a useful update on the graphite irradiation test facilities at the High Flux Reactor at Petten. He surveyed recent and ongoing work on testing graphites for high and very high temperature reactors and for the European fusion technology programme. He also showed how the NRG hot cell laboratories can be used to measure properties of graphite such as dimensional change, elastic modulus, strength, thermal expansion and diffusivity.

THE CONSEQUENCES OF GRAPHITE DETERIORATION: WHOLE CORE BEHAVIOUR

This session moved to the macro-length scale and addressed the behaviour of the whole core of the graphite moderated reactors. The session contained eleven papers which covered the UK Magnox and AGR generating plant together with the challenges associated with the proposed new generation of graphite moderated gas cooled reactor, the Pebble Bed Modular Reactor, that is being developed in South Africa. As pointed out by the speakers in the

Introductory Session, the whole core behaviour considerations seek to use predicted, measured or assumed component configurations, mechanical property data and stress analyses within models that investigate and describe the response of the graphite reactor cores to the service imposed environmental and loading conditions. Within this session, the papers divided broadly into the categories of (i) inspection and monitoring of the reactor cores, (ii) data handling and (iii) modelling of the plant to underwrite safe continued operation. Indeed within these categories the papers considered the challenges of graphite oxidation and shrinkage, differential thermal expansion and other factors that could lead to potential distortion and cracking within individual graphite reactor core bricks. However there is a noticeable difference in the incidence of cracking between the two types of gas cooled reactors operated in the UK.

Inspection and Monitoring

The following papers were in this category:

(i) Measurement of AGR Graphite Panel Brick Shrinkage and Channel Distortion, A Cole-Baker and J Reed (British Energy).
(ii) An Integrated Architecture for Graphite Core Data Analysis, G J John, S D J McArthur and J Reed (The University of Strathclyde and British Energy).
(iii) Graphite Core Condition Monitoring Through Intelligent Analysis of Fuel Grab Load Trace Data, CM West, GJ John, SDJ McArthur and J Reed (The University of Strathclyde and British Energy).

As the graphite reactor cores age the combined effects of irradiation and radiolytic oxidation causes the individual bricks to differentially expand and contract over the operating life. These have to be coupled with differential thermal expansion due to changes in both the coefficient of thermal expansion and dimensions arising from changes of temperature during start-up and shut-down to evaluate the potential for brick cracking of the type observed in the AGR reactors. As a consequence, the methods described to (i) monitor the dimensional changes using the Norebore channel bore diametrical and tilt system and (ii) to establish the existence of cracking that leads to core distortion in AGR cores from fuel grab load monitoring, provide an essential input to the overall safety arguments. It is clear that these data need to be interrogated correctly and the paper by John *et al.* seeks to provide an appropriate framework for undertaking such analyses.

Data Handling

There were two papers in this category:

(i) Statistical Models for Bore Cracking in AGR Graphite Cores, PR Maul and PC Robinson (Quintessa).
(ii) Application of Whole Core Modelling Methodology to Life Extension in AGR Reactor Graphite Cores, N McLachlan, DJ Shaw and MHS Salih (British Energy and AMEC NNC).

The first of these two papers take inspection data measured, to date, on AGR bricks observed to contain single and double cracks and uses statistical analysis procedures to evaluate the data and set a monitoring strategy. The statistical analysis methodology is based upon the principle of maximum likelihood and uses a multiple regression set in a Bayesian framework. At present, the output is limited by the sparseness of the data. However since the authors are seeking to use the output for limited prediction beyond the range of the data this approach can be challenged. Statistical analysis, however, should become of increasing value as the database is increased over time. The second approach offers an improved framework for addressing the safe continued operation of the AGR graphite reactor cores. It is based upon a methodology that is conceptually similar to that adopted for the R6 failure avoidance procedure, but involves Functionality Assessment Diagrams to establish safe operating margins. Both methodologies are at an early stage of development and, in part, are limited by the fact that the data are sparse, but both provide an innovative approach for the future.

Core Modelling

There were five papers contained in this category:

(i) Seismic Modelling of an AGR Nuclear Reactor Core, B Kralj, S Humphreys and B Duncan (Atkins Nuclear).

(ii) Investigation of Degradation of AGR Graphite Core Geometry using a Whole Core Scale Model, E Castro (AMEC NNC).

(iii) The Feasibility of Introducing Core Stiffening Devices and Diverse Shutdown Systems at Hinkley Point B/Hunterston B and Hartlepool/Heysham 1, MW Davies (AMEC NNC).

(iv) Development and Validation of Whole Core Modelling Methodology for AGR Graphite Cores. DJ Shaw, MHS Salih and N McLachlan (AMEC NNC British Energy).

(v) A Method of Determining if Control Rods and Fuel Strategies can be Inserted/Removed from Core, A Thomson, C Ablott, M Ripley and N McLachlan (AMEC NNC British Energy).

These papers, in general, described the development and use of scaled or full size section models of reactor cores to provide input data for, or validation of the predications from finite element or equivalent analytical models. Typically, an AGR core is a multi-layered arrangement of discrete graphite bricks located together using an arrangement of keys in keyways. The Magnox and AGR reactor cores have bricks of different sizes and geometry but both contain fuel and interstitial channels.

These theoretical and experimental models offer the opportunity to explore the response of the reactor cores to a range of service conditions and loadings including limited brick cracking and the overall seismic response. Some papers also addressed issues required for mitigation. For example, consideration was given to the possibility of introducing stiffening devices into specific AGR core channels which would then limit the potential for brick displacement and, or column bowing in cores with bricks that contain cracks. Here, the concept of internal sleeves was adopted. As an alternative to the engineering modifications to the reactor cores the challenge of safe shut-down was addressed by exploring the adequacy of the diverse shut-down systems and how these may be enhanced. Moreover one paper addressed, via modelling, the ability to continue to insert control rods into degraded cores to enhance the arguments for safe continued operation.

Finally the paper, PBMR Graphite Core Structures Fitness for Purpose Management Strategy, MA Davis, MN Mitchell and PJ Venter (RPS Engineering and Safety PBMR Co) provided an overview of how fitness for purpose is to be demonstrated throughout the planned life of this reactor type based upon (i) life assessment and qualification during the design phase; (ii) design of a suitable in service monitoring and inspection programme; and (iii) design and maintenance of the CSC including replacement. It was encouraging to note that experience gained from operating reactors with graphite cores is being used in these considerations.

CONCLUSIONS

It is clear from the presentations in this conference that there is increasing integration of the results and the data obtained from the nano- and micro length scale work described in the earlier sessions into the whole reactor core modelling. The experimental results obtained from the whole core array models, either full size or scaled, offer an important contribution for validating the theoretical model predictions. It is important to establish the underlying mechanisms that lead to the degradation of graphite during service so that (i) realistic whole core models can be developed and underwritten, (ii) optimum inspection and monitoring programmes can be established over the remaining life time of the Magnox and AGR reactor cores and (iii) safe operation can be underwritten with appropriate safety margins. It is surprising, in view of the potential for relaxation of internal stresses that none of the papers considered the role and implications of irradiation induced creep for the overall life evaluation of the graphite cores. Nevertheless, it is very clear that the story of nuclear graphite in the generation of civil nuclear power is far from over, both in terms of life extension for the UK's remaining fleet as well as the future reactor designs. Graphite remains an amazing material to study and a complex structural material, but functionally suitable for use in nuclear reactor cores!

Author Index

Subject Index (by Keywords)